KU-286-224

INVERTEBRATE ZOOLOGY

A Laboratory Manual

ITY

TEL. 0151 231 4022

WITHDRAWN

LIVERPOOL JMU LIBRARY

3 1111 00929 0691

LIVERPOOL
JOHN MOORES UNIVERSITY
AVRIL ROBARTS LRC
TEL. 0151 231 4022

INVERTEBRATE ZOOLOGY
A Laboratory Manual

5th
Edition

Robert L. Wallace
Ripon College

Walter K. Taylor
University of Central Florida

Prentice Hall
Upper Saddle River, NJ 07458

Library of Congress Cataloging-in-Publication Data
Wallace, Robert L.,   1948–
     Invertebrate zoology :   a laboratory manual
  / Robert L. Wallace, Walter K. Taylor. — 5th ed
          p.   cm.
     " This manual began as a revision of Beck and Braithwaite's
Invertebrate Laboratory Workbook"—Pref.
     Includes bibliographical references and index.
     ISBN 0-13-270026-3
     1. Invertebrates—Laboratory Manuals.   I. Taylor, Walter
Kingsley, 1939–   .   II. Beck, D. Elden, 1906–
III. Braithwaite, Lee F.   IV. Title.
     QL362.W33   1996
     592—dc21   96-35494
                                                                CIP

Acquisitions Editor: Teresa Ryu
Assistant Vice President and Director of Production: David W. Riccardi
Senior Production Editor: Jennifer Fischer
Production Editor: Kimberly Dellas
Special Projects Manager: Barbara A. Murray
Buyer: Benjamin D. Smith
Production Editorial/Composition: Innodata Corporation
Cover Design: Karen Salzbach

©1997 by Prentice-Hall Inc.
Upper Saddle River, New Jersey 07458

Earlier editions, © 1989 Macmillan Publishing Company,
a division of Macmillan, Inc.; © 1968, © 1962, © 1960
D Elden Beck and Lee F. Braithwaite

All rights reserved. No part of this book may be
reproduced in any form or by any means,
without permission in writing from the publisher.

Printed in the United States of America
10 9 8 7 6 5

ISBN 0-13-270026-3

Prentice-Hall International (UK) Limited, London
Prentice-Hall of Australia Pty. Limited, Sydney
Prentice-Hall Canada Inc., Toronto
Prentice-Hall Hispanoamericana, S.A., Mexico
Prentice-Hall of India Private Limited, New Delhi
Prentice-Hall of Japan, Inc., Tokyo
Prentice-Hall Asia Pte. Ltd., Singapore
Editora Prentice-Hall do Brasil, Ltda., Rio de Janeiro

CONTENTS

PREFACE

Invertebrate zoology is an enormous field. At least 96% of all animal species lack backbones, but this assessment may be too conservative. Recent estimates suggest that more than 10 million insect species in the Amazonian forests have yet to be described. If this prediction is true, the numerical dominance of invertebrate species will be approximately 99%. Nevertheless, humans seem preoccupied with organisms possessing vertebral columns, especially if an animal resembles, in any way, one of those cuddly toys made for young children. Invertebrates, on the other hand, are often viewed with disgust, evoking unwarranted fears and horrific screams of terror when encountered in a disused corner of a basement, in a half-eaten apple, or crawling on one's body. We do not mean to imply that invertebrates do not cause human suffering or seriously damage agricultural products. They do, and it is for these reasons and because as a group the invertebrates possess such diverse and rich biologies that they are worthy of intensive study.

We dedicate this manual to our teachers who introduced us to the invertebrate world, our families who put up with our interests, and new students of invertebrates who will discover additional wonders about animals without backbones.

This manual began as a revision of Beck and Braithwaite's *Invertebrate Zoology Laboratory Workbook* and therefore fulfills Dr. L.F. Braithwaite's dedication to the third edition: that the comprehensive laboratory study guide of invertebrate zoology begun by Dr. D Elden Beck should not become extinct. This continues to be our goal. Beck and Braithwaite's manual has been an excellent source of instruction for over three decades. Our purpose in continuing the revision was to retain the pedagogical approach of the manual while updating the text to take into account advances in the discipline and improving coverage of important groups. We believe that a study based on morphology is an excellent way to achieve a comprehensive understanding of invertebrates. We do not suppose that this is the only way in which invertebrates may be studied (e.g., one may do a study of systems across taxa, the functional approach). However, for the student who has had a course in introductory zoology or general biology, the approach of this manual will provide a solid conceptual framework for advanced work on behavior, ecology, physiology, and related subjects.

Although changes are evident in this edition, we have attempted to keep the original intent of the manual. "To excite the interest of the student to get acquainted with the world of invertebrates about them wherever they may live. It isn't necessary [or even desirable] to travel to far off forest or shore to find exciting animals to study. Wherever one lives, those animals usually listed as 'common' are often the least known and frequently make the best subjects for study and research" (Beck and Braithwaite 1968: i). Both live and preserved organisms are used throughout the manual and it is designed so that instructors may omit sections that are inappropriate to their particular interests and/or to substitute other studies.

The approach we continued in the fifth edition of the manual was to develop exercises using representatives of invertebrate phyla for which specimens are readily available from commercial suppliers. Therefore, some of the smaller phyla are not covered here because we have not found commercial material to support meaningful laboratory work. In the fifth edition all of the artwork was reworked (both figures and photographs). We have continued the practice of adding simple figures of the geological time scale for phyla that have a significant fossil record. While these figures do not indicate diversity throughout geologic time, they will help students appreciate the enormous length of time some phyla have been in existence (i.e., the concept of **Deep Time**). We also have kept the figures of fossils, parasitic forms, and larvae, the simple phylogenetic descriptive inserts (see Taxonomic Considerations and Evolution of the Invertebrates), the simple pronunciation guide, the etymons for each phylum name, and the use of **boldface type** for key concepts. We believe that these features will make the learning process easier for students. Thus, our basic idea for the laboratory manual continues to be one of supporting both instructors and students in their efforts in understanding the invertebrates. We believe that it is a flexible instrument so that it may be used successfully in a **learner-centered** teaching environment. Students can pick up the manual and, along with their text and the study material, learn the animals for themselves—first-hand discovery.

Many people have helped in the production of our two revisions, and to these colleagues we offer our most sincere thanks. Their efforts and expertise have added much to the project. We especially wish to thank Michele Johnson, who did all the original artwork; without her dedication this revision would not have been completed. We also thank Jan Richardson and Teresa Ryu, who offered their ideas, enthusiasm, and considerable logistic help in completing the revisions. Paul Conant, President of Triarch Inc., of Ripon, Wisconsin deserves special thanks for the preparation of several special order slides. Thanks also are due to over 150 instructors from around the country who took the time to fill out a preliminary questionnaire concerning their approach to laboratory instruction of invertebrate zoology and the types of features they wished to see in the manual. Those who acted as reviewers of the manuscript or who otherwise contributed to both the fourth and fifth edition were: John A. Allen, University Marine Biological Station Millport, Scotland (Bivalvia); Kenneth J. Boss, Museum of Comparative Zoology, Harvard University (Polyplacophora, Monoplacophora, Aplacophora, Scaphopoda); C. Bradford Calloway, Museum of Comparative Zoology, Harvard University (Priapulida); Ralph O. Brinkhurst, Ocean Ecology Laboratory, British Columbia (Oligochaeta); William S. Brooks, Ripon College (general editing); Robert Bullock, University of Rhode Island (general review); John L. Cisne, Cornell University (Trilobita); G. Arthur Cooper, Smithsonian Institution (Brachiopoda); Edward B. Cutler, Utica College, Syracuse University (Pogonophora, Sipuncula); William C. Dewel, Appalachian State University (Cnidaria); W.T. Edmondson, University of Washington (Rotifera); Stan Edwards, South Australian Museum (Echiura); Christian C. Emig, Station Marine D'Endourne et Centre D'Oceanographie (Phoronida); K. Fauchald, Invertebrate Zoology (Polychaeta); William McKay Fender, Soil Biology Associates, McMinnville, Oregon (Oligochaeta); Merrill W. Foster, Bradley University (Brachiopoda); Tom Frost, University of Wisconsin—Madison (Porifera); John J. Gilbert, Dartmouth College (Porifera, Rotifera); Mary Haskins, Rockhurst College (general review); K. Herrmann, Universität Erlangen-Nürnberg (Phoronida); Robert P. Higgins, National Museum of Natural History (Kinorhyncha); Meg Hummon, Ohio University (Gastrotricha); William Hummon, Ohio University (Gastrotricha); Alan Kohn, University of Washington (Gastropoda); Robin Leech, Northern Alberta Institute of Technology (Cephalopoda); Denis H. Lynn, University of Guelph (Protozoa); A.R. Maggenti, University of California—Davis (Nematoda); Frank Maturo, University of Florida (general); Janice Moore, Colorado State University (Acanthocephala); Brian Morton, University of Hong Kong (Bivalvia); Tom Near, Illinois Natural History Survey (Acanthocephala); Brent Nickol, University of Nebraska (Acanthocephala); Claus Nielsen, Zoologisk Museum, Denmark (Entoprocta); David L. Pawson, National Museum of Natural History (Echinodermata); Marian H. Pettibone, National Museum of Natural History (Polychaeta); Leland W. Pollock, Drew University (Tardigrada); Mary E. Rice, National Museum of Natural History (Sipuncula); Reinhard M. Rieger, Institut fur Zoologie der Universitat Innsbruck (Turbellaria), Ripon College students, especially: Wm. Brinkman, Mellisa Walden, Wm. Rochon, Trisha Schmitt, Erica Smith (general); Edward E. Ruppert, Clemson University (Gnathostomulida), J.S. Ryland, University College of Swansea (Bryozoa); Steward C. Schell, University of Idaho (Trematoda); Gerald Schmidt, University of Northern Colorado (Acanthocephala, Cestoda); J. Teague Self, The University of Oklahoma (Pentastomida); Nancy Shontz, Grand Valley State University (general); Tracy Simpson, University of Hartford (Porifera); R.W. Sims, British Museum (Natural History)

(Oligochaeta); Eve Southward, Marine Biological Laboratory, Plymouth, UK (Pogonophora); Peter L. Starkweather, University of Nevada–Las Vegas (Rotifera); Wolgang E. Sterrer, Bermuda Biological Station (Gnathostomulida); Margaret E. Stevens, Ripon College (general editing, artwork consultant); Julia Stuart, Trentham, New Zealand (Hirudinea); Sidney L. Tamm, Boston University, Marine Biology Laboratory, Woods Hole (Ctenophora); Bryn H. Tracy, Carolina Power and Light, New Hill, North Carolina (Entoprocta); Jean C. Tryon, Ripon College (Porifera, Acanthocephala, Gastrotricha, Tardigrada); Seth Tyler, University of Maine (Turbellaria); University of Central Florida students, especially: Zack Prusak, Ed Chicardi, Lisa Roberts, Susan Rallo (general); James W. Valentine, University of California (Phoronida); Gilbert L. Voss, Rosenstiel School of Marine and Atmospheric Science (Cephalopoda); Mary Wicksten, Texas A&M University (general review); and Craig Williamson, Lehigh University (Porifera). Thanks also are due to the anonymous reviewers who read and improved this and earlier versions of the manuscript. However, we take responsibility for any errors that remain and welcome constructive criticism from instructors and students of invertebrate zoology.

R.L.W.
W.K.T.

TAXONOMIC CONSIDERATIONS AND EVOLUTION OF THE INVERTEBRATES

This laboratory manual outlines procedures for the study of representatives of invertebrate phyla for which specimens are readily available from commercial suppliers. The way we chose to arrange the phyla in this manual generally reflects that of most of the current textbooks of invertebrate zoology and may be used to support any one of them. In the third edition of Beck and Braithwaite's workbook the authors offered a taxonomic list for the classification of invertebrates. They intended it as a workable outline that would be reviewed by the student before each taxon was studied in the laboratory. We believe that this list was useful, but lacked the impact necessary to achieve its purpose as a teaching aid. Therefore, we have replaced the taxonomic list with a series of Phylogenetic Descriptions of the major groups. Each time organisms with a distinct body plan are considered, a new phylogenetic description is presented, e.g., acoelomate worms, pseudocoelomates, mollusks, lophophorates, etc. Instructors may have their students use these descriptions as given or they may wish to augment them, perhaps by providing a phylogenetic tree to illustrate putative relationships among the invertebrates. No phylogenetic tree has been offered here, as we believe that consideration of the various evolutionary trees that have been proposed in the literature should be left to instructors. We hope that the addition of the phylogenetic descriptions will provide a structured framework so that halfway through the course students do not become overwhelmed and lost by the tremendous diversity of invertebrates.

■

Subkingdom Protozoa

Over 60,000 species of flagellates, amebas (or amoebas), opalinids, spore formers, and ciliates, often called collectively the protozoa, represent some of the most fascinating organisms to both novice and experienced microscopists. The tremendous number of individuals found in one drop of water can be overwhelming. Until recently, these mostly microscopic and primarily unicellular organisms were grouped in a single phylum Protozoa [PRO-to-ZO-ah; G., *proto*, first + G., *zoon*, animal]. Most protozoologists now agree that this grouping constituted a heterogenous assemblage of distantly related forms. Currently, many biologists place the protozoans within the subkingdom Protozoa (kingdom **Animalia**) containing at least six, probably polyphyletic phyla (Levine et al. 1980). Others prefer the term *protists* instead of *protozoa* and follow Whittaker (1969) in grouping the subkingdom under the kingdom **Protista**. Regardless of which grouping is followed, the protozoans play important roles in the web of life and are fascinating and challenging organisms to study.

Few places on the earth are devoid of protozoans. Free-living forms occur in aquatic habitats, moist soils, and decaying organic matter. Many protozoans are important parasites of plants, animals, and humans. Nearly one-half of the 60,000 protozoan species are fossils; some have been in existence at least since the **Precambrian** (Fig. 1.1).

The basic body plan is the single eukaryotic cell representing a functionally complete organism that performs all physiological processes found in multicellular animals or the **Metazoa** (G., *meta*, after or between + G., *zoa*, animal). Some biologists prefer to call protozoans acellular [G., *a*, without] organisms instead of unicellular, to emphasize the complete, functional organismal viewpoint.

Protozoan organelles tend to be more specialized than those found in the typical cell of metazoans. Some of the common organelles (together with their functions) are as follows: **food vacuole** or **phagosome** (digestion), **contractile vacuole** or water expulsion vesicle (water regulation), **myoneme** (contractile), **paraflagellar swelling** and **stigma** (sensory), **extrusome** (food getting and defense), and **pseudopodium, flagellum,** and **cilium** (food getting and locomotion). The traditional designation of protozoans as amebas, flagellates, spore formers, and ciliates was based primarily on their locomotory organelles.

Some species are colonial and can be seen with the unaided eye. However, there are no tissues, organs, or germ layers. Digestion is intracellular, occurring within food vacuoles. Respiration and excretion are accomplished by diffusion across the cell membrane. Reproduction is by **budding, fission, conjugation,** or **syngamy**. The motile form is often called a **trophozoite** (trophont), whereas the nonmotile form is called a **cyst**.

Selected representatives of the major phyla will be studied in the following exercises. Your instructor may provide additional examples for study.

Classification

1. **Phylum Sarcomastigophora.** Protozoans with a monomorphic nucleus. When present reproduction is

TIME (IN MILLIONS OF YEARS)

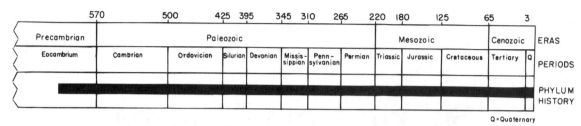

Figure 1.1. Geologic history of the subkingdom Protozoa.

by sexual means (syngamy). Locomotory organelles include flagellum, cilium, and pseudopodium.

a. Subphylum Mastigophora. Protozoans with flagella typically on the trophozoite. Reproduction is usually interkinetal (symmetrogenic) binary fission.

(1) Class Phytomastigophorea. Plantlike flagellates, typically with chloroplast and eyespots. Botanists call these flagellates algae.

(a) Order Dinoflagellida. Two flagella located in a cell with transverse and ventral grooves. Cell body typically covered with cellulosic plates. Examples: *Glenodinium, Gymnodinium*, and *Ceratium*.

(b) Order Euglenida. Cells with one to two flagella that arise anteriorly within the reservoir. Both chloroplast-containing and colorless forms present. Examples: *Euglena* and *Peranema*.

(c) Order Volvocida. Cells usually with two flagella, one chloroplast, and two contractile vacuoles. Solitary or colonial forms are known. Examples: *Chlamydomonas* and *Volvox*.

(2) Class Zoomastigophorea. Animal-like flagellates, typically lacking chloroplasts and eyespots.

(a) Order Choanoflagellida. Flagellates with a single anterior flagellum encircled by funnel-shaped collar. Both solitary and colonial forms are known. Choanoflagellates are similar to choanocytes of sponges (Phylum Porifera, Exercise 2). Examples: *Monosiga, Codosiga*, and *Proterospongia*.

(b) Order Trichomonadida. Flagellates parasitic in invertebrates and vertebrates. Trophozoite typically possess four to six flagella. Cysts usually lacking. Examples: *Trichomonas* and *Tritrichomonas*.

(c) Order Diplomonadida. Flagellates parasitic primarily in alimentary canal of host. Trophozoites with one or two karyomastigonts (e.g., complex of flagellum, nucleus, and associated organelles), each with one to four flagella. These protozoans are often bilaterally symmetrical. Examples: *Giardia* and *Hexamita*.

(d) Order Hypermastigida. Intestinal inhabitants of termites, cockroaches, and woodroaches, possessing many flagella. The motile cell usually is uninucleate. Example: *Trichonympha*.

(e) Order Kinetoplastida. Parasitic, endocommensal, and free-living protozoans with cells possessing one to two flagella. The kinetoplast is the largest extranuclear repository of DNA for any cell type. Their life cycle alternates between vertebrates and invertebrate hosts. Examples: *Leishmania* and *Trypanosoma*.

b. Subphylum Opalinata. Ovoid protozoans with numerous cilia in oblique rows over entire body. The cytosome and cytopharynx are absent. Nuclei are monomorphic. Reproduction is usually interkinetal (symmetrogenic) binary fission. Example: *Opalina*.

c. Subphylum Sarcodina. Protozoans with pseudopodia typically on trophozoite. Cell body is naked or with a test or internal skeleton. Reproduction is primarily by binary fission.

• **Superclass Rhizopodea.** Sarcodines that use lobopodia, filopodia, reticulopodia, or protoplasmic flow without pseudopodia for locomotion.

(1) Class Lobosea. Sarcodines with pseudopodia lobose to filiform in shape. Cells are usually uninucleate. The cell body either naked or in a test. Both free-living and parasitic forms are known. Examples: *Amoeba proteus*, *Chaos carolinense* (=*Pelomyxa carolinensis*), *Entamoeba histolytica*, *Arcella*, and *Difflugia*.

(2) **Class Filosea.** Filiform pseudopodia that often branch and sometimes anastomose. No flagellate stage known. Cell body is naked or in a test. Examples: *Lecythium* and *Euglypha*.

(3) **Class Granuloreticulosea.** Foraminiferidans (forams). Amebas primarily with delicate reticulopodia. The majority with a test of one to many chambers. Examples: *Globigerina*, *Elphidium*, and *Allogromia*.

• **Superclass Actinopodea.** Mostly planktonic amebas with radially arranged axopods. Capsular membrane between ectoplasm and endoplasm in most forms.

(1) **Class Acantharea.** Radiolarians. Skeleton of strontium sulfate with centrally joined radial spines. Mostly planktonic and all marine. Example: *Acanthometra*.

(2) **Class Polycystinea.** Radiolarians. Planktonic marine radiolarians with skeleton of silica. Example: *Collosphaera*.

(3) **Class Phaeodarea.** Radiolarians. Planktonic marine radiolarians with skeletons, when present, of silica and organic matter. Spines of skeleton hollow. Example: *Circoporus*.

(4) **Class Heliozoea.** Sun animalcules. Primarily freshwater amebas with a skeleton, when present, of siliceous material or chitinoid. Capsular membrane absent. Examples: *Actinosphaerium* and *Actinophrys*.

2. **Phylum Apicomplexa.** Entirely parasitic forms without cilia or flagella (except for flagellated microgametes). Nucleus is vesicular. Both asexual and sexual phases present in the life cycle.

(1) **Class Perkinsea.** Flagellated sporozoites with polar capsular tubules forming an incomplete cone. Example: *Perkinsus marinus* of oysters (only species).

(2) **Class Sporozoea.** Infective stage is a sporozoite, either naked or enclosed. Both asexual and sexual phases in complex life cycles. Examples: *Monocystis*, *Gregarina*, *Eimeria*, and *Plasmodium*.

3. **Phylum Microspora.** Intracellular parasites of mainly insects. Cells possesing polar filaments. Example: *Nosema*.

4. **Phylum Myxozoa.** Obligate, extracellular parasites of annelids and poikilothermic vertebrates (mainly fish) with usually two polar capsules. Examples: *Myxosoma* and *Triactinomyxon*.

5. **Phylum Ciliophora.** Cells typically with cilia and two nuclei types: macronucleus and micronucleus. Reproduction is primarily by transverse binary fission (homothetogenic) and conjugation.

(1) **Class Spirotrichea.** Somatic dikinetids present usually with postciliodesmata; usually with right and left oral ciliature. Paroral membranes may be present on right side of oral region. Examples: *Stentor*, *Spirostomum*, *Euplotes*, *Stylonychia*, *Kerona*, and *Blepharisma*.

(2) **Class Prostomatea.** Somatic ciliature of monokinetids with radial transverse ribbon. A ciliary crown of dikinetids surrounds the cytosome. A somatic ciliature of monokinetids is present. Cystostome either apical or subapical. Example: *Coleps*.

(3) **Class Litostomatea.** Somatic ciliature of monokinetids with tangential transverse ribbon. Oral cilia are simple. Example: *Didinium*.

(4) **Class Phyllopharyngea.** Somatic ciliature mostly as monokinetids with reduced transverse ribbon. The oral region bears radially arranged microtubular ribbons called phyllae. Examples: suctorians, *Ephelota* and *Acineta*.

(5) **Class Nassophorea.** Somatic ciliature as monokinetids, dikinetids, or polykinetids. The oral region bears microtubular rods called nematodesmata. Extrusomes, when present, are trichocysts. Example: *Paramecium*.

(6) **Class Oligohymenophorea.** Somatic ciliature of monokinetids with radial transverse ribbon. Oral apparatus, when present, of paroral dikinetid and usually three polykinetids. Examples: *Vorticella* and *Tetrahymena*.

(7) **Class Colpodea.** Somatic cilature as dikinetids with overlapping posterior transverse ribbons. Oral region with left and right ciliature. Resting cysts are common. Example: *Bursaria*.

A. Phylum Sarcomastigophora

Protozoans commonly called flagellates, opalinids, and amebas belong to the large phylum Sarcomastigophora [SAR-ko-MAS-ti-GOF-or-ah; G., *sarkos*, fleshly + G., *mastix*, whip + G., *phora*, to bear]. Members of the phylum move by **flagella** (flagellates), **cilia** (opalinids), and **pseudopodia** (amebas). Studies with the electron microscope show that the structural configuration of the flagellum and cilium is the same and that both arise from a **basal body** (**blepharoplast, kinetosome**). Three subphyla occur in Sarcomastigophora: (1) Mastigophora [MAS-ti-GOF-or-ah; G., *mastix*, whip + G., *phora*, to bear], the flagellates; (2) Opalinata [O-pa-LIN-a-tah; N.F., *opaline*, like a opal in appearance], the opalinids; and (3) Sarcodina [SAR-ko-DI-nah; G., *sarkos*, flesh + G., *ina*, belonging to], the amebas.

Subphylum Mastigophora

The most primitive protozoans belong to the subphylum Mastigophora. The flagellum is the primary locomotory organelle. Flagellates are mostly free living and often divided into plantlike and the animal-like forms.

Class Phytomastigophorea

These plantlike flagellates have a cell wall of cellulose, chloroplasts, and stigma. The usual type of nutrition is holophytic (autotrophic). Representatives of the following three orders will be emphasized: order Dinoflagellida (*Ceratium*, *Glenodinium*, and *Gymnodimium*), order Euglenida (*Euglena* and *Peranema*), and order Volvocida (*Chlamydomonas* and *Volvox*).

Order Dinoflagellida. Dinoflagellates are widely distributed in marine, brackish, and freshwater habitats. Most are free living. Dinoflagellates have an unequal pair of flagella. One is ribbon-like and typically lies in a transverse surface furrow called the **cingulum** or **girdle** (Fig. 1.2). The second flagellum is directed posteriorly and usually in a longitudinal furrow, the **sulcus**. Reproduction is by binary fission and syngamy.

The region above the cingulum in armored forms is the **epitheca** (epicone in unarmored ones); below the cingulum is the **hypotheca** (hypocone in unarmored ones). Unarmored or naked species (e.g., *Gymnodinium*) have a cell covering of membranes, whereas thecal plates of cellulose or other polysaccharides occur on armored species (e.g., *Ceratium* and *Glenodinium*). Many dinoflagellates (e.g., *Noctiluca*, *Gonyaulax*, and *Pyrocystis*) are bioluminescent and their bluish-green glow can be seen for some distance in the oceans. Certain genera, such as *Gonyaulax*

and *Gymnodinium*, may occur in hugh numbers; these blooms are often called **red tides**. Millions of cells per liter of water may be produced. Blooms of *Gymnodinium breve* produce toxins that directly kill fish and some invertebrates. The toxins of other species may be concentrated in the bodies of clams, oysters, and scallops, rendering these organisms unsafe for human consumption. Recent deaths of manatees in coastal waters of Florida have been blamed on dinoflagellate toxins.

■ **Observational Procedure: Order Dinoflagellida**

1. Make a small ring, using a thick suspension of methyl cellulose, on a glass slide. Add a drop of the culture in the center of the ring, cover with a coverslip, and observe under low power of a compound microscope. After the organism is located, change to high magnification.

2. Describe the locomotion of these protozoans. Attempt to locate the cingulum and sulcus in a stationary individual. Can you see flagella beating? What variations in the body shape, size, and structure do you see, especially in *Ceratium*, which has hornlike extensions (Fig. 1.2A)? What might cause these variations?

3. Examine prepared slides of several species of dinoflagellates (e.g., *Ceratium*, *Glenodinium*, and *Gymnodinium*). Note the nature of the body covering. *Ceratium* has one apical horn and one to three antapical horns. The nucleus should be evident. Locate the cingulum and the sulcus. ■

Order Euglenida. Euglenoid flagellates are primarily freshwater organisms. Over 800 species in at least 37 genera have been described. Most species are bilaterally symmetrical with bodies elongated, ovoid, spherical, or leaflike. Both chloroplast-containing and colorless forms are common. Chloroplasts (organelles with chlorophylls) vary in number and shape according to the species.

The anterior end of the cell is invaginated forming a flask-shaped **reservoir** (Fig. 1.3A). The narrow part of the reservoir is sometimes called the **cytopharnyx** and its opening the **cytostome** (cell mouth). Most euglenoid flagellates have two flagella (generally unequal in length) and a contractile vacuole near the reservoir. All species with an eyespot (**stigma**) have the photoreceptor or paraflagellar swelling near the base of the longer, emergent flagellum and opposite to the eyespot. Surrounding the cell is a **pellicle**, consisting of interlocking proteinaceous strips and microtubules beneath the cell membrane. Rapid changes of

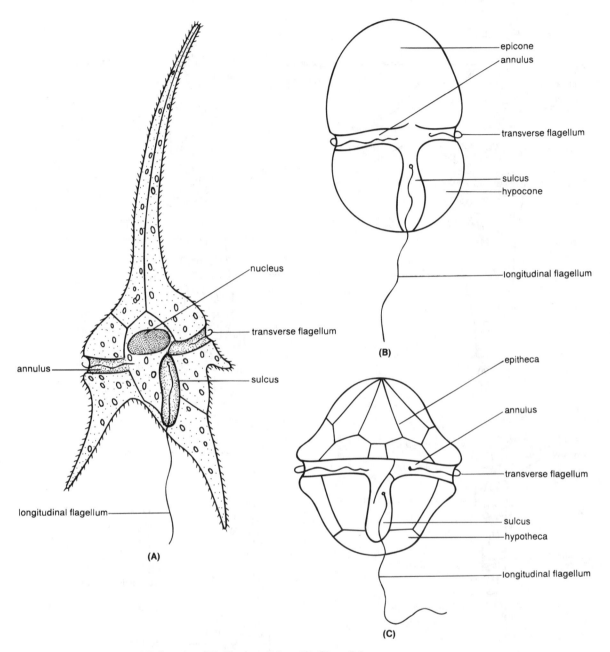

Figure 1.2. Dinoflagellates: (A) *Ceratium*. (B) *Gymmnodinium*. (C) *Glenodinium*.

body shape, called euglenoid movement, are seen in many euglenoids. A single nucleus with **endosome** (nucleolus) is present. Reproduction is primarily by longitudinal fission.

■ Observational Procedure: Order Euglenida

1. Add a drop of the *Euglena* culture in the center of a ring of methyl cellulose and cover with a coverslip.

2. Examine under low power of a compound microscope to locate the organism and then change to high power. Describe *Euglena's* movement and body shape. Does the cell rotate, creep, or glide? Do all individuals move in the same direction? Are any of the individuals showing euglenoid movement? Euglenoid movement in *E. viridis* is somewhat jerky. *Euglena gracilis* exhibits pronounced euglenoid movement, but not all *Euglena* species perform this movement.

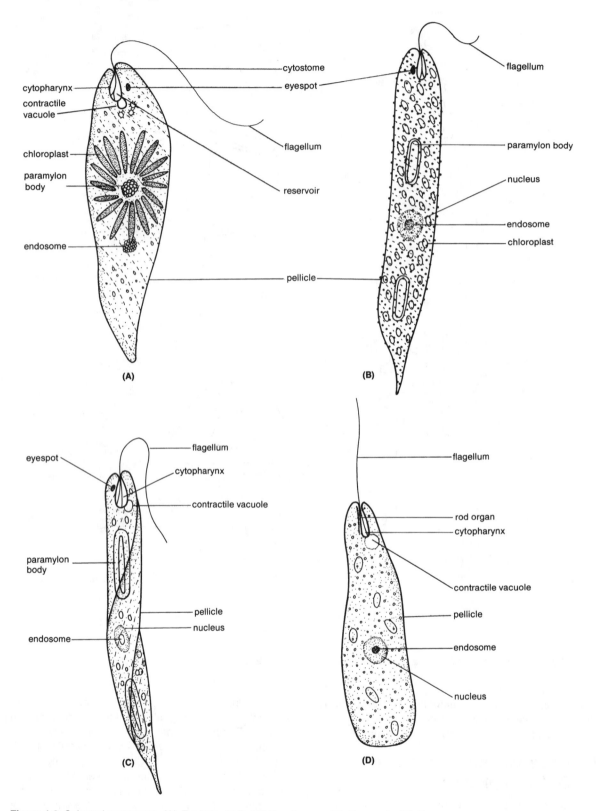

Figure 1.3. Selected protozoans. (A) *Euglena viridis*. (B) *E. spirogyra*. (C) *E. oxyuris*. (D) *Peranema*.

3. The outer covering of the organism is the pellicle. The forward end is generally less pointed than the trailing end. Carefully focus the forward end with the fine adjustment to see movement of the whiplike flagellum. You might want to make another wet mount and add a drop of Lugol's iodine solution before applying the coverslip. The solution will stain the flagellum, but will kill the organism. Near the reservoir is a contractile vacuole and a reddish eyespot (stigma). The eyespot is more easily seen in strains lacking pigment. The paraflagellar swelling at the base of the long flagellum will not be observed.

4. Much of the organism's cytoplasm contains chlorophyll-containing chloroplasts. Is your specimen greenish in color? The shape and position of the chloroplasts vary with the species. What shape are the chloroplasts in your species? In *E. viridis* the chloroplasts radiate from a common **paramylon body** (carbohydrate storage area). In the chloroplasts, there may be **pyrenolds**, proteinaceous bodies containing starch reserves. Although chlorophyll allows *Euglena* to manufacture its own food, this organism also has saprozoic nutrition; that is, it absorbs nutrients across the body surface in the absence of light. Holozoic nutrition, the ingesting of whole organisms, has never been demonstrated in *Euglena*; however, it does occur in *Peranema*, a relative of *Euglena*.

5. Attempt to locate the ovoid nucleus of the organism. This organelle is often obscured by chloroplasts and is best observed in colorless phytoflagellates. If a drop of acidified methyl-green stain is added to the culture, the nucleus will stain bright green. The nucleus can be observed in the prepared slide.

6. Examine a prepared slide of *Euglena* under high power of a compound microscope. Note the body shape. Locate the nucleus. Can you see the nucleolus (endosome) within the nucleus? Are chloroplasts evident? Carefully focus on the anterior end of a specimen. Attempt to see the flagellum, reservoir, eyespot, cytostome, and cytopharynx. These are more easily seen in the larger species of *Euglena*. *Euglena* reproduces by longitudinal binary fission. Are any individuals undergoing division?

Peranema. This freshwater, euglenoid phytoflagellate is similar to *Euglena*, but is colorless, lacks the eyespot, has a long trailing flagellum, is holozoic in nutrition, and

bears two rodlike structures (**rod organ**) in the reservoir for food-catching (Fig. 1.3).

■ Observational Procedure: *Peranema*

Mount a drop of the culture on a slide as before. Observe the shape and movement of *Peranema*. Can you see striations in the pellicle? *Peranema* undergoes pronounced euglenoid movement. Observe a prepared slide of *Peranema* and compare the organism with *Euglena*. ■

Order Volvocida. Most species occur in still freshwater ponds and ditches. The organisms usually have two flagella of equal length directed anteriorly, two contractile vacuoles, one large chloroplast, a red eyespot, and a centrally located nucleus (Figs. 1.4 and 1.5). The chloroplast is cup-shaped, usually grass green, and contains chlorophylls *a* and *b* and other pigments. Associated with the chloroplast is a single basal pyrenoid body with stored starch. Asexual and sexual means of reproduction are present.

Chlamydomonas. *Chlamydomonas* is a small (9–16 μm), solitary volvocid. Over 500 species, mostly freshwater, belong to this genus (Fig. 1.4).

■ Observational Procedure: *Chlamydomonas*

Prepare a wet mount of *Chlamydomonas* as before. Note the ellipsoid shape and bilateral symmetry. Examine a stained slide of *Chlamydomonas*. Not all structures in Fig. 1.4 will be observed. Near each flagellum is a contractile vacuole. Observe the single, large chloroplast and nucleus. Attempt to see the eyespot in the anterior half of the chloroplast. The single pyrenoid body probably will not be seen. ■

Volvox. *Volvox* is a green, spherical or ellipsoidal colony (**coenobium**) surrounded by a cellulose cell wall. The colony may reach 1.5 mm in diameter. The phytoflagellate is common in ponds, ditches, and pools. Colony formation in *Volvox* is similar to embryonic development in some metazoans.

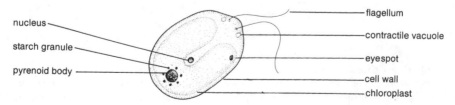

Figure 1.4. *Chlamydomonas*, a unicellular Volvocida.

The organism consists of many cells or **zooids** (500–60,000) embedded in a single surface layer in a gelatinous matrix (Fig. 1.5). The colony consists of mainly somatic (vegetative) and some reproductive zooids. Each somatic zooid is chlamydomonas-like with two flagella directed outward; the zooids are often attached laterally by protoplasmic connections. The few fertile zooids (**gonidia**) are somewhat larger than the somatic ones. Vegetative reproduction results in a hollow sphere, the **daughter colony**, formed by repeated mitotic division of a gonidium

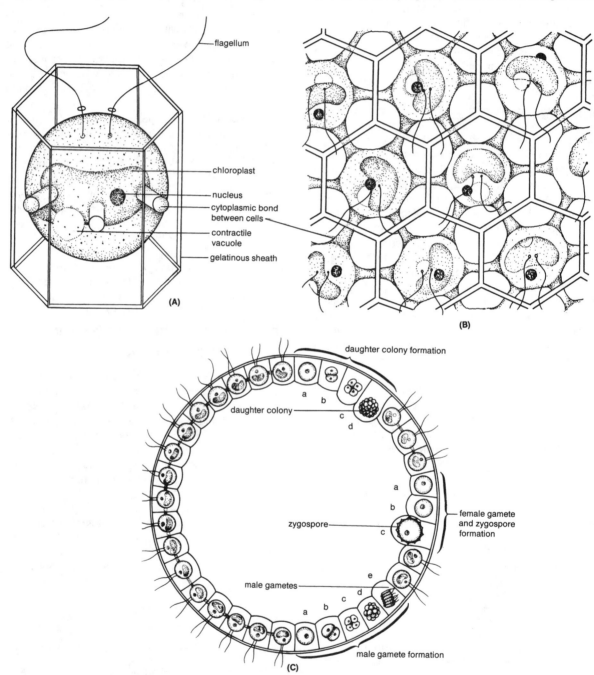

Figure 1.5. *Volvox.* (A) One zooid (cell). (B) Surface view of coenobium showing position of zooids in relation to cytoplasmic bonds. (C) Daughter colony formation, gametogenesis, and zygospore formation (diagrammatic). Lower case letters indicate sequence of development for three different processes: clockwise — daughter colony formation, female gamete and zygospore formation, and male gamete formation.

that has lost the flagella. When the parent dies, the daughter colony is released in the water where it forms a new colony. Colonies of *Volvox* are monoecious or dioecious. Sexual maturation involves formation of female and male gonidia. One egg (**macrogamete**) develops from the female gonidia. Those of the male result in sperm packets (**plakeas**) of 16 to 512 biflagellate **microgametes**. One sperm cell penetrates the colony and fertilizes the egg. A thick, protective wall surrounds the zygote, forming a **zygospore**. When the parent colony disintergrates, the zygospore is released. Inside the zygospore repeated division occurs and during the spring a new colony is formed.

■ Observational Procedure: *Volvox*

Prepare a wet mount of the *Volvox* culture using methyl cellulose. Raise the coverslip with clay or other supports. Study the organism under low power and then change to high power of a compound microscope. Describe movement of the colony. Observe the daughter colonies if present. How many are there? Do they move about inside the colony or remain stationary? Can you tell if the colony possesses sexual stages? The zooids are very small and details of structure will not be observed. Examine a prepared slide of *Volvox*. Observe the general anatomy of the colony and compare it with the living organism. ■

Class Zoomastigophorea

In contrast to the phytoflagellates, the animal-like zooflagellates belong to the class Zoomastigophorea [ZOO-mas-ti-GOF-or-e-ah; G., *zoon*, animal + G., *mastix*, whip + G., *phora*, to bear]. These organisms lack chromoplasts, eyespots, cellulose coverings, and other plantlike features. Four orders will be emphasized: order Trichomonadida (*Trichomonas*), order Diplomonadida (*Giardia*), order Hypermastigida (*Trichonympha*), and order Kinetoplastida (*Leishmania* and *Trypanosoma*).

Order Trichomonadida. These zooflagellates typically inhabit the digestive tracts of invertebrate and vertebrates (Fig. 1.6A). The motile cells reproduce only by binary fission. The key structural feature of trichomonadids is a mastigont system, a complex of flagellum and associated organelles (e.g., **kinetosome, undulating membrane, parabasal body** or Golgi apparatus, **axostyle**, and nucleus). Some organelles may be absent in certain species. The motile cell typically has four to six flagella; one that is directed backwards (recurrent) and is associated with the undulating membrane. The undulating membrane is a fold of the body surface's membrane coalesced with the flagellar membrane.

■ Observational Procedure: Order Trichomonadida

Examine living specimens or prepared slides of the human parasites, *Trichomonas vaginalis* or *T. tenax* and observe under oil immersion (Fig. 1.6A). *Trichomonas vaginalis*, the most common *Trichomonas* in humans, occurs in the vagina and male urethra and prostate gland; *T. tenax* occurs in the human mouth. Cysts are unknown and transmission occurs in the trophozoite stage.

Locate the four, anterior flagella and the single recurrent flagellum attached to the undulating membrane. An axial rod (axostyle) runs down the length of the parasite. Note the location of the nucleus, parabasal body, and cytostome. In well-stained slides a **rhizoplasts** (flagellar rootlet) may be seen extending from the kinetosomes. ■

Order Diplomonadida. These zooflagellates generally are found in the digestive system of their hosts. The best known species is *Giardia intestinalis* (= *lamblia*) from the human small intestine (Fig. 1.6B). Heavy infections cause severe diarrhea. Transmission is by cysts passed in the feces.

■ Observational Procedure: Order Diplomonadida

Examine a prepared slide of *Giardia intestinalis* under oil immersion of a compound microscope (Fig. 1.6B). The trophozoite is pear-shaped, with two parallel rods (axostyles) running longitudinally in the middle of the organism. Note the two large, ovoid nuclei with endosomes and eight flagella. *Giardia* has a large ventral **adhesive disc** for attachment to the host's intestinal mucosa. Are cysts present? They have distinct cell walls, usually four nuclei (two in young cysts), axostyles, and flagella.

Most students will not have an opportunity to observe live *Giardia*, but if you live in a region where this protozoan is endemic and if the local hospital is cooperative, you may be able to work with live samples. *Take great care to follow the instructions of your laboratory instructor.*

To observe the living organism, collect a very small amount of fresh feces on an applicator stick. Transfer the feces onto a glass slide. Add a drop of normal saline solution and Lugol's iodine solution to the sample. The iodine solution should be diluted 1:5 with distilled water. Mix well, spread out the sample, and cover with a coverslip. Locate the organism under low and then high magnification. Do not use oil immersion unless directed by the instructor. Compare the fresh trophozoite and cyst with those observed in the prepared slide. ■

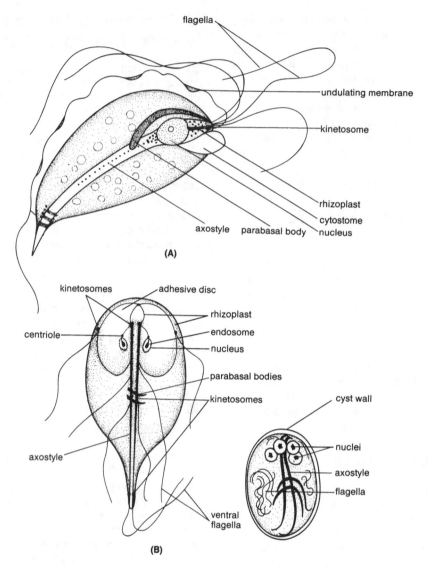

flagella

undulating membrane

kinetosome

rhizoplast

cytostome

nucleus

axostyle   parabasal body

(A)

kinetosomes   adhesive disc

centriole

rhizoplast

endosome

nucleus

parabasal bodies

kinetosomes

cyst wall

nuclei

axostyle

axostyle

flagella

ventral
flagella

(B)

Figure 1.6. Two parasitic protozoans. (A) *Trichomonas*. (B) *Giardia intestinalis*, trophozoite and cyst.

Order Hypermastigida. These zooflagellates live in the intestines of termites, cockroaches, and woodroaches. The motile cell has one nucleus, many flagella, and multiple parabasal bodies (Fig. 1.7A).

■ Observational Procedure: Order Hypermastigida

Obtain a slide of *Trichonympha*. Note the body shape, position of nucleus, and arrangement of flagella (Fig. 1.7A). Living specimens may be seen by examining the gut contents of a termite. Place the termite on a glass slide. Hold the anterior end of the termite's body with forceps. Pull the remaining body posteriorly with another pair of forceps until the gut contents emerge onto the slide. Spread the contents thinly, add a small drop of water, and examine under low and high powers with the compound microscope. There may be several species and genera represented. Note the shape of the body and arrangement of the flagella. *Calonympha* (order Trichomonadida) may be in the gut contents; it has an axostyle and several anterior nuclei which are lacking in *Trichonympha*. ■

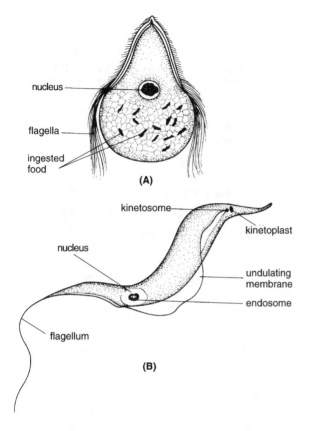

Figure 1.7. Two flagellated protozoans. (A) *Trichonympha* (After Cable). (B) Trypanosome.

Order Kinetoplastida. Parasites of blood and tissue parasites of vertebrates as well as free-living forms are in this order. Several species of *Trypanosoma* and *Leishmania* are dreaded human pathogens in tropical regions, causing **sleeping sickness, Chagas' disease**, and **Kala azar**. The motile cells have one or two flagella. A key feature of these organisms, for which the order is named, is the **kinetoplast** (Fig. 1.7B). This is a large stainable, mass of mitochondral DNA located near the base of the locomotory apparatus. The kinetoplast is usually larger than the kinetosome.

■ Observational Procedure: Order Kinetoplastida

Observe prepared slides of *Trypanosoma* and *Leishmania* under oil immersion. Note the body shape and single locomotory flagellum that is either free or attached to the undulating membrane, depending upon the species and the stage of the life cycle (Fig. 1.7B). Study the locational relationship of the nucleus, basal body, and kinetoplast in each genus.

Living *Trypanosoma lewisi* may be obtained from an infected rat by examining its blood. Put a drop of blood on a glass slide and quickly cover with a coverslip. Observe under low power of a compound microscope and search for movement. A moving trypanosome causes the red blood cells to move. Examine the parasite under high power. Do you see the undulating membrane moving? ■

Subphylum Opalinata

Members of subphylum Opalinata [O-pa-LIN-a-tah; N.F., *opaline*, like an opal in appearance] are found primarily in the intestines of amphibians; a few species occur in fish. Their flattened leaf-shaped bodies are covered with numerous cilia (Fig. 1.8). The fibrillar associates of the kinetosomes are not unlike those of ciliates. There are at least two monomorphic nuclei. A cytostome and cytopharynx are lacking; ingestion is by pincytosis and egestion is by exocytosis.

■ Observational Procedure: *Opalina*

Examine a prepared slide of *Opalina*. Note the body shape, body ciliation, and nuclei (Fig. 1.8). ■

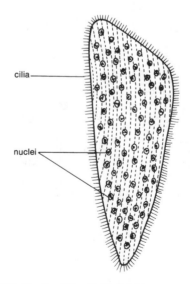

Figure 1.8. *Opalina* (After R. M. Cable).

Subphylum Sarcodina

The protozoans of subphylum Sarcodina [SAR-ko-DI-na; G., *sarkos*, flesh + G., *ina*, belonging to] use pseudopodia, protoplasmic extensions of the motile cell, as the primary means for locomotion and feeding. These organisms are commonly known as amebas. The body is naked (e.g., *Amoeba proteus*) or provided with a **test** or outer covering (e.g., *Difflugia*). Most species are free living. Reproduction by binary fission is common.

The subphylum is usually divided into two superclasses based primarily on the type of pseudopodia. Members of the superclass Rhizopodea [ri-ZOP-o-de-ah; G., *rhiza*, root + G., *podos*, foot] have pseudopodia called **lobopodia, filopodia**, and **reticulopodia** (granuloreticulopodium or rhizopoda), or by protoplasmic flow without production of pseudopodia (e.g., slime molds). Actinopodea [AK-tin-OP-o-de-ah; G., *aktinos*, ray + G., *podos*, foot] includes amebas with pseudopodia called **axopods**. Representatives from both superclasses will be studied.

Superclass Rhizopodea

This large group of amebas includes five classes differentiated primarily on the type of pseudopodia present. Included in the superclass Rhizopodea are slime molds (Class Mycetozoea), that have both animal and plant features. Slime molds will not be studied in this exercise.

Class Lobosea

The Lobosea [LO-bo-SE-ah; G., *lobos*, lobe] includes amebas that produce lobose to filiform pseudopods. Their bodies are naked or housed in a secreted proteinaceous covering (test). Habitats are varied: in fresh and salt waters, in moist soils, on plants, in manure, in sewage waters, and in guts of plants and animals. Some are parasitic (e.g., *Entamoeba histolytica*).

■ Observational Procedure: *Amoeba proteus* or *Chaos* (= *Pelomyxa*) *carolinense*

1. To prevent crushing the ameba, small pieces of a broken coverslip or bits of clay can be used to raise the coverslip used to cover the sample being examined. Place a drop of the ameba culture, taken from the bottom of the jar, on a glass slide and cover with a coverslip. Locate the organism under low power of a compound microscope and then change to high magnification.

2. Body size ranges from 100 µm to 500 µm, depending on whether the organism has assumed a round or extended shape. Describe the symmetry of your specimen. Observe movement of the organism. *Amoeba proteus* (Fig. 1.9) has large, conspicuous pseudopodia called lobopodia used in food-capture and locomotion.

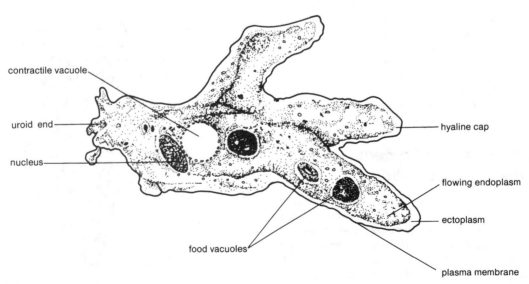

Figure 1.9. *Amoeba proteus.*

The body area opposite the advancing pseudopodia is the **uroid end**. Note the streaming of the cytoplasm within the body and as a pseudopodium is formed.

3. The cytoplasm consists of an outer, clear, firmer (**gel**), the ectoplasm, and an inner, granular, more liquid endoplasm (**sol**), the endoplasm, that contains the various organelles. These areas represent different colloidal states and have been used to explain in part amoeboid movement. Observe an extending pseudopodium. Note the large clear area of ectoplasm, the **hyaline cap**, at the end of the pseudopodium. This is the region of sol-to-gel conversion. Do other pseudopodia have the cap?

4. Do you see a large, clear spherical area in the endoplasm? This is the contractile vacuole, a water-regulatory organelle. The vacuole forms near the advancing end, but discharges its contents at the uroid end. At the time of expulsion, the vacuole is about 30 μm. Make an effort to see the vacuole discharge.

5. Scattered throughout the endoplasm are food vacuoles (5–20 μm) where intracellular digestion occurs. A food vacuole consists of food material and fluid surrounded by a membrane.

6. Examine a prepared slide showing whole mounts of the organism. Compare the shape and size of the stained amebas to that of the living organism. The nucleus should be evident. Are food vacuoles present? Can you identify any of the food items within the food vacuoles?

7. *Amoeba proteus* reproduces by binary fission. Observe prepared slides showing fission in *A. proteus*. Is the division equal?

8. *Chaos carolinense* is a large ameba (over 300 μm) that can be seen with the unaided eye, especially against a dark background. The ameba is found in swampy freshwaters where it feeds on protozoans and small invertebrates. Examine the living ameba first with the low power and then under high power of a compound microscope. Compare its size and shape to that of *A. proteus*. Describe the shape of the pseudopodia. Are hyaline caps present? Unlike *A. proteus*, *Chaos* has 1,000 or more nuclei and several contractile vacuoles that range in size from 30 μm to 40 μm.

9. Examine a prepared slide and observe the pseudopodia, ectoplasm, and endoplasm. Look for food vacuoles, contractile vacuoles, and nuclei. ■

Entamoeba histolytica. This widely distributed intestinal parasite of man, apes, monkeys, dogs, rats, cats, and other animals causes **amebic dysentery** or **amebiasis**. The parasite produces large abscesses in the intestinal lining. Sometimes the ameba causes death in its host when it penetrates the gut, enters the bloodstream, and invades vital organs such as the brain, liver, and spleen.

Other *Entamoeba s*
commensal in the
and *Entamoeba*
contamination
lacks cyst and
transmission a

■ Observation..
Entamoeba histolyti..

1. Examine prepared slides of *E. histolytica* containing trophozoites and cysts (Fig. 1.10). Locate the organism under high power and then under oil immersion of a compound microscope. The ameboid trophozoite is 15–40 μm long, depending on whether it is in the round or extended state. The ectoplasm is clear and the endoplasm is granular. There are usually four nuclei in the mature ameba. The structure of the nucleus is important in identifying *E. histolytica*. A nucleus stained with hematoxylin has fine, beadlike peripheral chromatin and a small (less than 1.0 μm) centrally placed endosome. In *E. coli* the trophozoite usually has eight nuclei, each with a large eccentric endosome and coarse peripheral chromatin. Both species lack contractile vacuoles. Food vacuoles are small (2–3 μm).

2. Examine the slide carefully for cysts. The mature cyst (10–20 μm) is round, smooth, and has four nuclei (3.5–6 μm). Eight amebas develop from a single cyst. In hematoxylin-stained specimens, dark-staining barlike rods may be seen. These are **chromatoid bodies** which represent an accumulated mass of RNA. The number varies from one to four. As the cyst ages, these bodies disappear. In contrast, *E. coli* has more chromatoid bodies that resemble a bundle of splinters. Reproduction is by binary fission.

3. To study the living organism, prepare a fecal sample as was done for the *Giardia*. *Be careful not to contaminate anything that may later come into contact with your hands or clothing; wash your hands thoroughly after completing this exercise.*

4. First locate the trophozoite or cyst with low power and then change to high power of a compound microscope. Do not use oil immersion unless directed by the instructor. Does the trophozoite move quickly? Describe the appearance of the nucleus. How does it differ from that observed in the stained slide? The chromatoid bodies appear as whitish refractile bodies instead of dark-stained structures observed in the prepared slide. Does your *E. histolytica* contain erythrocytes as a result from having recently fed on the intestinal mucosa? ■

Arcella vulgaris. Arcella is a beautiful testate ameba often found on aquatic plants (Fig. 1.11A, B). The rigid, chitinoid test is variable in shape and size depending

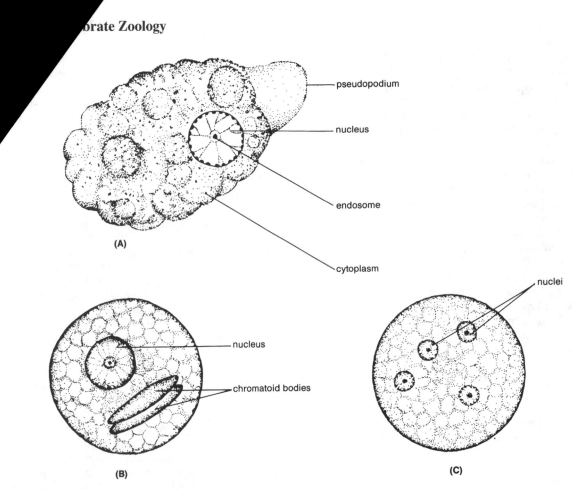

Figure 1.10. *Entamoeba histolytica*. (A) Trophozoite. (B) Immature cyst. (C) Mature cyst.

upon the species studied. The opening (**pseudostome**) into the test is on the ventral side and is rounded in shape.

■ Observational Procedure: *Arcella vulgaris*

Examine living specimens of *Arcella*. The organism is 100–150 µm in the round shape and 50–80 µm high. Describe the color of the test. *Arcella* is binucleate; each nucleus has an endosome (Fig. 1.11A, B). Can you see the lobose pseudopodia projecting through the pseudostome as the organism moves and feeds? Examine a prepared slide of *Arcella*. Locate the pseudopodia, nuclei, and pseudostome. ■

Difflugia. Difflugia is covered by a protective test (Fig. 1.11E). Members of the genus are beautiful organisms and are widely distributed on freshwater plants.

■ Observational Procedure: *Difflugia*

Prepare a sample from the culture as you did for *A. proteus*. Examine closely the ovoid test (Fig. 1.11E). Note its color and shape; attempt to determine of what materials the test is composed. Can you see through the test? Is the pseudostome circular? Locate the pseudopodia as they extend and retract. Is the ameba moving and feeding? Examine a prepared slide and compare it with the living organism. ■

Class Filosea

Most amebas in this class are herbivorous or bactivorous and have bodies protected in a test. The tests are often formed of siliceous scales with attached spines and other

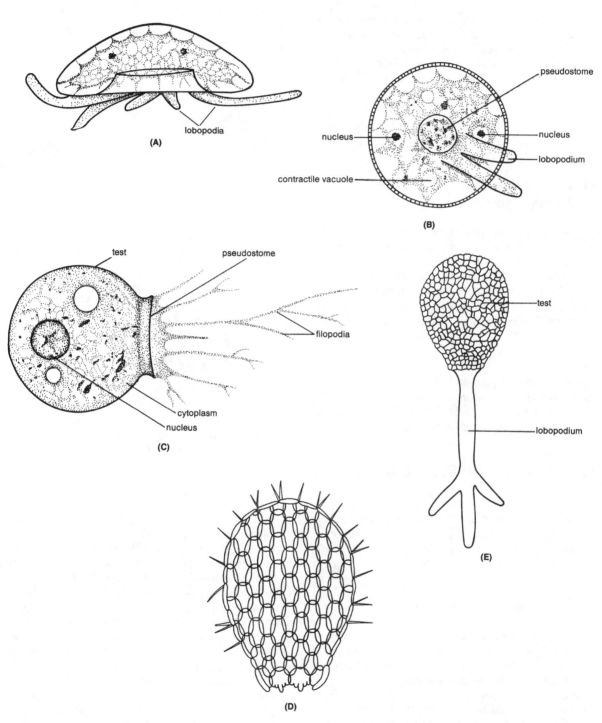

Figure 1.11. Testate ameba. (A) *Arcella*, side view. (B) Apical view (After Barnes). (C) *Lecythium*. (D) *Euglypha* test (After Barnes). (E) *Difflugia*.

debris. Scales may be lacking, but spines and other particles are attached to the test. Many species occur in freshwater, often with lobose amebas. A distinguishing feature of these amebas, from which the class name is derived, is the presence of a filopodium. This long pseudopodium is filamentous, generally without anastomoses and a central axial rod. It is used in food capture and locomotion.

Lecythium. This herbivorous ameba is found in freshwater. Individuals typically occur in clusters or groups.

■ Observational Procedure: *Lecythium*

Locate living specimens of *Lecythium* with the low power of a compound microscope and then change to high power. The organism is 30–40 μm in diameter and lives within a clear, ovoid test that has a round pseudostome (Fig. 1.11C). There is a single spherical nucleus. Note the clear filopodia, some of which are branched. Compare stained slide preparations with the living material. ■

Euglypha. Species in this large genus are often found living on freshwater plants. The test is covered with internally secreted scales that are transported to the outer surface of the ameba (Fig. 1.11D). The scales are in definite patterns and may or may not have spines.

■ Observational Procedure: *Euglypha*

Examine living *Euglypha* with a compound microscope. Note the shape and pattern of the scales forming the test (Fig. 1.11D). How would you describe their shapes? Observe the filopodia. ■

Class Granuloreticulosea

This large class typically contains amebas that have filamentous pseudopodia with branching anastomosing networks of protoplasm. This type of pseudopodium is a **reticulopodium** (granuloreticulopodium or rhizopodium). Most species have a single-chambered (**unilocular**) or multichambered (**multilocular**) test (Fig. 1.12). The first-formed chamber is the **proloculum**. Multichambered **foraminiferidans** are single individuals and not colonies. Perforate shells have many pores, whereas imperforate shells generally have one (e.g., the oral aperture) pore.

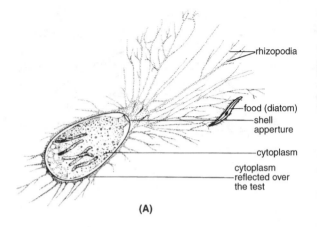

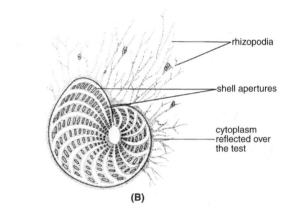

Figure 1.12. Two foraminiferidans. (A) *Allogromia* with unilocular shell (After Schulze). (B) *Elphidium* with multilocular shell (After Jahn).

Order Foraminiferidia. There are more described species of foraminiferidans (34,000) than any protozoan group. *Globigerina* is a predominate genus; much of the ocean's ooze consists of large deposits of the shells of *Globigerina* species. Foraminiferidans (forams) are ancient, primarily marine amebas. Forams are large in size (0.5 mm–1 cm), but larger ones (12 cm) are known from the fossil record. The composition, mostly of calcium carbonate, arrangement of chambers, and openings of the test vary. The test is covered by a portion of the cytoplasm (Fig. 1.12). In multilocular forams the cytoplasm and rhizopodia extend through surface pores as well as the oral aperture. In unilocular species, the pseudopodia extend from the oral opening of the shell. Multilocular species add chambers in definite patterns as they grow. The test is absent in a few species. The complex life cycle involves multiple division and alternation of haploid and diploid generations.

■ Observational Procedure: Foraminiferidans

Obtain prepared slides and/or unmounted specimens of several species of forams. Examine the organisms with the compound and dissection microscopes. Note the shapes of the specimens and how chamber formation has taken place. Are the surfaces of the tests adorned with ribs, keels, spines, or other structures? Are the tests of some of the forams perforated with holes? Is the perforate or imperforate test the more abundant? You might want to examine bulk specimens against a black background. Do any of these specimens coil? If so, determine the direction of coiling.

Return your microscope to low power and obtain a pair of polarizing filters. Notice how these filters stop all transmitted light when they are rotated 90° out of phase from one another. Orient the filters so that light pass through them and place one filter above the foram preparation (**analyzer**) and another on top of the light source of the microscope (**polarizer**). You will probably will have to refocus the microscope to bring the specimen into focus. While observing the forams, rotate the polarizer until it is out of phase with the analyzer. The background will be dark due to the exclusion of light by the crossed filters; however, the forams will appear to be bright. This phenomenon, called **birefringence**, occurs because the $CaCO_3$ of the test has two or more principal refractive indices (is highly ordered) and will rotate light themselves, thus permitting light to pass through the second filter. ■

Superclass Actinopodea

Four classes are recognized currently in the superclass Actinopodea [AK-tin-OP-o-de-ah; G., *aktinos* + G., *podos*, foot]. Members of the three marine classes (e.g., Acantharea, Polycystinea, and Phaeodorea) are commonly called **radiolarians**. The fourth class, Heliozoea, is primarily freshwater and commonly called the **sun animalcules**.

Members of all classes have a type of pseudopodium called an **axopodium** (actinopodium). This straight, slender pseudopodium possesses an axial rod (**axoneme**) composed of microtubules. Axopodia are radially arranged and used primarily in feeding rather than in locomotion. Most of these amebas have tests of silica.

Radiolarians

Radiolarians are among the oldest protozoans. All are marine and currently assigned to three classes. Most radiolarians are planktonic and pelagic. Their beautiful skeletons are composed primarily of silica and strontium

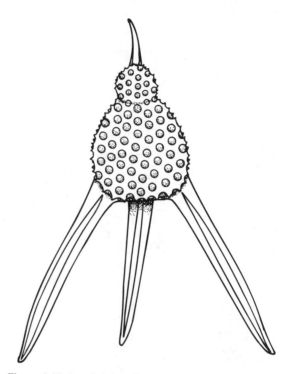

Figure 1.13. A radiolarian, *Podocytris.*

sulfate or calcium aluminum silicate, with radially arranged spines that extend outward from the center of the body (Figs. 1.13 and 1.14). A central capsule, a double or single pseudochitinous or mucinoid membrane, divides the body into an inner, granular endoplasm and an outer, frothy mass of ectoplasm (**calymma**). The central capsule has pores (**fusules**) that allow outward passage of the axial rods of the axopodia. One or more nuclei occur in the endoplasm. Axopodia arise from the capsule membrane and extend through the ectoplasm. They are sticky and capture prey that is carried by the streaming cytoplasm to the central capsule where digestion occurs within a food vacuole. The ectoplasm on one side of the axopodium moves toward its tip, while the other side moves toward the test. Binary fission, budding, and sporogenesis have been observed.

■ Observational Procedure: Radiolarians

Study prepared slides of radiolarian skeletons from several genera and species. Note the sizes and shapes of the skeletons as well as the varied ornamentations such as spines, thorns, and hooks. How do these organisms resemble the architectural structure known as a geodesic dome? How do they resemble the carbon form known as fullerene? Are the skeletal elements of any of the radiolarians specimens birefrigent? ■

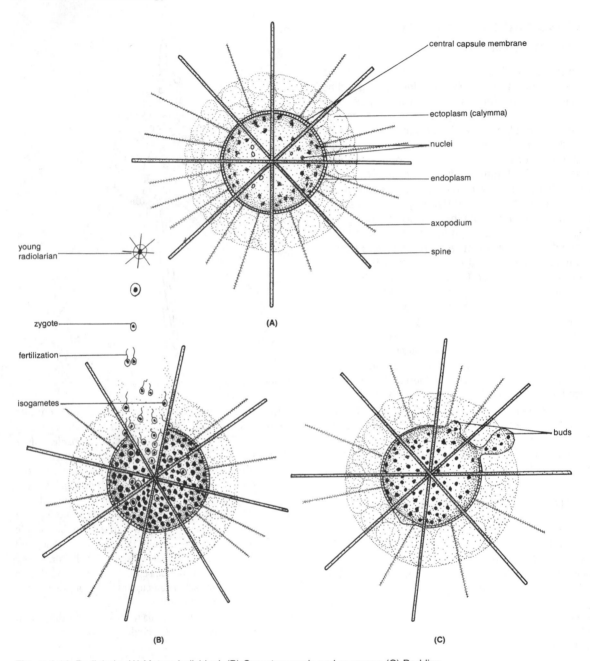

Figure 1.14. Radiolaria. (A) Mature individual. (B) Gametogenesis and syngamy. (C) Budding.

Class Heliozoea

Heliozoans [G., *helios*, sun + G., *zoon*, animal] are among the most beautiful protozoans. Most species are freshwater and generally benthic or attached to substrates such as rocks and plants. Heliozoans are spherical in shape and lack the central capsule of radiolarians, but have long, slender, granule-studded axopodia (Figs. 1.15 and 1.16).

■ Observational Procedure: Heliozoans

Examine living specimens of *Actinosphaerium* and *Actinophrys* in depression slides under low and high powers of the compound microscope (Figs. 1.15 and 1.16). Note their spherical shape, the lack of skeletal spicules, and the radiating axopodia. Observe the streaming ectoplasm as it moves over the axopods.

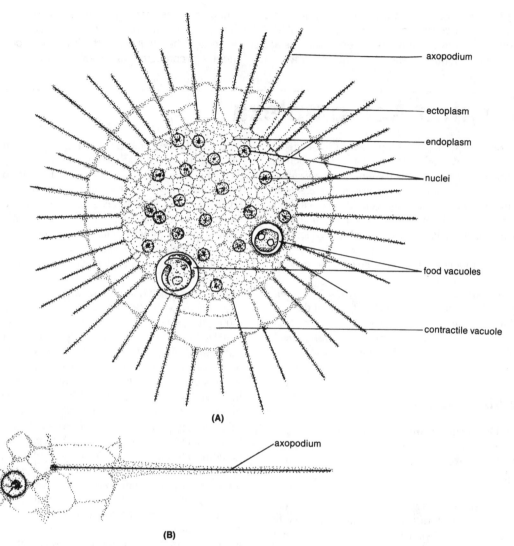

axopodium

ectoplasm

endoplasm

nuclei

food vacuoles

contractile vacuole

(A)

axopodium

(B)

Figure 1.15. *Actinosphaerium.* (A) Whole cell. (B) Enlarged axopodium.

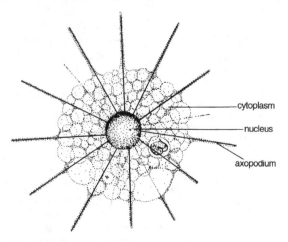

cytoplasm

nucleus

axopodium

Figure 1.16. *Actinophrys.*

Which of the two organisms is the largest? Note that the ectoplasm of *Actinosphaerium* is highly vacuolated with a frothy appearance, whereas the endoplasm is a dense, granular central mass. The multinucleated endoplasm has many axial rods originating from its outer portion. The endoplasm of *Actinophrys* is uninucleate and the axial rods attach to the nuclear membrane. Are contractile vacuoles present?

B. Phylum Apicomplexa

Members of the Apicomplexa [A-pi-com-PLEX-ah; (L., *apex*, tip + L., *complex*, twisted around] have a distinctive suite of structures called the **apical complex**. These structures, which are distinguishable only with the electron microscope, include a polar ring, conoid, micronemes, and rhopty. The somewhat simple-structured cell always has a vesicular nucleus. Cilia or flagella are lacking, except in flagellated microgametes in some groups. The phylum contains about 4,000 species and over 300 genera, all of which are parasitic.

Sporozoea [SPOR-o-ZO-e-ah; G., *sporos*, seed + *zoon*, animal] is the largest and most important class of the two present in Apicomplexa. The class contains the **gregarines** (Gregarinasina), **coccidians** (Coccidiasina), and **piroplasmids** (Piroplasmasina) as three separate subclasses.

Generalized Life Cycle

The life cycle is complex, usually involving multiple fission, gamete production, and one or more hosts (Fig. 1.17). All stages in the life cycle, except the 2n zygote, are haploid.

Gregarines. There are two types of gregarines. (1) **Aseptate** (acephaline) forms lack septa dividing the body into segments (Fig. 1.18). These gregarines are found especially in the coeloms of polychaetes, oligochaetes, sipunculids, nemerteans, and mollusks. (2) **Septate** (cephaline) gregarines have septa dividing the body (Fig. 1.19). The anteriormost segment, a modified holdfast organelle called an **epimerite**, usually breaks off when the trophozoite detaches from the host. The middle segment is the **protomerite** and the posterior **deutomerite** contains the nucleus. Septate gregarines are common in the guts of insects, millipedes, and crustaceans.

Monocystis: An Aseptate Gregarine. *Monocystis* is a large genus with about 72 species. The trophozoites (100–240 μm long) of most species live in the seminal vesicles of earthworms where they feed on developing earthworm sperm cells (Fig. 1.18). Tiny filaments, seen on the surface of the trophozoite, are the tails of disintegrated spermatozoa. The life cycle of *Monocystis* is similar to that of most gregarines (Fig. 1.18). The trophozoites pair off and unite together in a process called **syzygy** [siz-i-GE; G., *zygon*, yoke]. A cyst wall forms around the trophozoites. Each trophozoite, now a **gamont**, undergoes multiple fission producing isogametes that undergo syngamy with isogametes from the other gamont, so forming zygotes within a gametocyst. Each zygote is surrounded by its own wall and develops into a biconical **oocyst (sporocyst)**. Each oocyst (17–25 by 8–10 μm) forms eight **sporozoites** by meiosis and mitosis (sporogony). Oocysts are released when the earthworm dies and decays or perhaps via feces of the earthworm. Eight sporozoites are released in the next earthworm's gut after the oocyst is ingested. The sporozoites traverse the gut and enter the seminal vesicles where they mature into trophozoites.

■ Observational Procedure: *Monocystis*

Examine prepared slides (section and smear) of *Monocystis agilis* showing developmental stages in the seminal vesicle of earthworms. Identify as many stages as possible, using Fig. 1.18 as a guide. Is any one stage more abundant than the rest?

To obtain living *Monocystis*, put an earthworm in 7% alcohol for about 30 minutes. When the worm is fully anesthetized, take sharp-pointed scissors or

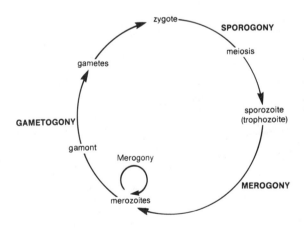

Figure 1.17. Generalized life cycle of Apicomplexa.

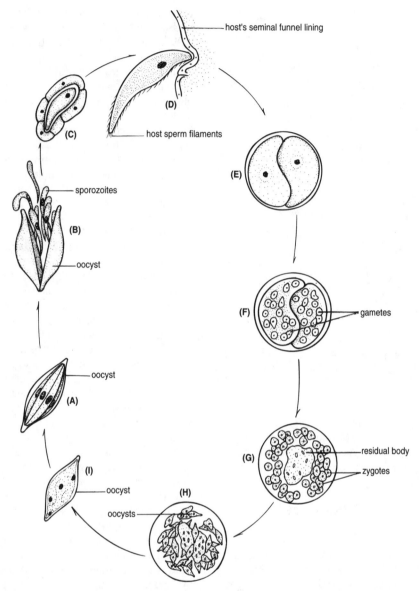

Figure 1.18. *Monocystis* life cycle. (A) Oocyst containing sporozoites. (B) Sporozoites released from oocyst. (C) Young trophozoite. (D) Mature trophozoite with sperm tails clustered about the outer surface. (E) Pair of encysted trophozoites. (F) Production of isogametes. (G) Zygotes within a gametocyst. (H) Oocysts developing. (I) One oocyst.

scalpel and make a shallow, dorsal cut in the body wall from segments 10 to 15. Extend two cuts each through both sides of the body wall. The cream-colored **seminal vesicles** should be evident. Excise a small part of the vesicle and place it on a clean slide. Add a drop of water to the piece of vesicle and tease apart with dissection needles to make a very thin smear. Cover with a coverslip and examine under low and high magnifications of the compound microscope. Do the stages differ from those observed in the prepared slides? ■

Gregarina: A Septate Gregarine. Gregarina is the largest genus of septate gregarines. The mature trophozoite (350–500 by 185–200 μm) parasitizes the guts of insects (Fig. 1.19). With maturity, the epimerite detaches from the body and two trophozoites unite a head-to-tail in syzygy. A wall forms about the two, but each trophozoite maintains its individuality. Each trophozoite develops into a spherical gamont within its own cyst wall. Multiple fission occurs inside each gametocyst, resulting in many gametes; the gametes are released with breakdown of the gametocyst wall. Syngamy results in

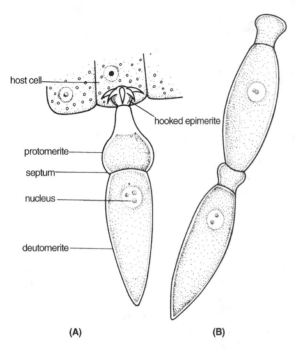

host cell

hooked epimerite

protomerite

septum

nucleus

deutomerite

(A)   (B)

Figure 1.19. Cephaline gregarine. (A) Attached to host. (B) Two gregarines in syzygy. Note the lack of the epimerite.

zygotes that develop into oocysts. Infection to a new host is by ingestion of oocysts. Meiosis inside the oocyst results in eight sporozoites. The sporozoites emerge in the host after the oocyst is ingested and soon develop into trophozoites.

■ Observational Procedure: *Gregarina*

Obtain a live insect such as a grasshopper or cockroach and cut off the head. Open the body with sharp-pointed scissors or scalpel. Locate the tube-shaped intestine in the ventral area of the body and remove carefully with forceps. Place the intestine on a glass slide, tease it open, and add a drop of water. Cover the gut contents with a coverslip and examine with low power of a compound microscope. Change to high magnification once the organism has been located. Locate the immature trophozoites. Can you distinguish the epimerite, protomerite, and deutomerite? Based upon your study of *Monocystis*, can you identify other stages of the life cycle of *Gregarina*? Study prepared slides and compare with the living organism. ■

Malarial Coccidians. The coccidians are perhaps the most important infectious disease-causing parasites of humans. The suborder Haemospororina (subclass Coccidiasina) contains the malaria-causing parasites (*Plasmodium*) and allies (e.g., *Haemoproteus*). Over 3.5 million humans die annually from malaria; some biologists believe that malaria is responsible for over one-half of all human deaths worldwide. The Haemospororina contains blood parasites of vertebrates, transmitted by a blood-sucking dipteran such as a mosquito, blackfly, or midge. The vector of *Plasmodium* in humans is the female *Anopheles* mosquito. Four species of *Plasmodium* cause malaria in humans (*P. falciparum, P. malariae, P. ovale*, and *P. vivax*) and these species differ in details related to frequency of multiple fission, morphology of schizonts (merozoites), and gametocytes (gamonts). A brief study of *P. vivax* will be made (Fig. 1.20).

Plasmodium vivax. Of the four malarial diseases in humans, the type caused by *Plasmodium vivax* is the most common and widespread, but not the most severe. When an infected female *Anopheles* mosquito takes a blood meal (only females bite), the spindle-shaped sporozoites leave her salivary glands and enter a victim's bloodstream. Within a few minutes, the sporozoites enter primarily liver cells. Here they grow and multiply by several schizogonous divisions, producing **merozoites**. These merozoites are sometimes called metacryptozoites; however, they are morphologically similar to other merozoites. The merozoites enter the bloodstream, invade the erythrocytes, and transform into **trophozoites**. A young trophozoite inside an erythrocyte resembles a signet ring: the single nucleus is ruby red and is perched on one side of a blue ring of cytoplasm. This is often called the **ring stage**. Inside the erythrocytes, the ring-shaped trophozoites grow and undergo several nuclear divisions. The erythrocytes increase in size to accommodate the growing trophozoites. The cytoplasm of an infected erythrocyte contains fine spots called **Schuffner's dots**. In older trophozoites a brownish iron-containing residue from metabolized hemoglobin may be observed. The multinucleated trophozoite is an immature **schizont** (merozoite) distinguished by having many nuclei throughout a continuous cytoplasm. With maturity, the cytoplasm of the schizont divides into distinct smaller bodies each with a nucleus. This mature schizont is called a **segmenter**. When the erythrocyte ruptures the small nucleated merozoites are released into the bloodstream. Chills and fevers, so characteristic of malaria, occur when toxic wastes are released in the bloodstream with rupture of erythrocytes. The released merozoites invade other erythrocytes. Some merozoites develop into male and female **gametocytes** (gamonts). The male **microgamete** (microgamont) has a pale cytoplasm and a large nucleus with irregular granules. The female **macrogamete** (macrogamont) has a dense cytoplasm, a small compact nucleus, and a dark red nucleolus. A mature gametocyte nearly fills the erythrocyte. The gametocytes circulate in the blood and further

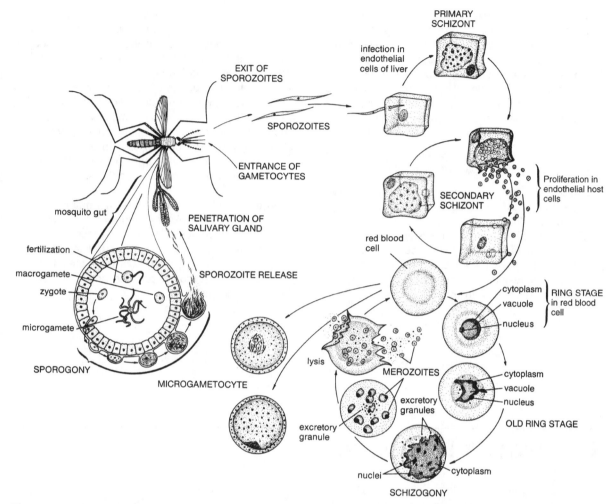

Figure 1.20. Life cycle of *Plasmodium*. (A) Infective stage (sporozoites) enter human, invade liver cells, undergo multiple fission (merogony), and become merozoites. (B) Some merozoites instead of reinvading liver cells, enter erythrocytes. These merozoites increase in numbers by multiple fission. (C) Some merozoites become microgametocytes (potential male gametes) and macrogametocytes (potential female gametes). (D) Gametocytes enter mosquito with a blood meal and become gametes. (E) Fertilization of macrogamete and microgamete results in a motile zygote (ookinete). The ookinete encysts on the mosquito's stomach wall and becomes an oocyst. Inside the oocyst the sporozoites develop. (F) With rupture of the oocyst, the released sporozoites enter the salivary glands of the mosquito and remain there until a blood meal is taken.

development ceases until a mosquito takes a blood meal which includes the gametocytes.

Upon entering the mosquito's stomach, each gametocyte undergoes gametogenesis. One microgametocyte undergoes three nuclear divisions to produce eight flagellated microgametes, the process of **exflagellation**. In the stomach of the mosquito, each microgamete fertilizes a female macrogamete. The resulting motile zygote is called an **ookinete** that lodges on the wall of the mosquito's stomach. Here the ookinete develops into an oocyst. Growth of the oocyst occurs and inside it, numerous **sporoblasts** (future sporozoites) form by meiosis and multiple fission. With rupture of a mature oocyst, 10,000

or more sporozoites are released and migrate to and enter the salivary gland. When the mosquito takes a blood meal, the infective sporozoites enter the host and the cycle repeats.

■ Observational Procedure:
Plasmodium vivax

Microscope slides with sections of infected human liver, human blood smears, and preparations showing mosquito salivary glands and stomach are necessary to study developmental stages in the life cycle of

Plasmodium vivax. Some of the stages may be set up as special demonstrations. Additional preparations representing other *Plasmodium* species may be provided. Do not expect to find all stages on one slide preparation. Examine the slides under high power of the compound microscope and then under oil immersion. Extreme care must be exercised when using oil immersion to prevent breaking the slide and causing damage to the lens. Refer to Fig. 1.20 as you make your observations.

Liver Section. Find the large schizonts among the liver cells. Note the size of the parasite with that of a liver cell.

Blood Smear. Find a ring stage of a young trophozoite.
1. Note the single ruby red nucleus perched on a bluish ring of cytoplasm.
2. Find trophozoites in various stages of growth. Observe Schuffner's dots in the cell's cytoplasm. Distinguish the parasite's nuclei from the brownish, iron-containing residue from metabolized hemoglobin.
3. Locate a schizont with many nuclei in a uniform

cytoplasm and segmenters with merozoites. The merozoites may not be as distinct as they appear in drawings. Attempt to locate the free spindle-shaped merozoites in the blood. These minute bodies are often difficult to find in most smears.
4. Look for microgametocytes and macrogametocytes. You may experience difficulty differentiating between the two. The microgametocyte has a large nucleus with irregularly distributed granules and a pale blue cytoplasm. The blue cytoplasm of the macrogametocyte appears dense and the smaller compact nucleus may show a dark red nucleolus.

Mosquito Sections. Observe a preparation showing exflagellation of the microgametocyte.
1. How many tail-like structures are present?
2. Examine a stomach preparation of the mosquito showing oocysts. How many oocyst are present?
3. Observe a preparation showing sporozoites. Describe their shape. ■

C. Phylum Myxozoa

These are obligate, extracellular parasites in body cavities, in gills, and various tissues of poikilothermic vertebrates (especially fish) and annelids. Entire fish populations may be destroyed by the parasite. Unlike the Apicomplexa and microsporans (Phylum Microspora), a spore with one to two ameboid **sporoplasms** and one to six (usually two) **polar capsules** and **coiled filaments** are present (Fig. 1.21). The polar filaments are holdfast structures. Each uninucleate ameboid sporoplasm enters the host when a spore is ingested and grows into a multinucleate trophozoite. The trophozoite feeds on the host's tissues and undergoes multiple fission (schizogony), thereby increasing the infection. Eventually sporogony occurs and one or more spores are produced.

Observational Procedure:
Myxosoma

Examine prepared slides of *Myxosoma* or other myxozoans. Specimens can often be found in the gills of certain fish. Note the shape and size of the parasite

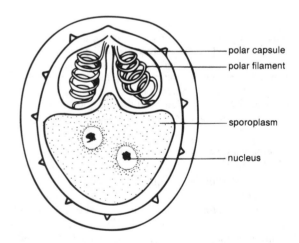

Figure 1.21. *Myxosoma.*

(Fig. 1.21). If spores are available, locate the polar filaments, polar capsules, sporoplasms, and nuclei.

D. Phylum Ciliophora

The approximate 8,000 species of Ciliophora [SIL-e-OF-or-ah; G., *cilium*, eyelash + G., *phora*, to bear] occur in all major ecological habitats. Most are free living and phagotrophic. Ciliates are the most structurally complex group of protozoans. Unique features include cilia, **dikaryotic nuclei**, complex cortex and associated organelles, and modes of reproduction. Most ciliates have a cytostome and cytopharynx for ingestion of food, at least one fixed contractile vacuole to regulate cytoplasmic water content and ion concentration, and a **cytoproct** (**cytopyge**) for egestion of digested food from food vacuoles. A variety of organelles collectively called **extrusomes** release materials from the cell. These include **mucocysts** that coat the body and help form cysts, **toxicysts** that eject toxins to capture prey, **haptocysts** for prey-capture, and **trichocysts** that are nontoxic explosive organelles whose exact function is unknown.

The dikaryotic nuclei distinguish ciliates from other protozoans. The small, compact **micronucleus** is primarily involved in genetic and sexual recombination, undergoing mitosis and meiosis. The larger, polyploid **macronucleus** regulates the metabolic processes of the cell. The macronucleus develops from the micronucleus, but can usually replicate once formed.

Certain ciliates have **membranelles** and **cirri**, specialized ciliary organelles, used in locomotion and food getting (Fig. 1.22). Membranelles, composed of two or three rows of short cilia, appear as a thin membrane, whereas cirri are ciliary tufts that function as a unit. Cilia usually cover most of the cell. The outer portion of the ciliate body is called the **cortex**. This includes the outer plasma membrane and the membrane-lined alveoli (together called the **pellicle**) and the **infraciliature** (Fig. 1.23). The latter is a complex network consisting of **kinetosomes** and associated microtubules and fine, striated kinetodesmal fibrils. The kinetosomes plus associated fibrils form kinetids, which may be arranged in a row called a **kinety**. Kinetids may have one (monokinetid), two (dikinetid), or more (polykinetid) kinetosomes. Kinetid distribution and specializations over the ciliate body are used in taxonomy.

Reproduction is primarily by transverse binary fission; the plane of division is across the long axis of the body (homothetogenic fission). This is unlike symmetrogenic fission of flagellates, where the longitudinal furrow forms between rows of kinetosomes, if present. Budding occurs in some ciliates. Free gametes are not formed during the sexual process rather micronuclei are exchanged between two sexually compatible individuals in a process called conjugation.

The classification scheme of Small and Lynn (in Lee et al. 1985) published in *An Illustrated Guide to the Protozoa* by the Society of Protozoologists (1985)

divides the phylum Ciliophora into eight classes: Karyorelictea, Spirotrichea, Prostomatea, Litostomatea, Phyllopharyngea, Nassophorea, Oligohymenophorea, and Colpodea. Representatives of six classes will be studied in this exercise.

Subclass Heterotrichia (Class Spirotrichea)

These usually large, contractile species have conspicuous oral ciliature, well-defined myonemes, and body cilia uniformly distributed. The body shape is variable.

Stentor coeruleus. *Stentor coeruleus* is a large (1–2 mm long), freshwater ciliate with a conspicuous adoral zone of membranelles around the ciliated peristome. The body is vase or trumpet in shape.

■ Observational Procedure: *Stentor coeruleus*

Place a drop of the *Stentor* culture in a ring of methyl cellulose and cover with a coverslip. Observe with both low and high magnifications of the compound microscope. Note the size and shape of this protozoan (Fig. 1.22A). The bluish color is derived from a cytoplasmic pigment called **stentorin**. Which end moves forward in a swimming specimen? Locate an individual that has become attached. Observe the active-moving adoral membranelles around the peristome. In what direction do these fused rows of cilia beat? *Stentor* is a suspension feeder. Can you see the area where the membranelles lead to a spiral-shaped buccal cavity (oral cavity)? Your specimen should extend and contract its body. Longitudinal contractile filaments called **myonemes** shorten the body; microtubular arrays lengthen the body by sliding. Myonemes lie in the pellicle and may be observed in stained preparations. Locate the contractile vacuole with its long collecting canal. Is its position constant or variable? Attempt to see the moniliform-shaped (beadlike) macronucleus. The macronucleus and other structures can be seen in the prepared slide. ■

Spirostomum ambiguum. This freshwater ciliate is slender, somewhat flattened, and very elongated. It may reach a length of 3.5 mm. Body ciliation is short and arranged in longitudinal rows.

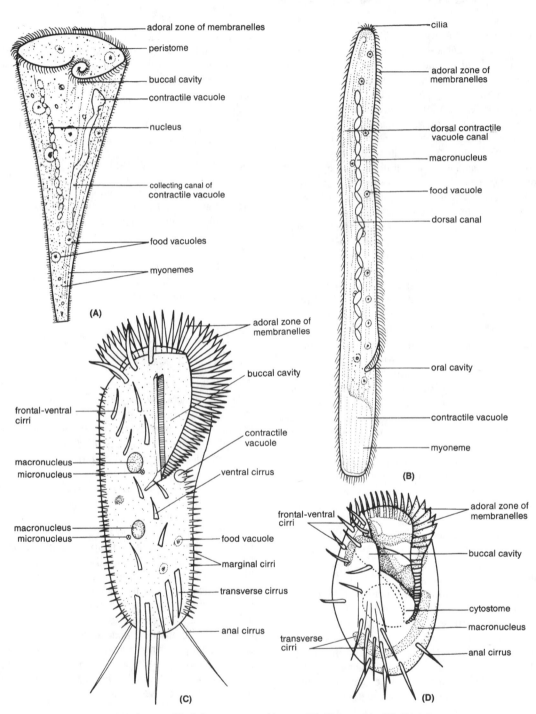

Figure 1.22. Selected ciliates. (A) *Stentor*. (B) *Spirostomum ambiguum*. (C) *Stylonychia*. (D) *Euplotes*.

■ Observational Procedure: *Spirostomum ambiguum*

Examine living specimens under low and high magnifications of a compound microscope and note their size and shape (Fig. 1.22B). Observe beating of the cilia. Do cilia cover the entire body? Note that the posterior end is more blunt than the anterior end. Which end usually moves forward? Like *Stentor, Spirostomum* contracts its body with myonemes. Attempt to see at the posterior end the large contractile vacuole with a long dorsal canal. The peristome, bordered by short adoral membranelles, occupies an area about two-thirds of the body length. Study a prepared slide and observe the dark-stained macronucleus. Describe its shape. Is the micronucleus evident? ■

Subclass Stichotrichia (Class Spirotrichea)

The body shape of these ciliates is usually ovate and dorsoventrally flattened. Oral membranelles for food capture and cirri for locomotion are characteristic of the group.

Stylonychia. *Stylonychia* is a common freshwater genus, but does occur in marine waters.

■ Observational Procedure: *Stylonychia*

Examine living specimens under low and high magnifications of a compound microscope. Describe its size and shape. Observe how the specimen darts about. Locate the three anal (caudal) cirri, eight frontal-ventral cirri, five ventral cirri, five transverse cirri, and several marginal cirri on the ventral surface of the body (Fig. 1.22C). These fused tufts of cilia serve as leglike structures. Can you see membranelles around the wide buccal (oral) cavity? Attempt to see the contractile vacuole near the membranelles. *Stylonychia* has two macronuclei and two micronuclei. Observe these along with the structures described above, in the prepared slide.

Examine specimens of *Kerona* and *Oxytricha* of the same subclass. Note similarities and differences between these and *Stylonychia*. Some species of *Kerona* are ectocommensal on the cnidarian hydra. *Kerona's* ventral cirri are arranged in five oblique rows and caudal cirri are lacking. ■

Subclass Hypotrichia (Class Spirotrichea)

These ciliates have ovoid bodies more or less dorsoventrally flattented. Somatic or body ciliation is reduced, usually as cirri; oral ciliature is conspicuous.

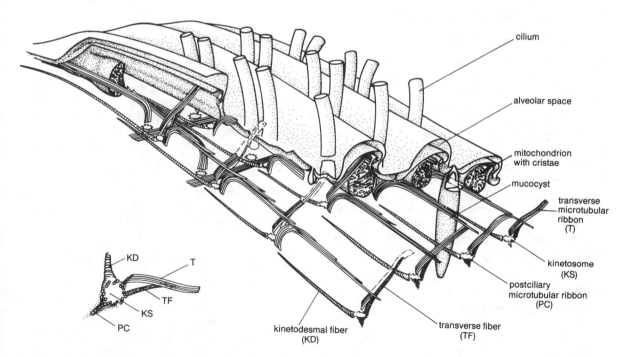

Figure 1.23. Structure of cortex and associated infraciliature of a ciliate (After Peck 1977).

Euplotes. *Euplotes* (100–195 µm) occur in fresh and brackish waters. Symbiotic zoochlorellae are sometimes present in their bodies.

■ Observational Procedure: *Euplotes*

Observe living *Euplotes*; note its size, body shape, and way of movement (Fig. 1.22D). Does the organism creep, crawl, or dart about? Observe the 9–10 frontal-ventral cirri, the five transverse cirri, the four anal (caudal) cirri, and adoral zone of membranelles. Are these involved in locomotion? Examine a stained slide and locate the large C-shaped macronucleus as well as the membranelles and cirri. The buccal cavity is broadly triangular in shape. Is a posterior contractile vacuole present? ■

Subclass Haptoria (Class Litostomatea)

These ciliates usually have uniformly ciliated bodies. The cytopharynx inverts on ingestion and the **proboscis** or oral dome, when present, is supported by microtubules.

Didinium nasutum. This species (125 µm long) is a freshwater, raptorial ciliate feeding almost exclusively on *Paramecium*. Several didiniums may attack and kill one paramecium with discharged toxic trichocysts.

■ Observational Procedure: *Didinium nasutum*

Place together on a depression slide a small drop each from cultures of *Didinium* and *Paramecium*. Cover with a coverslip and observe under low and high powers of the compound microscope. Note the barrel-shaped body with the posterior end round and the cone-shaped anterior end (Fig. 1.24A). The cytostome, located at the tip of the proboscis, can be greatly expanded to swallow a paramecium. Describe the feeding process if *Didinium* is seen taking a paramecium. Is a contractile vacuole evident? Note the girdle of cilia around the middle of the body and another circle around the anterior end. These cilia can be seen in a prepared slide along with the macronucleus and contractile vacuole. Describe the shape of the macronucleus. ■

Subclass Trichostomatia (Class Litostomatea)

Most members of this subclass are endosymbionts in vertebrates. The oral region forms a densely ciliated depression called a **vestibulum**.

Balantidium coli. This medium-sized ciliate (50–100 µm long) is a common commensal in the cecum and large intestine of pigs. In humans, *Balantidium coli* is a pathogenic parasite that invades the intestinal mucosa.

■ Observational Procedure: *Balantidium coli*

Obtain a prepared slide of *Balantidium coli* and locate the ovoid trophozoite under low and then high power (Fig. 1.24B). The anterior end is more pointed than the posterior end. Body cilia occur in oblique, longitudinal rows. Attempt to locate the cytopyge (cell anus) at the posterior end and the cytostome and long vestibulum in an apicoventral position near the anterior end. Note the large, sausage-shaped macronucleus and two contractile vacuoles, one at the posterior and one at the anterior end. The vesicular micronucleus may not be evident.

Balantidium coli is transmitted via contaminated food or water containing the ovoid cysts (40–60 µm in diameter) passed in the host's feces. Cysts often stain poorly and the large macronucleus and one contractile vacuole are the most conspicuous organelles observed. Look for cysts in a prepared slide. Note the size, shape, and structure of the cyst wall. Locate the large macronucleus and contractile vacuole. The micronucleus will probably not be observed. Are food vacuoles evident? ■

Subclass Suctoria (Class Phyllopharyngea)

Suctorians are carnivorous ciliates found in both marine and freshwater habitats. Suctorians get their name from the way they obtain their food. Mature suctorians are usually sessile, stalked, lack cilia, and have one or more tentacles. Immatures are ciliated and lack the stalk and tentacles.

Ephelota. *Ephelota* (250 by 220 µm, with a stalk of up to 1.5 mm) is a marine suctorian and is often found attached to the marine hydroid *Obelia*.

■ Observational Procedure: *Ephelota*

Examine living specimens or prepared slides with a compound microscope. Note the short, thick, flat-tipped tentacles and their arrangement (Fig. 1.24D). The stalk enlarges distally. Do the tentacles and stalk extend or contract in the living organism? Do you see myonemes

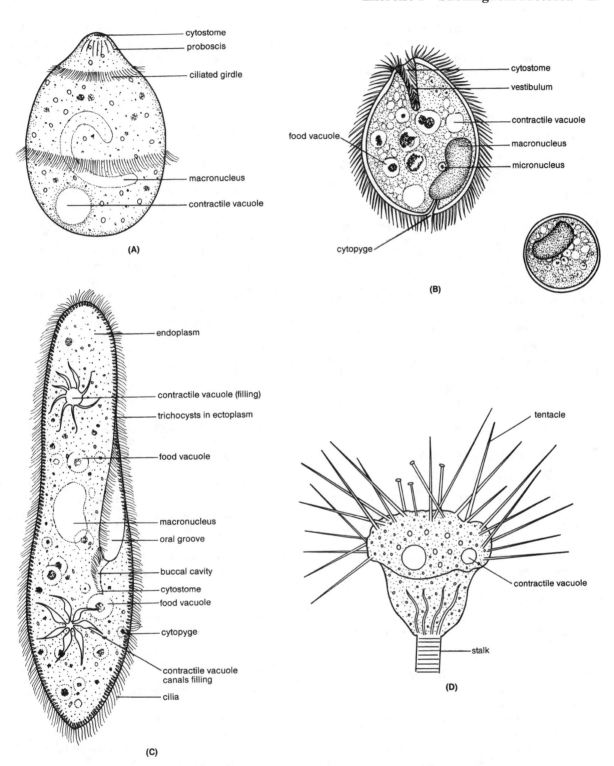

Figure 1.24. Ciliates and suctoria. (A) *Didinium nasutum.* (B) *Balantidium coli*, trophozoite and cyst (After Kudo). (C) *Paramecium caudatum.* (D) *Ephelota.*

LIVERPOOL JOHN MOORES UNIVERSITY
LEARNING SERVICES

or striations in the stalk? Suctorians have **haptocysts** at the tip of the tentacles used in feeding. Can you see a contractile vacuole? Examine a prepared slide and locate the curved, elongated macronucleus. ■

Subclass Nassophoria (Class Nassophorea)

These ciliates have an oral cavity that is either shallow or deep. The body ciliation is typically uniform and trichocysts are present.

Paramecium caudatum. This is the most widely distributed species of paramecium. It is 150–300 μm long, has a single large macronucleus, and has one compact micronucleus (Fig. 1.24C). The species occurs in freshwater streams and ponds containing plants and decaying organic matter. *Paramecium caudatum* is holozoic, living on bacteria, algae, and small animals.

   Paramecium aurelia with two micronuclei and *P. multimicronucleatum* with four or more micronuclei and three to seven contractile vacuoles are similar species to *P. caudatum. Paramecium busaria*, unlike the above-mentioned species, has unicellular **zoochlorellae** as symbionts in its body.

■ Observational Procedure: *Paramecium caudatum*

1. Place a drop of the paramecium culture on a glass slide containing a ring of methyl cellulose. Cover with a coverslip and locate an individual under low power of the compound microscope.
2. Observe locomotion in the paramecium. Note the organism's forward spiral movement as it slowly rotates on its long axis. Does the organism change its direction of movement by reversing the beat of its cilia? Can the organism bend its body?
3. Locate an individual that has ceased moving rapidly and observe under high power. The anterior end is narrower and more blunt than the broad, pointed posterior end (Fig. 1.24C). Can you see beating of the cilia?
4. Locate the oblique depression on the right side of the body. This is the **oral groove**; it gives the organism an asymmetrical shape.
5. The oral groove leads to a buccal cavity (oral cavity).
6. Note the longitudinally arranged cilia covering the entire body and lining the oral groove. The buccal cavity has on its inner surface numerous cilia arranged in two major bands. These cannot be observed with the light microscope. Food trapped in the oral groove trav-

els through the buccal cavity, then through the cytostome, where a food vacuole is formed. Formation and circulation of food vacuoles can be observed by dusting a few grains of powered carmine on the under surface of a coverslip before it is placed over the culture sample.
7. The outermost membrane layers of the body are called the pellicle. If a fresh drop of culture is allowed to dry on the slide uncovered, a hexagonal pattern on the pellicular surface can be seen with the compound microscope.
8. In the ectoplasm are numerous rodlike trichocysts that alternate with the bases of the cilia. Each trichocyst is discharged through a pellicular pore and is not reused once discharged. Trichocysts are quickly regenerated. An electron micrograph of a discharged trichocyst resembles a golfer's tee attached to a long, striated shaft. If a small drop of acetic acid, methylene blue, or Dahlia stain is added to a paramecium sample, the organisms should discharge numerous trichocysts.
9. Behind the posterior end of the oral groove is a small permanent opening through which solid wastes are voided. This is the cytoproct (cytopyge); it is difficult to observe in the living organism.
10. A contractile vacuole with radiating canals occurs at each end of the paramecium. These are in fixed positions and open to the outside via a pellicular pore. If another preparation is made and some of the water removed by placing a blotter or piece of filter paper near the edge of the coverslip, the vacuoles can be observed.
11. The macronucleus and micronucleus are located near the center of the paramecium. Observed these in a prepared stained slide since they are difficult to see in the living organism. The larger macronucleus is kidney-shaped and the smaller micronucleus lies in a depression of the macronucleus.
12. Paramecium reproduces asexually by transverse binary fission (across the kineties) while sex occurs by conjugation. Examine prepared slides showing fission and conjugation in *Paramecium.* ■

Subclass Peritrichia (Class Oligohymenophorea)

The oral apparatus, when present, is distinct. The oral structures are located in a ventral oral cavity or deeper infundibulum. The oral region can be contracted. Sessile forms are usually stalked.

Vorticella. *Vorticella* is a solitary, but often gregarious ciliate with a long stalk (up to 4.2 mm long). The stalk can retract and extend because of a large fiber called a myoneme (**spasmoneme**). Members of the genus occur in both fresh and marine waters. *Vorticella* is a suspension feeder on bacteria.

■ Observational Procedure: *Vorticella*

Place a drop of culture containing *Vorticella* in a ring of methyl cellulose. Cover with a coverslip and observe under low power of a compound microscope. Note the inverted bell-shaped body and stalk (Fig. 1.25). The stalk can appear like a coiled spring as it retracts and extends. Do you see popping movements? Observe beating of the membranelles that lie in a peristomal groove around the bell. There is one inner set of cilia and one row that projects outward like a shelf. Do all rows beat in the same direction? What function might this movement serve? Are food and contractile vacuoles evident? Look for the elongate sausage- or horseshoe-shaped macronucleus. It can be observed in a prepared slide. ■

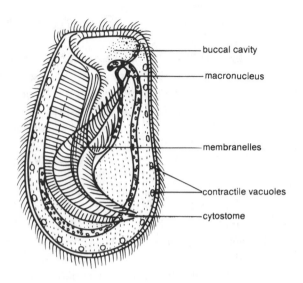

Figure 1.26. *Bursaria.*

Order Bursariomorphida (Class Colpodea)

These ciliates have an apical mouth and a large, funnel-shaped buccal (oral) cavity with membranelles on its left and right walls.

Bursaria truncatella. This large (500–1,000 μm long), broad, carnivorous ciliate has uniform body ciliation. The species is found in freshwater.

■ Observational Procedure: *Bursaria truncatella*

Make a wet mount from the *Bursaria* culture and observe under low and then with high power of a compound microscope. Describe the organism's movement. *Bursaria* is ovoid with the anterior end truncate and the posterior end round (Fig. 1.26). The ventral surface is flat, but the dorsal surface is convex. Note the funnel-shaped buccal cavity beginning at the anterior end and extending to the end of the body where it gives rise to the cytostome. A lengthwise fold divides the peristome into two chambers. Along the lateral and posterior edges of the body are many contractile vacuoles. Attempt to see the long, band-shaped macronucleus. *Bursaria* has several micronuclei. Compare what you see in the living organism with that in a prepared slide. ■

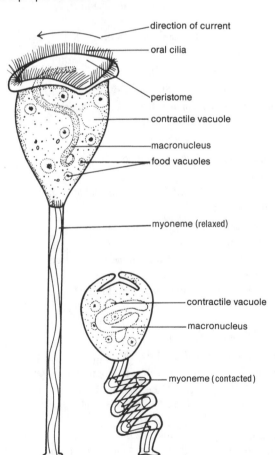

Figure 1.25. *Vorticella.*

Supplemental Readings

Anderson, O. R. 1983. Radiolaria. Springer, New York.

Bannister, D. H. 1972. The structure of trichocyst in *Paramecium caudatum*. J. Cell Sci. 11: 899–929.

Bonner, J. T. 1983. Chemical signals of social Amoebae. Sci. Am. 248:114–120.

Borror, A. C. 1973. Protozoa: Ciliophora. Marine flora and fauna of the northeastern U.S. NOAA Technical Report. NMFS circular 378. U.S. Printing Office, Washington, DC.

Bovee, E. C. 1953. Morphological identification of free-living Amoebida. Proc. Iowa Acad. Sci. 60:599–615.

Bovee, E. C., and T. L. Jahn. 1973. Locomotion and behavior of Amoebae. *In*: Jeon, K.W. (ed.). Biology of *Amoeba*. Academic Press, New York, pp. 249–290.

Bovee, E. C., and T. K. Sawyer. 1979. Protozoa: Sarcodina: Amoebae. Marine flora and fauna of the northeastern U.S. NOAA Technical Report. NMFS Circular 419. U.S. Printing Office, Washington, DC.

Buetow, D. E. (ed.). 1967. The Biology of *Euglena*. Vols. 1 and 2. Academic Press, New York.

Buetow, D. E. (ed.). 1982. The Biology of *Euglena*. Vol. 3. Academic Press, New York.

Capriulo, G. M. (ed.). 1990. Ecology of Marine Protozoa. Oxford Univ. Press, New York.

Chen, Y. T. 1950. Investigations of the biology of *Peranema trichophorum* (Euglenineae). Quart. J. Microsc. Sci. 91:279–308.

Coleman, A. W., and P. Heywood, 1981. Structure of the chloroplast and its DNA in chloromonadophycean algae. J. Cell Sci. 49:401–409.

Corliss, J. O. 1979. The Ciliated Protozoa: Characterization, Classification, and Guide to the Literature, 2nd ed. Pergamon Press, London.

Corliss, J. O. 1983. Consequences of creating new kingdoms of organisms. Bioscience 33:314–318.

Corliss, J. O. 1984. The kingdom Protista and its 45 phyla. Biosystems 17:87–126.

Edds, K. T. 1981. Cytoplasmic streaming in a heliozoan. Biosystems 14:371–376.

Farmer, J. N. 1980. The Protozoa: Introduction to Protozoology. C.V. Mosby Co., St. Louis, MO.

Giese, A. C. 1973. *Blepharisma*: The Biology of a Light-sensitive Protozoan. Stanford Univ. Press., Stanford, CA.

Grell, K. G. 1973. Protozoology. Springer-Verlag, New York.

Hall, R. P. 1953. Protozoology. Prentice-Hall, New York.

Hammond, D. M., and P. L. Long (eds.). 1973. The Coccidia. University Park Press, Baltimore, MD.

Haynes, J. R. 1981. Foraminifera. John Wiley and Sons, New York.

Jahn, T. L., E. C. Bovee, and F. F. Jahn. 1979. How to Know the Protozoa. 2nd ed. W.C. Brown, Dubuque, IA.

Jeon, K. W. (ed.). 1973. The Biology of *Amoeba*. Academic Press, New York.

Katz, M., D. D. Despommier, and R. Gwadz. 1983. Parasitic Diseases. Springer-Verlag, New York.

Kudo, R. R. 1966. Protozoology. 5th ed. Charles C Thomas, Springfield, IL.

Leadbeater, B. S. C. 1983. Observations on the life history and ultrastructure of the marine choanoflagellate *Proterospongia choanojuncta*. J. Mar. Biol. Assoc. U. K. 63:135–160.

Lee, J. J., S. H. Hutner, and E. C. Bovee (eds.). 1985. An Illustrated Guide to the Protozoa. Soc. of Protozoologists, Lawrence, KS.

Leedale, G. F. 1967. Euglenoid Flagellates. Prentice-Hall, Englewood Cliffs, NJ.

Leedale, G. F. 1974. How many are the kingdoms of organisms? Taxon 23:261–270.

Leedale, G. F. 1978. Phylogenetic criteria in euglenoid flagellates. Biosystems 10:183–187.

Levine, N. D. 1973. Protozoan parasites of domestic animals and man, 2nd ed. Burgess Publ. Co., Minneapolis, MN.

Levine, N. D., J. O. Corliss, F. E. G. Cox, G. Deroux, J. Grain, B. M. Honigberg, G. F. Leedale, A. R. Loeblich III, J. Lom, D. Lynn, E. G. Merinfeld, F. C. Page, G. Poljansky, V. Sprague, J. Vavra, and F. G. Wallace. 1980. A newly revised classification of the Protozoa. J. Protozool. 27:37–58.

Long, P. L. (ed.). 1982. The Biology of the Coccidia. University Park Press, Baltimore, MD.

Margulis, L. 1974. Five-kingdom classification and the evolution and origin of cells. Evol. Biol. 7:45–78.

Molyneux, D., and R. W. Ashford. 1983. The Biology of *Trypanosoma* and *Leishmania*, Parasites of Man and Domestic Animals. Taylor and Francis, London.

Nanney, D. L. 1980. Experimental Ciliatology. John Wiley, New York.

Noble, E. R., and G. A. Noble. 1983. Parasitology. The Biology of Animal Parasites, 5th ed. Lea & Febinger, Philadelphia.

Ogden, C. G., and R. H. Hedley. 1980. An Atlas of Freshwater Testate Amoebae. Oxford University Press, Oxford.

Ormerod, W. E. 1982. The life cycle of the sleeping sickness trypanosome compared with the malaria life cycle. *In*: Canning, E. U. (ed.). Parasitological Topics. A Presentation Volume to P.C.C. Garnham, F. R. S. on the Occasion of his 80th Birthday 1981. Allen Press, Lawrence, KS, pp. 191–199.

Peck, R. K. 1977. Cortical ultrastructure of the scuticociliates *Dexiotricha media* and *Dexiotricha colpidiopsis* (Hymenostomata). J. Protozool. 24:122–134.

Pitelka, D. R. 1963. Electron-microscopic structure of Protozoa. Macmillan, New York.

Pitelka, D. R. 1970. Ciliate ultrastructure; Some problems in cell biology. J. Protozool. 17:1–10.

Poag, C. W. 1981. Ecologic Atlas of Benthic Foraminifera of the Gulf of Mexico. Hutchinson Ross, Stroudsburg, PA.

Rudzinska, M. A. 1973. Do suctoria really feed by suction? Bioscience 23:87–94.

Sadun, E. H., and A. P. Moon (eds.). 1972. Basic Research in Malaria. Proc. Helminthol. Soc. Washington. 39 (special issue).

Sarjeant, W. A. S. 1974. Fossil and Living Dinoflagellates. Academic Press, London.

Silva, P. C. 1980. Names of classes and families of living algae. Regnum Veg. 103:1–156.

Sleigh, M. A. 1962. The Biology of Cilia and Flagella. Macmillan, New York.

Sleigh, M. A. 1973. The Biology of Protozoa. Elsevier, New York.

Steidinger, K. A., and E. R. Cox. 1980. Free-living dinoflagellates. *In*: Cox, E. R. (ed.). Dev. Mar. Biol 2, Phytoflagellates. Elsevier, New York, pp. 407–432.

Sweeney, B. M. 1979. The bioluminescence of dinoflagellates. *In*: Levandowsky, M., and S. H. Hutner (eds.). Biochemistry and Physiology of Protozoa. Academic Press, New York, pp. 288–306.

Tartar, V. 1961. The Biology of *Stentor*. Pergamon Press, Oxford.

Taylor, F. J. R. (ed.). 1987. The Biology of Dinoflagellates. Blackwell, Oxford.

Van Wagtendonk, W. J. (ed.). 1974. Paramecium; A Current Survey. Elsevier, New York.

Vickerman, K., and F. E. G. Cox. 1967. The Protozoa. Houghton Mifflin, Boston.

Whittaker, R. H. 1969. New concepts of kingdoms of organisms. Science 163:150–160.

Wichterman, R. 1953. The Biology of *Paramecium*. McGraw-Hill, New York.

THE METAZOA

Metazoans, unlike protozoans, are composed of numerous, specialized cells arranged in layers. In primitive metazoans the outer layer serves for protection and for sensory response, while the inner layer functions in digestion. Cells between these layers function in support, conduction of nutrients, reproduction, and muscular contraction. There is an increasing complexity in cellular organization from primitive to advanced metazoans. In the least complex metazoans, cells having similar functions form aggregations known as tissues (tissue grade). In more advanced metazoans, two or more tissues combine to form organs (organ grade). In the most advanced forms, combinations of specific organs form specialized organ systems (system grade). Although these grades do not have taxonomic status, they may be thought of as a logical way of organizing the metazoan phyla.

Branch Parazoa [G., *para*, beside + G., *zoon*, animal] includes two phyla, **Porifera** and **Placozoa**. Both phyla have a cellular or poorly defined tissue grade of construction. In the Porifera the body has numerous pores, canals, and chambers lined with specialized, flagellated feeding cells. The phylum Placozoa contains one species, *Trichoplax adhaerens*. This enigmatic organism resembles a microscopic pancake and is composed of a loose mesenchyme surrounded by layers of epithelial cells on the upper and lower sides (Pearse 1989; Pearse et al. 1994).

Branch Mesozoa [G., *meso*, middle] contains a single phylum of the same name. This enigmatic phylum contains minute parasitic and commensal marine animals composed of a single, outer cell layer and an inner, reproductive layer of one or more cells. Mesozoa has been thought of as a link between protozoans and metazoans, but some workers believe that their primitive characteristics are secondary, resulting from a long evolution as obligate parasites (Hochberg 1982; Lapan and Morowitz 1972).

Branch Eumetazoa [G., *eu*, good] includes the remainder of the metazoan phyla. These approximately 30 phyla are of the tissue, organ, or system grade of construction. Recently several research groups have attempted to build a phylogeny of the Metazoa. Some of those studies are based on morphology alone while others have emphasized molecular data such as the sequence of the 18S ribosomal subunit (Ax 1989; Field et al. 1988; Ghiselin 1989; Hillis 1987; Inglis 1985; Lake 1990; Meglitsch and Schram 1991; Nielsen 1985, 1994; Patterson 1989; Raff et al. 1994; Schram 1991; Wainright et al. 1993).

Ax, P. 1989. Basic phylogenetic systematization of the Metazoa. *In*: Fernholm, B., K. Bremer, and H. Jörnall (eds.). The Hierarchy of Life. Elsevier, Amsterdam, pp. 229–245.

Field, K. G., G. J. Olsen, D. J. Lane, S. J. Giovannoni, M. T. Ghislin, E. C. Raff, N. R. Pace, and R. A. Raff. 1988. Molecular phylogeny of the Animal Kingdom. Science 239: 748–753.

Ghiselin, M. T. 1989. Summary of our present knowledge of metazoan phylogeny. *In*: Fernholm, B., K. Bremer, and H. Jörnall (eds.). The Hierarchy of Life. Elsevier, Amsterdam, pp. 261–272.

Hillis, D. M. 1987. Molecular versus morphological approaches to systematics. Annu. Rev. Ecol. Syst. 18: 23–42.

Hochberg, F. G., 1982. The "kidneys" of cephalopods: a unique habitat for parasites. Malacologia 23:121–134.

Inglis, W. G. 1985. Evolutionary waves: patterns in the origin of animal phyla. Aust. J. Zool. 33: 153–178.

Lake, J. A. 1990. Origin of the Metazoa. Proc. Natl. Acad. Sci. USA 87: 763–766.

Lapan, E. A. and H. Morowitz. 1972. The Mesozoa. Sci. Am. 227: 94–101.

Meglitsch, P. A. and F. R. Schram. 1991. Invertebrate Zoology. Oxford University Press, New York, 623 pp.

Nielsen, C. 1985. Animal phylogeny in the light of the trochaea theory. Biol. J. Linn. Soc. 25: 243–299.

Nielsen, C. 1994. Larval and adult characters in animal phylogeny. Amer. Zool. 34: 492–501.

Patterson, C. 1989. Phylogenetic relations of major groups: conclusions and prospects. *In*: Fernholm, B., K. Bremer, and H. Jörnall, (eds.). The Hierarchy of Life. Elsevier, Amsterdam, pp. 471–488.

Pearse, V. B. 1989. Growth and behavior of *Trichoplax adhaerens*: first record of the phylum Placozoa in Hawaii. Pac. Sci. 43: 117–121.

Pearse, V. B., T. Uehara, and R. L. Miller. 1994. Birefringent granules in Placozoans (*Trichoplax adhaerens*). Trans. Am. Microsc. Soc. 113(3): 385–389.

Raff, R. A., C. R. Marshall, and J. M. Turbeville. 1994. Using DNA sequences to unravel the Cambrian radiation of the animal phyla. Annu. Rev. Ecol. Syst. 25: 351–375.

Schram, F. R. 1991. Cladistic analysis of metazoan phyla and the placement of fossil problematica. *In*: The Early Evolution of Metazoa and the Significance of Problematic Taxa. Simonetta, A. M. and S. Conway Morris, (eds.). Cambridge University Press, Cambridge, pp. 35–46.

Wainright, P. O. G. Hinkle, M. L. Sogin, and S. K. Stickel. 1993. Monophyletic origins of the metazoa: an evolutionary link with fungi. Science 260: 340–342.

■

Phylum Porifera

Phylum Porifera [po-RIF-er-a; L., *pori*, pore + L., *ferre*, bearing], or sponges, comprises a group of simple, sessile metazoans considered to have a cellular grade of construction. They are an ancient group whose geologic history extends back to before the **Cambrian Period** (Fig. 2.1). Sponges are either radially symmetrical or asymmetrical and have a body consistency varying from hard and stony to friable, rubbery, or gelatinous, depending upon nature and arrangement of their skeletal elements. They range in size from a few millimeters to about 1 m tall and are economically important as members of the marine-fouling community and because of the continued commercial importance of bath sponges. Lately the significance of sponges has increased because they are viewed as potential sources of unique biochemicals that may be important in fighting disease.

In their simplest form, sponges resemble a tube closed at one end surrounding an inner cavity called the **spongocoel** or **atrium** (Fig. 2.2). The tube wall consists of two layers of cells arranged as concentric cylinders separated by a thin, noncellular gelatinous layer, the **mesohyl**. The outer cell layer (**pinacoderm**) is made up of flattened cells called **pinacocytes** that do not secrete a basement membrane. The inner layer consists of flagellated feeding cells called **choanocytes** or **collar cells** (Fig. 2.2). Sponges are supported internally by a calcareous or siliceous skeleton of **spicules**, or by protein **spongin** fibers. Some sponges have a combination of siliceous spicules and spongin. A large variety of different sizes (**mega-** and **microscleres**) and shapes of spicules has been described (Fig. 2.3). Spicules are absent in some sponges and a massive or reticulate, calcareous skeleton is present.

TIME (IN MILLIONS OF YEARS)

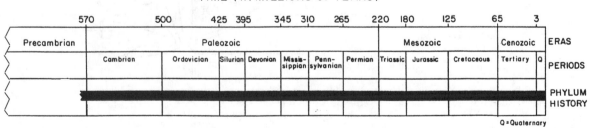

Figure 2.1. Geologic history of the phylum Porifera.

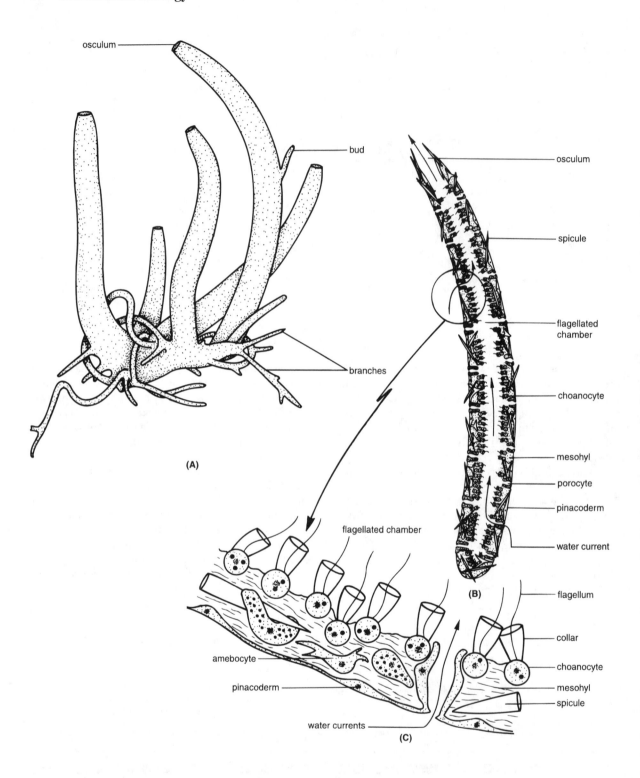

Figure 2.2. An asconoid sponge, *Leucosolenia*, and its canal system. A small portion of a *Leucosolenia* colony (A) and a diagramatic longitudinal section of *Leucosolenia* (B) with an enlargement of the body wall are shown (C).

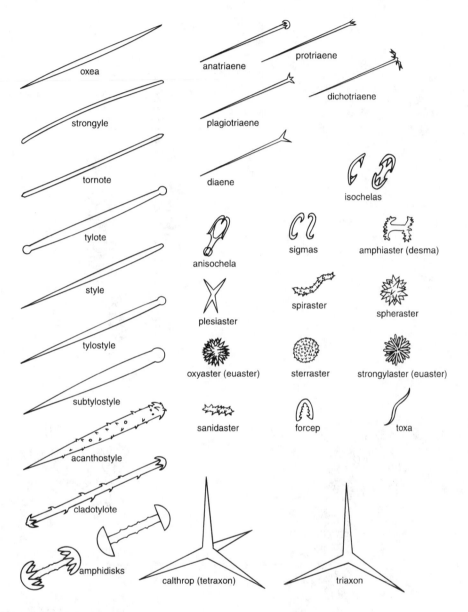

Figure 2.3. Common spicule types.

The name "pore bearer" refers to the fact that sponges are perforated by numerous, minute, incurrent pores (**ostium**, plural **ostia**) and by one or more large excurrent pores (**osculum**, plural **oscula**). Water currents, generated by flagellar action of choanocytes, enter the sponge through ostia, pass into choanocyte (flagellated) chambers, and exit the sponge via the osculum. Viewed under high magnification of a light microscope, choanocytes appear as globular cells with long flagella surrounded at the base by a collar-like extension (actually a ring of microvilli) of the cell membrane (Fig. 2.2).

Sponges feed by screening incurrent water through a progressively finer series of filters (canals to choanocyte chambers) until the finest particles are removed by the choanocytes.

The structure of sponges may be divided into three body types, based on anatomy of choanocyte chambers and incurrent and excurrent canals. These types, in increasing order of complexity, are as follows: **asconoid, syconoid,** and **leuconoid** (Figs. 2.2, 2.4, and 2.5).

Both asexual and sexual reproduction occur in sponges. Asexual reproduction is accomplished by frag-

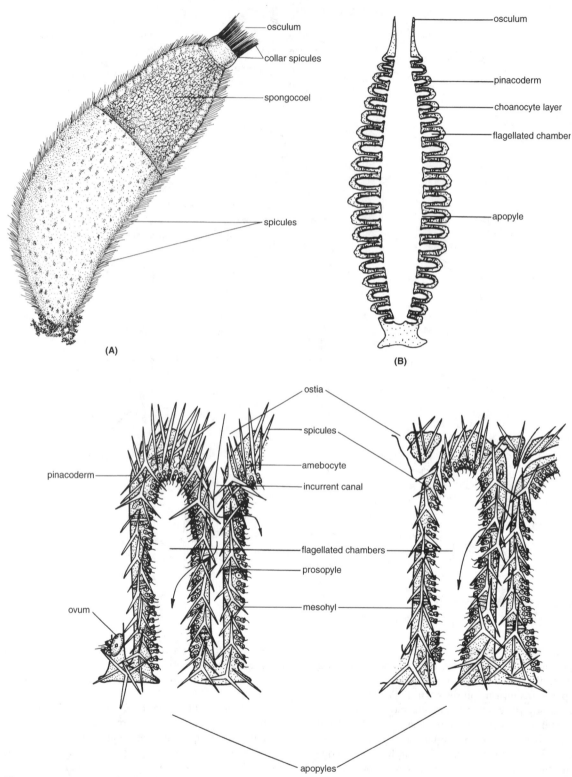

Figure 2.4. The syconoid sponge body plan of *Sycon*. A whole specimen with a portion of the body wall removed (A), a longitudinal (B), and two cross sections (C) are shown.

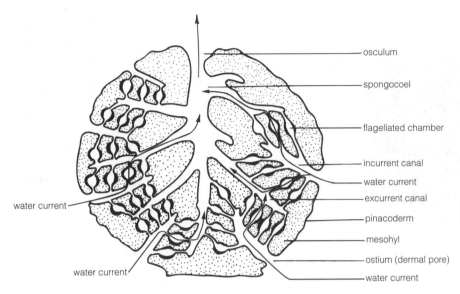

Figure 2.5. Diagrammatic view of a leuconoid sponge.

mentation and budding, and in freshwater and a few marine sponges, by production of resistant structures called **gemmules** (Fig. 2.6). Some sponges have remarkable powers of regeneration and reaggregation. Apparently all sponges are capable of sexual reproduction. Both hermaphroditic (**monoecious**) and **dioecious** species occur in the phylum. Fertilization and development are internal, except for some species that release fertilized eggs. Two different larvae occur in sponges. Most sponges have a **parenchymella** larva and a few an **amphiblastula** (Fig. 2.7). After a mobile phase of a few hours to several days, larvae attach to a suitable surface and metamorphose into young sponges.

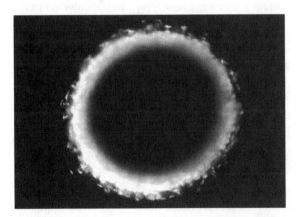

Figure 2.6. Gemmules of *Spongilla*, a freshwater sponge.

Classification

Sponge taxonomy is based primarily on the chemical nature and morphology of inorganic and organic skeletal elements. However, reproductive features, comparative biochemistry, and ultrastructural histology also are important. Four classes of sponges are recognized. Reiswig and Mackie (1983) separate sponges into two subphyla: **Cellularia** (having discrete cells) comprise the three main classes; **Symplasma** (having a syncytium) comprise one class (Hexactinellida) (cf. Vacelet 1985 and Bergquist 1985). Most of the 5,000 or more species of sponges are marine; about 150 species inhabit freshwaters.

1. **Class Calcarea**. Sponges with spicules of calcium carbonate; massive skeletal structures also are present. Microsclere spicules lacking; entirely marine, usually located in shallow water. All three body types occur. Examples include *Leucosolenia* and *Sycon*.

2. **Class Demospongiae**. Approximately 90% of all extant poriferans; siliceous spicules (megascleres and microscleres), spongin fibers, or both; one family, Oscarellidae, lacks any skeleton. Freshwater sponges (family **Spongillidae**) are members of this class. Only the leuconoid body type occurs. Examples include the freshwater genera *Ephydatia* and *Spongilla*, and the marine genera *Cliona* (boring sponge), *Spheciospongi* (loggerhead sponge), and *Spongia* (bath sponge).

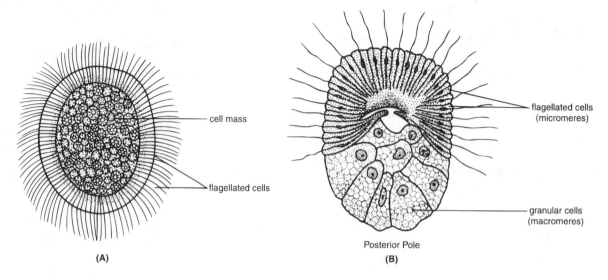

Figure 2.7. Parenchymella and amphiblastula larvae (After Hyman).

3. **Class Sclerospongiae**. Tropical to subtropical sponges usually found in cryptic habitats, possessing massive or reticulate calcareous skeletons. Only the leuconoid body type occurs in this class. Examples are *Calcifibrospongia* and *Merlia*. Vacelet (1985) suggests that these few species (ca. 15) are members of the Calcarea and Demospongiea.

4. **Class Hexactinellida**. Deep-water marine sponges with siliceous six-pointed spicules and long siliceous fibers forming a fused lattice structure. Some hexacts have a syconoid-like body type. A prominent example is *Euplectella* or Venus's flower basket. Because of their syncytial organization, it has been argued that hexacts should be a distinct phylum, Symplasma (Bergquist 1985).

■ Observational Procedure:

Caution: Please note that when you touch a sponge the tiny spicules may rub off on your hands. Therefore, never touch your eyes immediately after handling a sponge!

Class Calcarea

Leucosolenia. An Asconoid Sponge. Using a dissection microscope observe a preserved specimen of *Leucosolenia*. Keep the specimen immersed in fluid in a small shallow dish while making your observations. This small, marine sponge lives attached to rocks in the lower intertidal zone as a group of clustered, tubular structures in varying stages of growth (Fig. 2.2). The clusters result from budding. The body wall is quite thin and the surface has a rough texture due to spicules that project through the pinacoderm. Locate the osculum, flagellated chamber, and buds. Two types of spicules may be seen in *Leucosolenia*. Place a portion of a specimen on a glass slide, tease apart with dissection needles, and apply a small drop of bleach to the sponge. Cover the preparation with a cover slip and examine. The bleach will dissolve organic material and only spicules will remain, allowing identification of spicule types under high power (Fig. 2.3).

Return your microscope to low power and obtain a pair of polarizing filters. Notice how these filters stop all transmitted light when they are rotated 90° out of phase from one another. Rotate the filters so that light passes through them and place one filter above your spicule preparation (**analyzer**) and another on top of the light source of the microscope (**polarizer**). You probably will have to refocus the microscope. While observing the spicules, rotate the polarizer until it is out of phase with the analyzer. The background will be dark due to exclusion of light by the crossed filters, however, the spicules will appear to be bright. This phenomenon, called **birefringence**, occurs because $CaCO_3$ spicules have two or more principal refractive indices and will rotate light themselves, thus permitting light to pass through the analyzer. What does this indicate about these spicules?

Sycon (=Scypha). A Syconoid Sponge. *Sycon* is also a marine sponge. Small buds occasionally appear at the base of mature specimens and may separate from the parent. Examine a specimen of *Sycon* using a dissection microscope and locate the osculum and the large monaxial spicules arranged around the opening as a collar (Fig. 2.4A). Also note that the surface is interrupted by numerous, small projections (Fig. 2.4B). These are the flagellated chambers (radial canals or choanocyte chambers). Using a sharp scalpel, bisect the animal along its longitudinal axis and carefully examine the spongocoel. The numerous, inner openings are **apopyles** that communicate with the choanocyte chambers.

Examine a prepared slide of a cross and/or longitudinal section of *Sycon* and identify the following structures: (1) pinacoderm, (2) choanocyte chambers, (3) choanocytes, (4) incurrent canal, (5) apopyle, (6) mesohyl, (7) spicules, (8) ostia, and (9) spongocoel (Fig. 2.4C). The prosopyles will be indistinct. Choanocytes will not be as neatly arranged as depicted in Fig. 2.4C. In a prepared cross section, note the arrangement of spicules within the sponge. Ova or embryos may be present in some preparations. How is *Sycon* more complex than *Leucosolenia*?

Make a spicule preparation of *Sycon* as described above and identify the spicule types seen (Fig. 2.3). Are the spicules of *Sycon* birefringent?

Class Demospongiae

Spongilla. A Leuconoid Sponge. Examine a preserved specimen of *Spongilla* under a dissection microscope. Remember to keep the specimen covered with fluid. You should see numerous spicules projecting through the pinacoderm. Obtain a prepared thin section of *Spongilla* and observe the distribution of spicules in the sponge mass. Note the concentrations of spicules, called **fascicles**, that project through the pinacoderm (Fig. 2.8). The pinacoderm also should be visible. Choanocyte chambers are not located at the outer periphery of the sponge mass, but rather beneath the **subdermal cavity** (Fig. 2.8). The choanocytes are very small and it will require great care to identify them. Compare the body form of *Spongilla* to the other species studied.

Observe a preserved specimen of *Spongilla* with gemmules. How large are these structures? Examine a prepared whole mount-slide of gemmules under high magnification, noting the tough spongin coat in which are embedded numerous spicules (Fig. 2.6). If live gemmules are available, your instructor will provide directions for observation of their hatching and subsequent early sponge development.

Make a preparation of *Spongilla* spicules as described above and identify the spicule types seen (Fig. 2.3). Are the siliceous spicules of *Spongilla* birefringent? Why does it make sense that these spicules,

which are composed of silica, have this response to the test for birefringence?

Bath Sponge. Although you may have used a natural bath sponge before, examine one this time with a critical eye. Note how light and flexible it is. Use a hand lens or dissection microscope and observe the minute, interwoven spongin fibers. This is all that remains after the tissues have decayed. Bath sponges are noted for the amount of water they can hold. Because the fibers themselves cannot imbibe significant amounts of water, how can a sponge hold water? To answer this question do the following simple exercise. Obtain a dry sponge and weigh it to the nearest tenth of a gram. Wet the sponge, wring it out, and reweigh it. How much water is the damp sponge capable of holding? Calculate how many milliliters of water are held per gram dry weight of sponge? Under a dissection microscope re-examine the spongin fibers and note the film of water that is held by capillary action between adjacent fibers.

Cliona. The Boring Sponge. Examine a mollusk shell that has been colonized by *Cliona*. What does this sponge have to do to colonize a shell composed of calcium carbonate? Are the holes regularly spaced throughout the shell? Are they all the same diameter? Observe a shell that has a broken edge permitting a view of the extensive network of cavities within it. What do the cavities and holes represent in terms of the biology of the sponge?

Compare several shells with various levels of damage? Are the number of holes perforating the inside of the shell equal to the number perforating the outside? What might this indicate? Would you say that this sponge is parasitic? What do your observations indicate about the importance of *Cliona* in marine systems?

Class Hexactinellida

Euplectella. *Euplectella* (Venus's flower basket) is a marine glass sponge found in waters >200 m deep. Most people who have seen this sponge have viewed only the delicate skeleton. Extreme care should be taken when handling an unprotected specimen.

Observe the following aspects of general morphology (Fig. 2.9). The body of *Euplectella* is tubular and slightly curved. Internally, note the large spongocoel. Are remains of the commensal shrimp (*Spongicola*) present in the spongocoel of your specimen? Note that the osculum is covered by a **sieve plate** of fused spicules that adds additional strength to the skeleton. Carefully examine the skeleton using a dissection microscope. Note that the skeletal framework of *Euplectella* is composed of spicules united by crossbars of silica. Will these spicules be birefringent? Describe the shape and arrangement of spicules at the

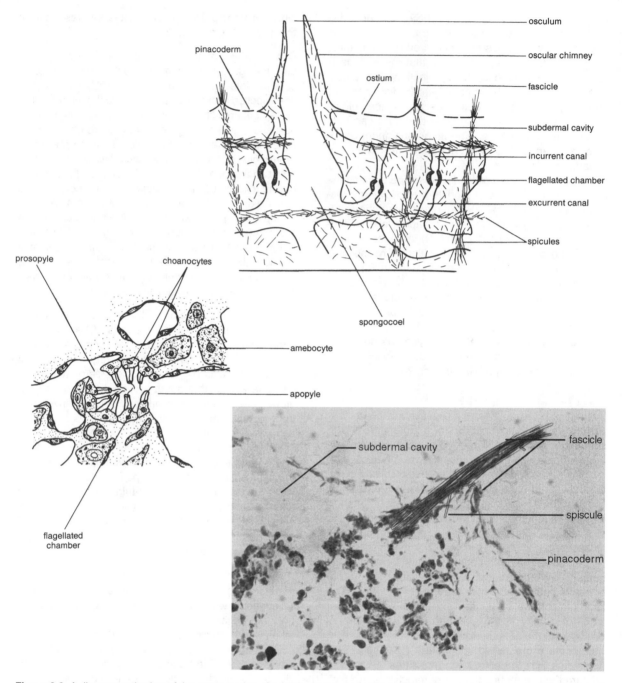

Figure 2.8. A diagrammatic view of the cross section of a freshwater sponge (*Spongilla*) with a photomicrograph of the region of the subdermal cavity (Drawings after Pennak).

base of the sponge. Examine them using a dissection microscope. Is their surface smooth? What function might the basal spicules serve?

Other Specimens. Observe other specimens your instructor may provide. Compare the size, shape, and skeletal elements of these specimens to the others you studied in more detail. Determine whether their spicules are birefringent. If specimens of Class Sclerospongiae are available, compare their general structure to the other specimens examined in this exercise. ■

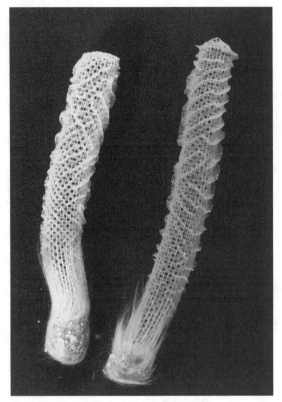

Figure 2.9. *Euplectella*, Venus's flower basket.

Fossil Forms

Many zoologists believe that sponges arose in the **Precambrian Era** from a group of protozoans similar to the modern choanoflagellates (Fig. 2.1). By the **Cambrian Period**, sponges had developed an extensive fauna. However, in spite of their auspicious start these colonial organisms are unique and difficult to relate to other phyla.

■ Observational Procedure:

Because of the delicate nature of poriferan tissue, sponges do not preserve well and most fossil sponges are not easily recognizable. However, *Cliona* is easily recognizable as a fossil and fossil mollusk shells are often found riddled with cavities similar to those produced by this genus in modern shells. Compare fossil and modern shells that have been colonized by *Cliona*. Are there any significant differences?

Another interesting fossil organism is the sponge-like phylum **Archaeocyatha**. This group of pore-bearing organisms was unlike other sponges. **Stromatoporoids**, classified by some as colonies of fossil cyanobacteria (bluegreen algae) or cnidarians, may be another extinct sponge-like group. Examine specimens of archaeocyathids and other fossil sponges (Fig. 2.10). What evidence do you see in these specimens that would lead you to conclude that they are really sponges? ■

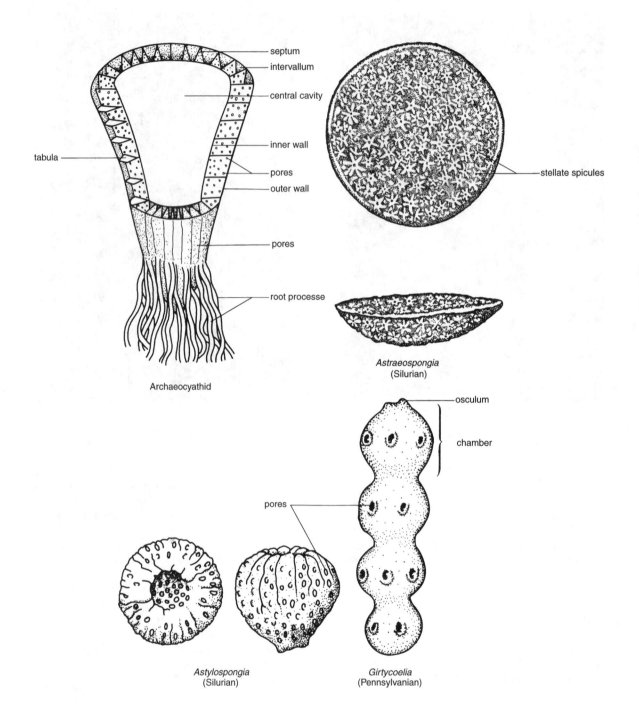

Figure 2.10. Examples of fossil sponges. An archaeocyathid with section cut away to view internal walls (After Moret, from Bergquist); *Astraeospongia* and *Astylospongia*, Silurian sponges; *Girtycoelia* showing repeating chambers, Pennsylvanian sponge.

Supplemental Readings

Bergquist, P. R. 1978. Sponges. University of California Press, Berkeley, CA.

Bergquist, P. R. 1985. Poriferan relationships. *In*: S. Conway Morris, J. D. George, R. Gibson, and H.M. Platt (eds.), The Origins and Relationships of Lower Invertebrates. Systematics Association special volume 28, Clarendon Press, Oxford, pp. 14–27.

Cobb, W.R. 1969. Penetration of calcium carbonate substrates by the boring sponge, *Cliona*. Am. Zool. 9:783–790.

De Vos, L. Rützler, K., Boury-Esnault, N., Donadey, C., Vacelet, J. 1991. Atlas of Sponge Morphology. Smithsonian Institution Press, Washington, DC.

Fell, P. E. 1995. Deep diapause and the influence of low temperature on the hatching of the gemmules of *Spongilla lacustris* (L.) and Eunapius fragilis (Leidy). Invert. Biol. 114: 3–8.

Finks, R. M. 1970. The evolution and ecologic history of sponges during Palaeozoic times. Symp. Zool. Soc. Lond. 25: 3–22.

Frost, T. M. 1978. In situ measurements of clearance rates of the freshwater sponge *Spongilla lacustris*. Limnol. Oceanogr. 23: 1034–1039.

Frost, T. M. 1991. Porifera. *In*: Thorpe, J. H. and A. P. Covich (eds.), Ecology and Classification of North American Freshwater Invertebrates. Academic Press, New York, pp. 95–124.

Gilbert, J. J. 1975. Field experiments on gemmulation in the freshwater sponge *Spongilla lacustris*. Trans. Am. Microsc. Soc. 94: 347–356.

Gilbert, J. J., and H. L. Allen. 1973. Chlorophyll and primary productivity of some green, freshwater sponges. Int. Revue ges. Hydrobiol. 58: 633–658.

Harrison, F. W. and J. A. Westfall (eds.), 1991. Microscopic Anatomy of Invertebrates, vol. 2: Placozoa, Porifera, Cnidaria, and Ctenophora. Wiley-Liss, New York.

Hartman, W. D., J. W. Wendt, and F. Wiedenmayer. 1980. Living and fossil sponges. Notes for a short course. Sedimenta VIII. Comparative Sedimentology Laboratory, Division of Marine Geology and Geophysics, Rosenstiel School of Marine and Atmospheric Science, University of Miami, Miami, FL.

Humphreys, T. 1970. Species specific aggregation of dissociated sponge cells. Nature 228: 685–686.

Lawn, I. D., G. O. Mackie, and G. Silver. 1981. Conduction system in a sponge. Science 211: 1169–1171.

Palumbi, S. R. 1984. Tactics of acclimation: morphological changes of sponges in an unpredictable environment. Science 225: 1478–1480.

Rasmont, R. 1970. Some new aspects of the physiology of freshwater sponges. Symp. Zool. Soc. Lond. 25: 415–422.

Reiswig, H. M., and G. O. Mackie. 1983. Studies on hexactinellid sponges. III. The taxonomic status of Hexactinellida within the Porifera. Phil. Trans. R. Soc. B301: 419–428.

Rützler, K. and G. Rieger. 1973. Sponge burrowing: fine structure of *Cliona lampa* penetrating calcareous substrata. Mar. Biol. 21: 144–162.

Simpson, T. L. 1968. The structure and function of sponge cells. Bull. Peabody Mus. Nat. Hist. 25:1–141.

Simpson, T. L. 1984. The Cell Biology of Sponges. Springer-Verlag, New York.

Vacelet, J. 1985. Coralline sponges and the evolution of Porifera. *In*: S. Conway Morris, J. D. George, R. Gibson, and H.M. Platt (eds.), The Origins and Relationships of Lower Invertebrates. Systematics Association special volume 28, Clarendon Press, Oxford, pp. 1–13.

THE RADIATA

The Eumetazoa is divided into two groups, one with radial symmetry called the **Radiata** and one with bilateral symmetry called the **Bilateria**. Radiata includes the phyla Cnidaria (Coelenterata) and Ctenophora. These phyla have well-developed tissues that have the beginnings of organ functions, but they have not developed well-defined organs systems. The radiate body consists of a body wall composed of an outer epidermis, an inner gastrodermis, and a thin to thick gelatinous layer between called the mesoglea. The latter may be a noncellular, gelatinous matrix or possess amoebocytes, connective tissues, and muscle cells. The body cavity (gastrovascular cavity or coelenteron) ranges from a simple sac in the lower Cnidaria to a complex canal system in the Ctenophora, with pharynx, stomach, and other canals. In Cnidaria, the body cavity opens to the exterior through only a mouth, while in the Ctenophora both a mouth and anal pores are present.

Harrison, F. W., and J. A. Westfall (eds.). 1991. Placozoa, Porifera, Cnidaria, and Ctenophora, vol. 2: Microscopic Anatomy of Invertebrates. Wiley-Liss, New York.

Muscatine, L., and H. M. Lenoff (eds.). 1974. Coelenterate Biology. Academic Press, New York.

■

Phylum Cnidaria (Coelenterata)

Hydras, jellyfishes, anemones, and corals are members of the phylum Cnidaria [ni-DAR-e-a; G., *cnide*, nettle] or Coelenterata [se-LEN-ter-a-ta; G., *koilos*, hollow + G., *enteron*, gut]. With the exception of freshwater hydras and a few jellyfishes, the approximately 9,000 species of cnidarians are marine. These carnivorous, tentaculate, multicellular animals represent the tissue level of organization; that is, organs or organ systems are lacking. This ancient phylum (Fig. 3.1) contains some of the most beautiful, interesting, and perhaps most dangerous of all marine invertebrates. The sea wasp (*Chironex fleckeri*), a cubomedusa of the Indo-Pacific waters, may be the most venomous invertebrate. Human fatalities have been reported from the noxious stings of this animal. The Portuguese man-of-war (*Physalia*) is less dangerous, but often becomes a nuisance when currents carry large numbers into shallow waters.

Cnidarians possess stinging organelles called **nematocysts**, the most distinguishing feature of the phylum. Each nematocyst arises from a specialized cell, the **cnidocyte** (Fig. 3.2A, B). Unlike sponges, cnidarians have two basic metazoan features, a **mouth** and a **gastrovascular cavity** (coelenteron or enteron) (Fig. 3.3). There is no anus and all materials enter and exit via the mouth. The gastrovascular cavity is a large, internal space for digestion and distribution of nutrients and other materials.

Two basic types of body forms are found in cnidarians: (1) **polyp (hydroid)** and (2) **medusa (jellyfish)**. Both forms are saclike and contain a gastrovascular cavity (Fig. 3.3). The sessile polyp is tubular with a mouth in the **hypostome** at the distal end that is usually surrounded by tentacles. The medusa is typically free-swimming and umbrella-shaped, with tentacles extending from the outer margin of the bell. The mouth is located at the end of the **manubrium**, which hangs from the center of the subumbrellar surface somewhat like a clapper hanging from a bell. The manubrium is comparable to the hypostome of the polyp. Both the polyp and medusa are composed of the same body layers and are radially symmetrical. The medusa possesses a much thicker mesoglea than does the polyp.

Two types of medusae are found in cnidarians based upon presence (**craspedote** medusa) or absence (**acraspedote** medusa) of the **velum**, an inward projecting membrane around the margin of the bell. Rhythmic contractions of the velar musculature propel the jellyfish through the water.

The cnidarian body (polyp and medusa) is composed of three layers: (1) an outer, protective dermal epithelium (epidermis); (2) a middle mesoglea; and (3) an inner, gastrodermis (endoderm) that serves in digestion and absorption. The mesoglea varies from a thin, noncellular layer to a thick, fibrous jelly-like mass with or without wandering amoebocytes. Some authors prefer the term **mesolamella** for the middle layer when it is thin and noncellular and the term **collenchyma** when the mesoglea is cellular. Regardless, the mesoglea is derived from both the epidermis and gastrodermis.

Polymorphism [G., *poly*, many + G., *morph*, shape] is common in cnidarians and is often associated with their colonial existence. Various zooids function in food procurement, reproduction, defense, or protection of the

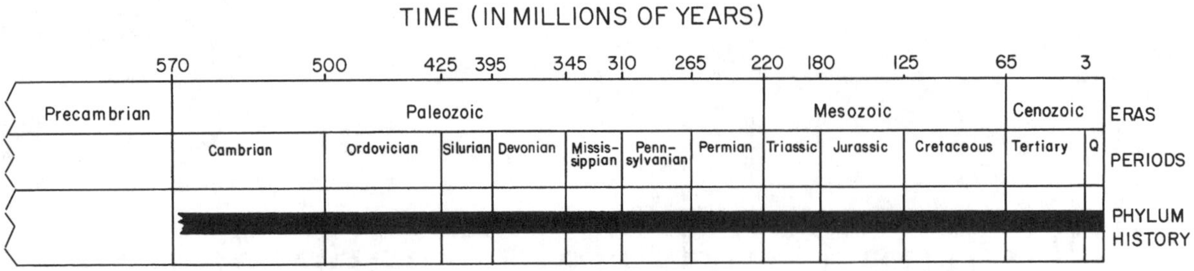

TIME (IN MILLIONS OF YEARS)
570
500
425
395
345
310
265
220
180
125
65
3
Precambrian
Paleozoic
Mesozoic
Cenozoic
ERAS
Cambrian
Ordovician
Silurian
Devonian
Missis- sippian
Penn- sylvanian
Permian
Triassic
Jurassic
Cretaceous
Tertiary
Q
PERIODS
PHYLUM HISTORY

Figure 3.1 Geological history of the phylum Cnidaria.

cnidocil
filament
nematocyst
nucleus
cnidocyte
(A)
filament
stylet
butt
capsule
cnidocyte
(B)
capsule
stylet
butt
spines
filament
operculum
(C)
(D)
(E)

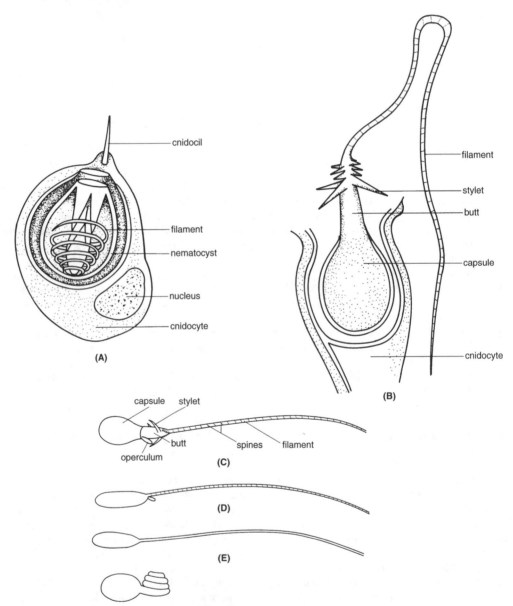

Figure 3.2 Cnidarian nematocysts: (A) undischarged, (B) discharged, (C) penetrant (stenotele), (D) streptoline glutinant (holotrichous isorhiza), (E) stereoline glutinant (atrichous isorhiza), and (F) volvent (desmoneme).

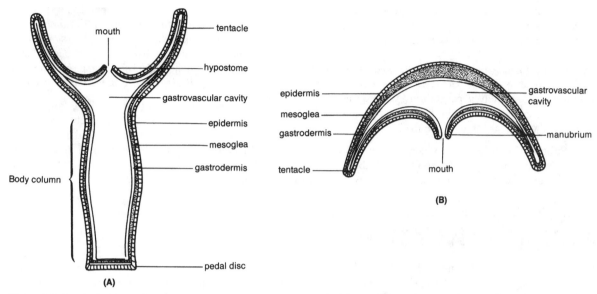

Figure 3.3 Two body forms of cnidarians: (A) polyp (hydroid) and (B) medusa (jellyfish).

colony. A large vocabulary has been developed to describe the structure and function of zooids.

Cnidarians are monoecious or dioecious. Reproduction is both asexual and sexual. Asexual budding or fission is generally associated with the polyp form, whereas sexual reproduction occurs in the medusa and in some polyps (e.g., *Hydra*). In the life cycle of many cnidarians, asexual reproduction in the polyp alternates with sexual reproduction in the medusa, a phenomenon termed **alternation of generations** or **metagenesis** [G., *meta*, back again + G., *genesis*, origin].

Sexual reproduction typically produces a free-swimming, ciliated larva called the **planula** (Fig. 3.8E). This nonfeeding, radially symmetrical larva is composed of a ciliated epidermis, mesoglea, and gastrodermis. It swims about for a time and then settles on a substrate and metamorphoses into a new individual or colony.

Classification

The phylum Cnidaria is divided into four classes.

1. **Class Hydrozoa.** Cnidarians whose life cycle includes both polyp and medusa. Medusa with velum. The mesoglea is noncellular. Gonads are epidermal. Both marine and freshwater forms are known. Seven orders are recognized, but only four are examined here.

   a. **Hydroida**. Polyp stage dominant. Medusa present or absent. Solitary and colonial. Examples: *Craspedacusta* (freshwater jellyfish), *Gonionemus*, *Tubularia*, *Clava*, *Halocordyle* (= *Pennaria*),
   Hydractinia, Polyorchis, freshwater hydras, *Obelia, Campanularia,* and *Plumularia.*

   b. **Chondrophora**. Polymorphic, pelagic, and polypoid colonies. Multichambered pneumatophore. Examples: *Velella* and *Porpita.*

   c. **Siphonophora**. Polymorphic, pelagic, and polypoid and medusoid individuals. Pneumatophore not multichambered. Colonies with float or swimming bells. Examples: *Physalia* (Portuguese man-of-war) and *Stephalia.*

   d. **Hydrocorallina**. Colonial polypoids with calcium carbonate skeletons. Examples: *Millepora* (stinging coral) and *Stylaster.*

2. **Class Scyphozoa.** Cnidarians with the medusa stage dominant. No velum. The mesoglea is cellular. Gonads are gastrodermal. All species are marine. Three orders are recognized.

   a. **Stauromedusae** (Lucernariida). Sessile with a trumpet-shaped body attached to the substratum by a stalk. Examples: *Haliclystus* and *Lucernaria.*

   b. **Semaeostomae**. Medusae bowl- or saucer-shaped with scalloped margins. Rhopalia present. Oral tentacles extend from manubrium. Examples: *Aurelia, Cyanea,* and *Pelagia.*

   c. **Rhizostomae**. Tentacles lacking around the bell. Oral arms of manubrium branched and fused. Filter feeders. Examples: *Cassiopeia, Stomalophus,* and *Rhizostoma.*

3. **Class Cubomedusae** (Cubozoa). Medusa having a cuboidal shape with unscalloped margins, four rhopalia, and velum. Examples: *Chironex* and *Carybdea*.

4. **Class Anthozoa**. Cnidarians with the medusa absent. The mesoglea is cellular. Gonads are endodermal. The gastrovascular cavity is partitioned by septa (mesenteries). All marine with both solitary or colonial forms. Two subclasses are recognized.

   a. **Subclass Zoantharia** (Hexacorallia). Polyps with more than eight, paired mesenteries. Unbranched tentacles. Solitary or colonial. Five orders of sea anemones (solitary polyps without a hard skeleton). Examples: *Metridium, Calliactis, Cerianthus*, and *Ceriantheopsis*. Two orders of hard or stony corals (solitary or colonial polyps with calcareous skeletons and without siphonoglyphs). Examples: *Astrangia* (eyed coral), *Porites*, and *Acropora* (staghorn coral).

   b. **Subclass Alcyonaria** (Octocorallia). Polyps with eight pinnate tentacles and eight, complete, unpaired mesenteries. Mostly colonial. Examples: *Tubipora* (organ pipe coral), *Clavularia, Gorgonia* (sea fan), *Leptogorgia* (sea whip), *Stylatula* (sea pen), and *Renilla* (sea pansy).

A. Class Hydrozoa

The simpler cnidarians belong to class Hydrozoa [hydro-ZO-a; G., *hydra*, water serpent + G., *zoon*, animal]. Seven representatives of the class are described in the following exercises. One or more of these may be omitted from your study to conserve time.

Hydra

Hydras are cosmopolitan inhabitants of freshwaters, often attached to the underside of plants and twigs in the water. Three common North American species are *Hydra littoralis, H. oligactis* (= *H. fusca* and *Pelmatohydra oligactis*), and *H. viridis* (= *Chlorohydra viridissima*). The green color of *H. viridis* is due to numbers of the symbiotic alga, **zoochlorella** (*Chlorella*), living within its gastrodermis. The description below will apply to any of these three hydras.

Hydras are small animals (0.3–5 cm long) with four major body regions: (1) **hypostome**, (2) **tentacles**, (3) **body column**, and (4) **pedal disc** (Fig. 3.4). The hypostome, a conical elevation of cells at the distal end, surrounds the mouth opening. Numerous mucus-secreting gland cells in the gastrodermis of the hypostome aid the hydra in swallowing food such as water fleas (*Daphnia*), cyclops (*Cyclops*), small annelids, and insect larvae. When the mouth is closed and at rest, it is stellate-shaped, but can be greatly distended when food is swallowed.

The hypostome is surrounded by 4–10 tentacles. Tentacles are hollow and can be extended more than twice the length of the body column. Numerous clumps or batteries of cnidocytes with nematocysts inside impart a bumpy appearance to the outer surface of each tentacle. Four kinds of nematocysts occur in hydra (Fig. 3.2): **penetrants, streptoline** and **stereoline glutinants**, and **volvents**.

The main part of a hydra is the body column, a cylindrical tube that can be greatly expanded or contracted. In *H. oligactis* the column can be divided into a large gastric region and a proximal, narrow, light-colored stalk (**peduncle**). The gastrodermis of the gastric region contains flagellated epithelionutritive cells and enzyme-secreting gland cells. Most digestion occurs in this region. The gastrodermis of the stalk is composed of vacuolated cells that do not produce enzymes. In *Hydra*, but not all hydrozoans, the gastrovascular cavity extends into the tentacles. Because the cavity is filled with water, it serves as a **hydrostatic skeleton** as well as in digestion and circulation.

The base of the pedal disc forms the proximal (or aboral) end of the body. Tall epidermal cells of the disc secrete a sticky mucus for attachment and form a gas bubble that allows the animal to float.

■ Observational Procedure: *Hydra*

Living Organism. A small dish containing hydras will be provided. Examine the specimen with a dissection microscope. A hydra in the extended state may reach 5 cm in length. Locate the tentacles, hypostome, body column, and basal disc (Fig. 3.4). Count the number of tentacles on your specimen. Can you see the mouth? Offer the hydra a small prey (e.g., microcrustacean) and observe prey capture. If your hydra has a lateral growth from the body, it is most likely an asexual **bud**.

Observe your specimen extending and contracting. Do the tentacles extend and contract at the same time? If disturbed, the organism will contract and remain in a nonextended state for several minutes. When contracted, a hydra resembles a jelly-like ball.

Microscopic Study

1. With a pipette, carefully remove a hydra from the dish and place it in a depression slide, adding a drop of

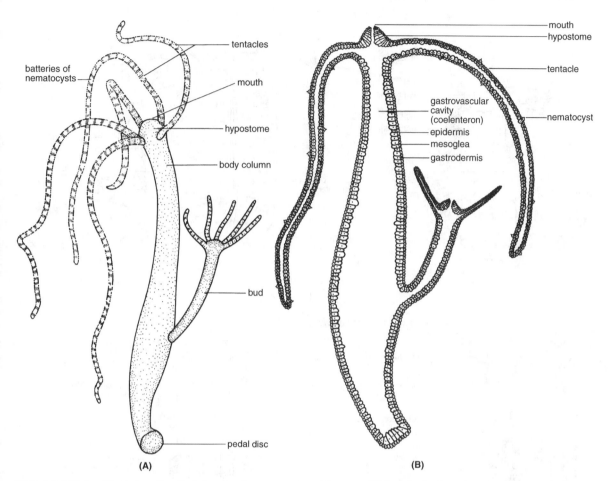

Figure 3.4 *Hydra.* (A) *Hydra* with a bud showing the gross external features. (B) Longitudinal section of *Hydra.*

pond water if needed. Your instructor will provide some hints on how this may be done without damaging the specimen. Cover the preparation with a cover slip and observe under low power of the compound microscope.

2. Locate the central, light-colored gastrovascular cavity and trace the cavity as it extends into the tentacles.

3. Compare the external morphology of the tentacles with that of the body column. Locate the batteries of cnidocytes along the tentacles. Can you determine whether the cnidocytes are more numerous on the tentacles than on the body column? Are any cnidocytes located on the pedal disc? Why does this distribution make sense?

4. Place a small drop of weak acetic acid or methylene blue near the edge of the cover slip and observe what happens to the cnidocytes when the solution contacts the specimen. Can you see discharged nematocysts? Carefully focus on the nematocysts using high power and attempt to identify the different types by referring to Fig. 3.2.

5. Examine prepared longitudinal and cross sections of a hydra under high power of the compound microscope. Locate the three body layers: outer **epidermis**, middle **mesoglea** (mesolamella), and inner **gastrodermis** (Fig. 3.5). The mesoglea appears as a line wedged between the epidermis and gastrodermis. The noncellular mesoglea is secreted by both epidermis and gastrodermis; it is thinnest in the tentacles and thickest in the body column.

Study the cell types and their distribution in the epidermis and gastrodermis by examining the slide preparations under oil immersion. *If you have never used oil immersion optics before, ask your instructor for help.* With the description below and using Fig. 3.5 as a guide, locate the following cells and understand their functions.

a. Epitheliomuscular cells. Most of the epidermis is composed of these cuboidal to columnar-shaped cells. The basal region is expanded longitudinally and contains contractile fibers. When contracted these fibers shorten the animal; therefore, they function as longitu-

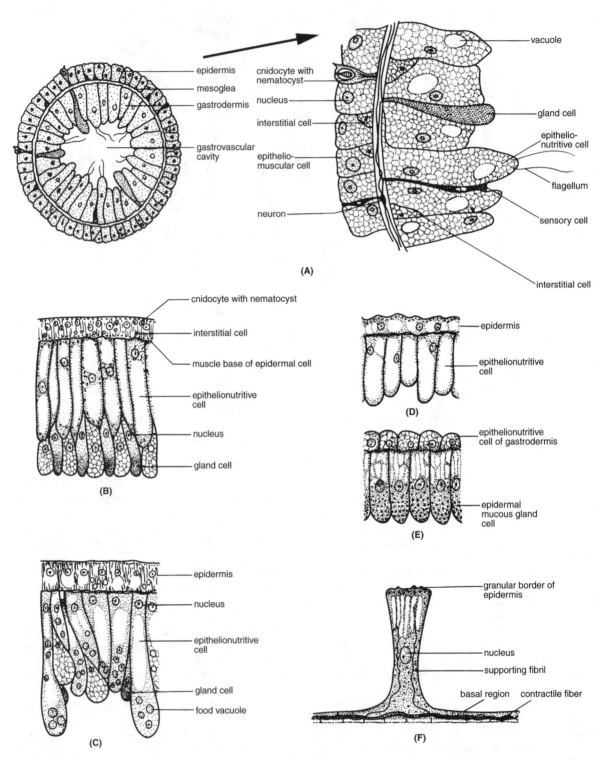

Figure 3.5 *Hydra*. (A) *Hydra* cross section. (B) Section of body wall in the region of the hypostome. (C) Section of the stomach region. (D) Section of the stalk region. (E) Section of the pedal disc. (F) Epitheliomuscular cell. (Figures B–F are after Hyman.)

dinal muscles. These cells also support and provide protection for the body.

b. Cnidocytes. These cells contain nematocysts. They occur in groups or batteries and are most abundant on the tentacles. A few cnidocytes occur in the stalk of *H. oligactis*, but are rare in the pedal disc. None occur in the gastrodermis.

c. Interstitial cells. These **I-cells**, as they are often called, are found primarily in the epidermis of the gastric region, wedged between epitheliomuscular cells. I-cells occur singly or in clusters (**nests**) and have dark-stained nuclei. These cells are precursors or stem cells of all other cell types in hydras.

d. Epithelionutritive (nutritive or nutritive-muscular) cells. These tall, columnar digestive cells with food vacuoles make up most of the gastrodermis. Like their counterparts in the epidermis (i.e., epitheliomuscular cells), the base of each nutritive cell is elongated and contains contractile fibers. The fibers are oriented circularly at right angles to those of the epitheliomuscular cells. Contraction of the nutritive cells elongates the hydra and reduces the size of its gastrovascular cavity. Unlike the epitheliomuscular cells, epithelionutritive cells bear two flagella that extend into the gastrovascular cavity.

e. Gland cells. These large, vacuolated cells secrete digestive enzymes and mucus. They are commonly found in the gastrodermis and hypostome.

f. Neurons and sensory cells. These form the nerve net plexus of both epidermis and gastrodermis. They will be difficult to see.

g. Mucous gland cells. These tall epidermal cells do not produce enzymes. The cells are especially abundant in the pedal disc and hypostome. Those in the pedal disc produce mucus for attachment.

Reproduction. Asexual (budding) and sexual reproduction occur in hydras. Dividing epitheliomuscular and nutritive cells in the stalk initiate budding. The young polyp or bud grows until it detaches from the parent hydra in 2 to 4 days. Observe budding in living hydras or from prepared slides (Fig. 3.4). What is the general shape of the bud; is it identical to the parent?

Most species of *Hydra* are dioecious. The ovary and testis arise from aggregated interstitial cells in the epidermis. Unlike most cnidarians, hydras lack a medusa and a planula larva.

Examine living hydra or prepared slides (whole mounts and cross sections) showing the testis and ovary. Distinguish between the two reproductive structures. Are the testis and ovary true organs such as found in higher animals? To be organs, what would these structures need to possess? ■

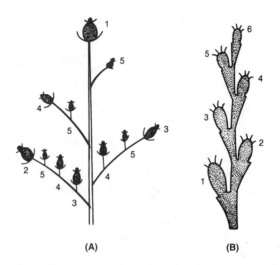

(A)   **(B)**

Figure 3.6 Monopodial (A) and sympodial (B) growth patterns (diagrammatic). Age of the polyps indicated by relative size and numbers with the original polyp numbered 1.

Pennaria

Pennaria is a colonial, shallow-water, marine hydroid that forms a minute plantlike growth on pilings, rocks, and seaweed. The pinnate colony, which may reach 15 cm in height, is attached to the surface of objects by hollow, rootlike **hydrorhiza** (stolons or rhizomes). Stemlike structures called **hydrocauli** grow upright from the hydrorhiza. The hydrocaulus buds off alternating side branches called **hydrocladia** from which are budded the terminal feeding polyps called **hydranths** (gastrozooids). The type of growth in *Pennaria* is **monopodial** because the original or primary hydrocaulus elongates throughout the colony's life. Therefore, the oldest polyp occupies the tip of the original hydrocaulus (Fig. 3.6A).

■ Observational Procedure: *Pennaria*

1. Obtain specimens of a preserved or living colony and prepared slides (whole mounts and cross sections) for study. Using dissection and compound microscopes, confirm that the growth of *Pennaria* is monopodial and not sympodial (Fig. 3.6).

2. Locate a hydranth and observe the large, conical-shaped hypostome containing the mouth (Fig. 3.7). Note the **capitate tentacles** on the hypostome and the ring of **filiform tentacles** that circles the base of the gastral region. Examine both types of tentacles for cnidocytes. Are they uniformly distributed? How might this distribution be effective in prey capture?

3. Find the **coenosarc**, a living hollow tube that extends throughout all branches of the colony (Fig. 3.7). The coenosarc consists of an outer epidermis, a

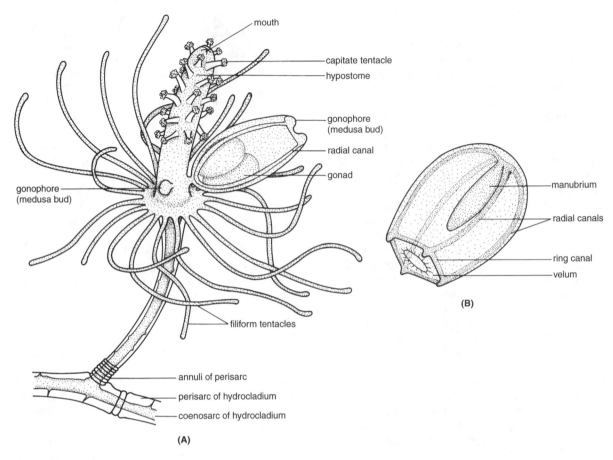

Figure 3.7 *Pennaria.* (A) Enlarged hydranth with medusa bud. (B) Medusa.

middle mesoglea (mesolamella), and an inner flagellated gastrodermis that lines the gastrovascular cavity. The cavity is continuous throughout the colony, at the base of the hydranth it enlarges as a type of stomach region. The epidermis of the coenosarc of the hydrocaulus and hydrocladium secretes an outer, protective chitinous covering, the **perisarc** (periderm).

4. In *Pennaria*, the perisarc adheres closely to the outer surface of the colony and is annulated at the base of each stem. What function(s) might these annulations serve? Carefully trace the extent of the perisarc; does it cover the entire polyp? *Pennaria* is an **athecate** (**gymnoblastic**) hydrozoan because the hydranths and gonophores (medusa buds) lack the perisarc covering.

5. Locate the small, thimble-shaped gonophores that arise by budding off the hydranth just distal to the filiform tentacles. The medusa lacks a mouth, but has a velum and four short tentacles (Fig. 3.7). *Pennaria* medusa are dioecious. ■

Obelia

Obelia is another colonial marine hydroid that forms a minute plantlike growth on rocks, pilings, and other substrates. *Obelia* is similar to *Pennaria* in a number of ways; however, there are important differences between the two hydrozoans.

■ Observational Procedure: *Obelia*

1. Obtain preserved or living specimens of the *Obelia* colony for study using dissection and compound microscopes. Determine the growth form of *Obelia*; is it monopodial or sympodial (Fig. 3.6)? Can you distinguish the youngest polyps from the oldest ones in the colony?

2. Locate the two kinds of polyps occurring on the *Obelia* colony: **hydranth** (**gastrozooid**) and **gonangium** (**gonozooid**) (Fig. 3.8). Both polyps are

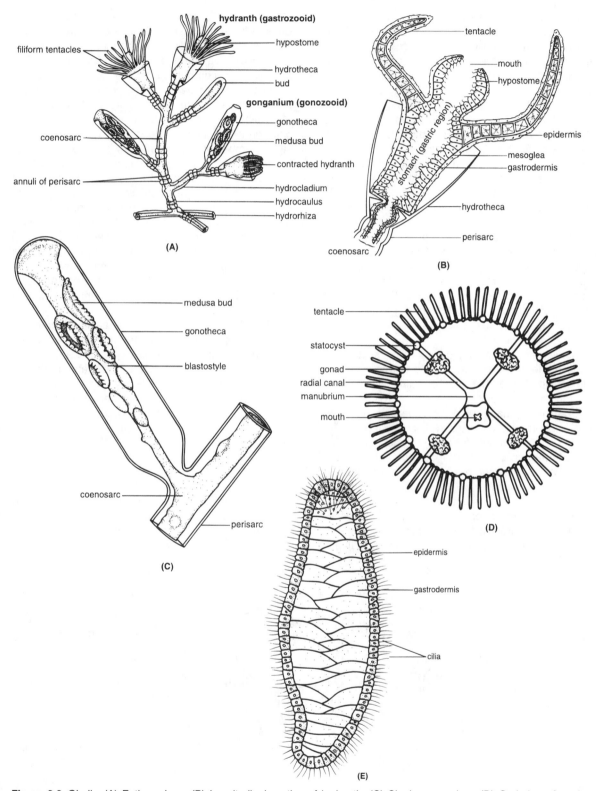

Figure 3.8 *Obelia*. (A) Entire colony, (B) Longitudinal section of hydranth. (C) Single gonangium. (D) Oral view of medusa. (**E**) Planula larva.

supported by upright, stemlike, branched hydrocauli and rootlike hydrorhiza attached to a substrate. Where are the two polyp types located in the colony?

3. Observe that the entire colony is covered with a nonliving, chitinous perisarc. The perisarc covering each hydranth is bell-shaped and called the hydrotheca, whereas the perisarc of the gonangium is the gonotheca. Because the perisarc extends onto the polyps, *Obelia* is a **thecate** (calyptoblastic) hydroid, unlike *Pennaria* which is **athecate** (gymnoblastic). The perisarc of the hydrocauli is annulate at various places which allows flexibility of movement to the colony. Like *Pennaria*, the living coenosarc consists of epidermis, mesoglea, and gastrodermis; the hollow gastrovascular cavity is continuous throughout the colony.

4. Examine a hydranth (feeding polyp) and locate its hypostome that bears the mouth. The filiform tentacles of *Obelia* are solid and provided with cnidocytes. The tentacles can be retracted into the hydrotheca. Note that the tentacle arrangement in *Obelia* is different from that of *Pennaria* and that capitate tentacles are absent. How are the cnidocytes distributed on the tentacles in a polyp of *Obelia*?

5. Study prepared cross and longitudinal sections of *Obelia* made in the regions of the hypostome and hydrocaulus. How do these sections differ with comparable regions of hydras?

6. Examine a **gonangium** or reproductive polyp. Note that it lacks tentacles, a mouth, and hypostome. Observe the asexually produced discoid medusa buds (**gonophores**) on the club-shaped **blastostyle** of the gonangium. The blastostyle is a modified polyp and represents a continuation of the coenosarc into the gonangium. The medusa, termed a **leptomedusa**, escapes into the water from the blastostyle via the gonopore at the terminal end of the gonangium. A pelagic existence occurs before the medusa reaches sexual maturity.

7. Observe a prepared slide of a medusa of *Obelia* (Figs. 3.8D and 3.9). Note its bell or umbrella shape. The upper, convex surface is the **exumbrella** and the lower, concave surface is the **subumbrella**. The **manubrium**, comparable to the hypostome of the hydranth, hangs down from the center of the subumbrella surface. Locate the mouth in the manubrium?

8. Observe the solid tentacles around the bell's margin. Count the tentacles. Does each hydranth have the same number of tentacles? How are the cnidocytes distributed on the tentacles in a medusa of *Obelia*?

9. Locate the four **radial canals** and four **gonads** attached to them. Radial canals meet the **ring canal** that encircles the margin of the bell. Eight **statocysts** are located on the margin of the bell in the interradial areas. The gastrovascular cavity of the medusa is lined with gastrodermal cells. The cavity includes the spaces of the manubrium, radial canals, and the ring canal. *Obelia's* medusa (Fig. 3.9) is atypical, compared to most hydromedusa such as *Gonionemus* because a velum is lacking (Fig. 3.11).

10. Locate the four gonads of the animal. Medusa of *Obelia* are dioecious. Gametes are fertilized in the water and the zygote develops into a planula larva which settles onto a substrate and metamorphoses into a sessile colony. *Obelia* is a good example for understanding metagenesis, the alternation of asexual and sexual reproduction (Fig. 3.8). ∎

Plumularia

Plumularia is a marine, calyptoblastic colonial hydroid with creeping hydrorhiza. Arising from the hydrorhiza are hydrocauli with attached plumelike branches (hydrocladia), each bearing on one side a series of bell-shaped

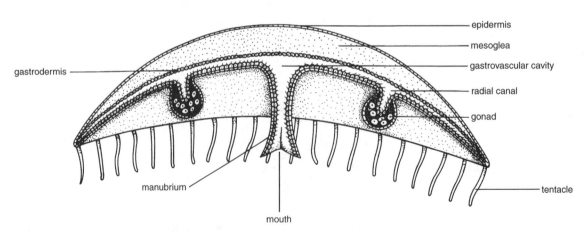

Figure 3.9 Oral-aboral section of *Obelia* medusa.

hydrothecae containing the gastrozooids. The sessile hydrotheca is so small that the gastrozooid with its filiform tentacles cannot be retracted within the encasement.

■ Observational Procedure: *Plumularia*

Examine a prepared slide or preserved or living specimen of *Plumularia* (Fig. 3.10). Note the small dactylozooids called **nematophores** (sarcostyles) that are characteristic of *Plumularia* and its relatives. Nematophores arise from tiny thecae, the **nematothecae**, located on the hydrocaulus and on the hydrotheca of the gastrozooid. There are usually three nematophores to each gastrozooid. The nematophores are protective polyps. They have both cnidocytes and a long amoeboid process that engulfs polyps, diatoms, protozoa, and larval stages of epizoic organisms. Gonangia extend from the hydrocaulus at the point where each hydrocladium arises. No special modifications for the gonotheca exist in *Plumularia*. Compare the colony of *Plumularia* with that of *Obelia* and *Pennaria*. Can you determine the growth pattern of *Plumularia*? ■

Gonionemus

Gonionemus, containing several species, is a marine hydrozoan commonly found in shallow water attached to seaweed. The small (2 cm) medusa is the dominant stage, whereas the athecate polyp is only about 1 mm in length. The solitary polyp with its four, solid tentacles resembles a small hydra. Nonciliated, planula-like bodies called **frustules** bud from the polyp and grow into additional solitary polyps. Seasonally, medusa buds (gonophores) are formed on the column of the polyp; the buds grow into medusae.

Gonionemus is dioecious. Gametes from the gonads, which are attached to the radial canals, are shed in the water where they unite. The zygote develops into a planula larva which settles onto a substrate and grows into the small polyp.

■ Observational Procedure: *Gonionemus*

Study a medusa of *Gonionemus* in a small dish with water using the dissection microscope. Locate the following structures: (1) exumbrella, (2) subumbrella, (3) velum, (4) short manubrium with its four frilled lobes, (5) tentacles, (6) four radial canals with their ruffled gonads, and (7) the ring canal around the margin of the bell (Fig. 3.11). Can you find the mouth in the manubrium? Count the number of hollow tentacles around the bell's margin. Closely examine a tentacle, noting that the cnidocytes with their nematocysts are arranged in a ringlike fashion (batteries) around the tentacles. Can you see the **adhesive sucker** near the distal end of each tentacle? What function might the sucker serve? At the base of each tentacle is a prominent sensory swelling called a **tentacular bulb**. Between the bulbs are balancing structures called statocysts. ■

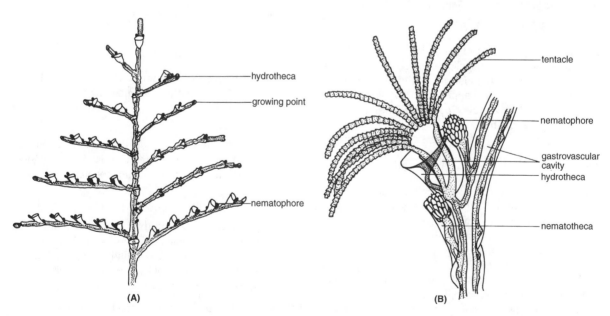

(A)

(B)

Figure 3.10 (A) Colony of *Plumularia* showing monopodial growth. (B) Hydranth of *Plumularia* showing two of three nematophores.

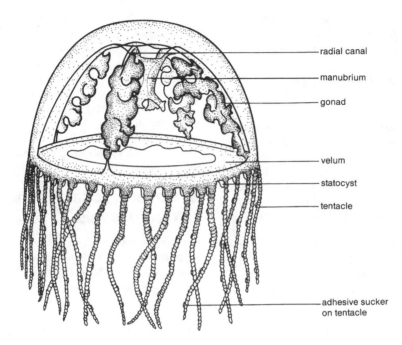

radial canal

manubrium

gonad

velum

statocyst

tentacle

adhesive sucker
on tentacle

Figure 3.11 *Gonionemus* medusa.

Craspedacusta

Craspedacusta sowerbyi is a freshwater hydrozoan whose medusa is the dominant form (Fig. 3.12). The athecate, solitary polyp is very small (2 mm) and lacks tentacles.

■ Observational Procedure: *Craspedacusta*

Examine a medusa of *Craspedacusta* in a small dish with water using the dissection microscope. Compare the anatomy of the medusa to that of *Gonionemus*. How are these two organisms similar? Do you see any differences? If available, compare the polyp of *Craspedacusta* to a hydra. ■

Siphonophora

Members of the order Siphonophora, such as the Portuguese man-of-war (*Physalia*), are perhaps the most complex and specialized of hydrozoans. These pelagic, colonial marine animals attain the maximum development of polymorphism found in cnidarians.

The siphonophore colony originates from a planula larva. The colony consists of both modified medusoid and polypoid individuals (zooids) specialized to perform different functions. The polypoid individuals are modified for feeding, capturing prey, and producing the repro-

ductive zooids. The medusoid individuals are concerned with locomotion, flotation, and reproduction. The different types of zooids that may occur in siphonophores are listed below.

1. **Polypoid zooids**

   a. **Gastrozooid (trophozooid).** A feeding polyp with a mouth and one long, contractile, hollow tentacle with cnidocytes. The tentacle originates near the base of the polyp.

   b. **Dactylozooid (palpon).** This tactile polypoid resembles a gastrozooid in having a long tentacle, but a mouth is lacking. The tentacle is never branched.

   c. **Gonozooid.** This reproductive polyp lacks a tentacle. A mouth may be present as in *Velella*. A gonozoid usually bears several gonophores (see below).

2. **Medusoid zooids**

   a. **Nectophore (swimming bell).** This modified, very muscular medusa is used for flotation. The nectophore has a bell, velum, four radial canals, and a ring canal. It lacks a mouth, manubrium, tentacles, and sense organs.

   b. **Phyllozooid (bract).** This leaflike zooid bears cnidocytes for protection.

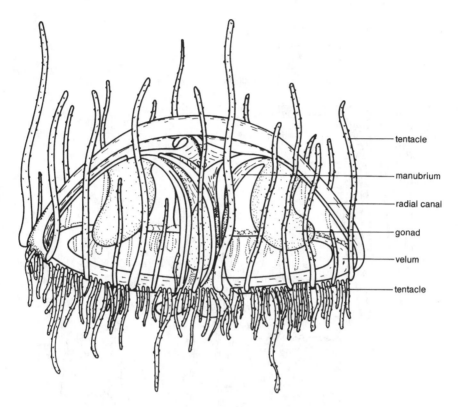

Figure 3.12 *Craspedacusta* medusa.

c. **Gonophore**. This medusoid occurs on gonozooids or branched stalks called **gonodendra**. Gonophores lack a mouth, tentacles, and sense organs. They have a bell, velum, manubrium, and radial canals. Gonophores occur singly or in grape-like clusters. Some are saclike such as the sporosacs of some hydroids. Gonophores are dioecious even though the colony is monoecious. Female gonophores are medusoid and may swim free; male gonophores are reduced.

d. **Pneumatophore**. This is a float that may or may not have a pore to regulate the gas it contains. Some workers believe that the pneumatophore is a modified polyp instead of a medusa.

Physalia. The bluish pneumatophore of *Physalia* is derived from an aboral evagination from the planula larva. The float is highly muscular, double-walled, lacks mesoglea, and lined with a chitinous epidermal secretion. The float has its greatest development in *Physalia* where it is an ovate bladder 30 cm in length. On the floor inside the pneumatophore is a modified glandular epithelium called the **gas gland** that produces gas (mainly nitrogen and carbon dioxide) contained inside the float. *Physalia*

unlike some species, lacks a pore to regulate the amount of gas within the float. Attached to the pneumatophore is an erectile, sail-like **crest** that can be lowered and raised. In the living organism the crest is blue bordered with pink.

The various zooids of *Physalia* are budded from a ventral disc area beneath the pneumatophore. The gastrovascular cavity and the coenosarc of the zooids and pneumatophore are continuous. Each group of zooids extending from the disc beneath the float is called a **cormidium**. *Physalia* possesses several cormidia, each consisting of a gastrozooid, male and female gonozooids, and one dactylozooid with an extremely long tentacle (10 m). The dactylozooids from several cormidia form a suspended driftlike net. Contact with the tentacles results in discharge of the nematocysts. The tentacles contract and pull the prey to the gastrozooids.

■ Observational Procedure: *Physalia*

In a dish containing water, carefully examine the specimen provided. Locate the pneumatophore, crest, and gas gland (Fig. 3.13). The latter may be observed within the floor of the pneumatophore by holding the pneu-

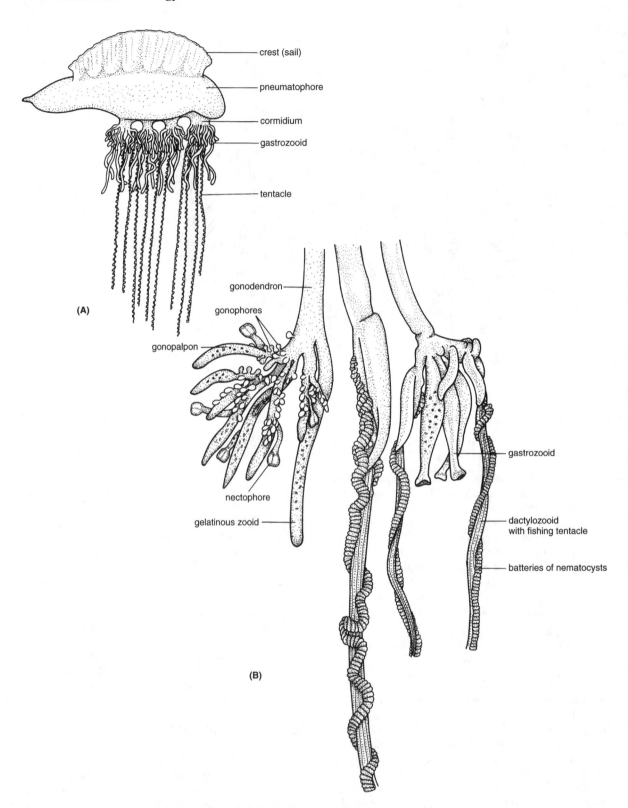

Figure 3.13 *Physalia*. (A) Entire colony. (B) Portion of a colony showing the types of zooids.

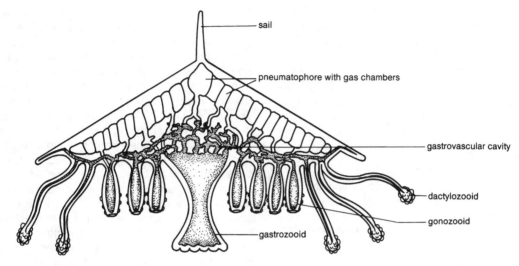

Figure 3.14 *Velella* colony.

matophore up to a bright light. Study one cormidium and identify the gastrozooid, with its mouth and long tentacle; the gonozooid, with small, grapelike, sessile gonophores; and the dactylozooid, with its long tentacle. The stemlike, branched structure bearing the gonozooid is called a **gonodendron**. A gonodendron may bear **gonopalpons** (modified dactylozooids lacking a mouth), **gelatinous zooids**, and **nectophores** (modified medusae). Each nectophore consists of a bell, velum, four radial canals, and a ring canal, but it lacks a mouth, manubrium, tentacles, and sense organs. ■

Velella. Velella, commonly called By-the-Wind Sailor, is a pelagic, polymorphic chondrophoran. The ovoid body is flattened with the pneumatophore on the aboral surface as a chitinous disc (Fig. 3.14). Arising from the aboral surface of the pneumatophore is a thin, chitinous, triangular **sail**. The living organism is bluish in color and the sail is edged with purple. *Velella* lacks control over the sail and is at the mercy of the ocean currents for movement.

Beneath the pneumatophore are three types of zooids forming the bulk of the colony (Fig. 3.14). Unlike *Physalia,* zooids of *Velella* are not greatly elongated. In the center of the colony is a single, large gastrozooid that contains the mouth at its free end. The mouth leads to a large gastrovascular cavity. On both sides of the gastrozooid are several gonozooids with small, grapelike gonophores that escape into the water as free-swimming medusae. Around the margin of the disc is a fringe composed of several dactylozooids.

The cavities of gastrozooids and gonozooids communicate with the cavity of the pneumatophore via ectodermal tubes. These tubes are extensions of the multichambered pneumatophore. *Velella* lacks swimming bells.

■ Observational Procedure: *Velella*

In a dish containing water, carefully examine a specimen of *Velella.* Locate the sail, pneumatophore (with gas chambers), gastrozooid, gonozooids (with gonophores), and the dactylozooids. ■

B. Class Scyphozoa

The approximately 200 species in the class Scyphozoa [sy-fo-ZO-a; G., *skyphos,* cup + G., *zoon,* animal] are often called true jellyfish. The dominant body form is an acraspedote medusa (velum lacking) having a cellular mesoglea or collenchyma. The medusa may attain a diameter of 1 m such as in *Cyanea capillata.* Scyphozoans are carnivorous, feeding on a variety of marine animals: fish, annelids, ctenophores, crustaceans, and other planktonic forms. Most scyphozoan medusae are active swimmers propelled through the water by rhythmic pulsations of the bell; however, some members are sluggish, bottom-dwellers.

Two examples of the class will be studied: (1) *Aurelia aurita*, an active pelagic scyphozoan in the medusa stage; (2) *Cassiopeia*, a sluggish jellyfish that spends most of its life on the bottom of shallow seas.

Aurelia aurita

The moon jelly, *Aurelia aurita*, is the most common jellyfish on both American and European coasts. The tetramerous medusa is relatively flat and ranges from 7 to 10 cm in diameter (Fig. 3.15A). Four **subgenital pits** occur on the subumbrella surface. Cilia cover both the exumbrella and subumbrella surfaces.

■ Observational Procedure: *Aurelia aurita*

1. Obtain an adult *Aurelia* in a bowl containing water and examine the specimen with a dissection microscope. Find the centrally located mouth, at the free end of the manubrium, surrounded by four long, frilly **oral arms** bearing short tentacles (Fig. 3.15A). The oral arms are extensions of the manubrium. Food trapped in mucus is transferred to the mouth by **lappets** located around the margin of the bell.
2. Follow the mouth into the central stomach that connects to four short canals, each leading to four radially arranged **gastric pouches**. The gastrodermis of the pouches bears tentacle-like projections called **gastric filaments**, chief sites of enzyme secretion and intracellular digestion. Numerous cnidocytes with nematocysts cover the filaments.
3. Observe the many radiating canals in the bell; these connect with the ring canal located around the bell's margin. The fluid-filled canals have special names depending upon their location and function: **adradial, interradial**, and **perradial** (Fig. 3.15A). All canals plus the stomach constitute the gastrovascular cavity.
4. Observe the fringe of numerous, short tentacles hanging around the margin of the bell. Where the perradial canal meets the ring canal is a ciliated **rhopalium** located between two rhopalial lappets (Figs. 3.15F and 3.16). *Aurelia* has eight rhopalia and sixteen rhopalial lappets. The rhopalium consists of sensory cells, sensory pits, and a statocyst, an organ of equilibrium. Unlike many of its relatives, *Aurelia* has a light-perceiving eyespot (**ocellus**) on the rhopalium.

5. *Aurelia* is dioecious. Find the four, interradial gastrodermal gonads and four well-developed genital pits. Fertilization and development to the planula larva occur in the folds of the oral arms. If small opaque patches are seen on the inner surfaces of the oral arms, carefully dissect out and open some of them and examine under high power of a dissection microscope. Do you find planulae? If so, examine them using the compound microscope. The mature, ciliated planula leaves the adult and metamorphose into an **actinula** or **hydratuba** stage (Fig. 3.15C). This stubby, hydra-like body with solid tentacles settles down and develops into a polypoid stage, the **scyphistoma**.
6. Examine a cross section of a scyphistoma and observe the four septa that partition off the gastrovascular cavity. Like the medusa, the scyphistoma is tetramerous. The scyphistoma can produce additional scyphistoma by asexual budding which occurs in autumn and winter. In late winter and spring, the scyphistoma undergoes transverse fission (**strobilation**), which results in a **strobila**, a polyp resembling a stack of saucers. The saucer-shaped bodies are **ephyrae** or immature medusae (Fig. 3.15F). Each ephyra breaks off the strobila one at a time and eventually develops into an adult medusa. Examine prepared slides showing all stages of development in *Aurelia*. Compare the stages with Fig. 3.15, paying particular attention to comparison of the ephyra to the adult *Aurelia*.

The life cycle of *Aurelia* involves alternation of asexual reproduction with sexual reproduction (metagenesis). Distinguish the asexual phase from the sexual. What advantage might there be to the organism in having such reproductive cycles? ■

Cassiopeia

Cassiopeia (also spelled *Cassiopea*), was named from a mythological queen of Ethiopia. This common rhizostome scyphozoan lives in calm, shallow-water lagoons of Florida and the West Indies. The adult medusa may reach 26 cm in diameter and is unusual in its habit of living upside-down on the bottom of the seafloor. *Cassiopeia* attaches to the sea floor by having a raised circular zone around the margin of the bell thus creating a suckerlike action on the bell's aboral surface. Pulsation of the bell creates water currents that bring oxygen and food to the animal as well as carry wastes away. Small planktonic organisms are swept over the frilly oral arms, are paralyzed by discharged nematocysts, trapped in mucus strands, and then carried into the secondary, porelike mouths. Symbiotic **zooxanthellae** also live within the mesoglea of *Cassiopeia*. If disturbed, *Cassiopeia* rises off the sea floor, swims a short distance, and settles down again.

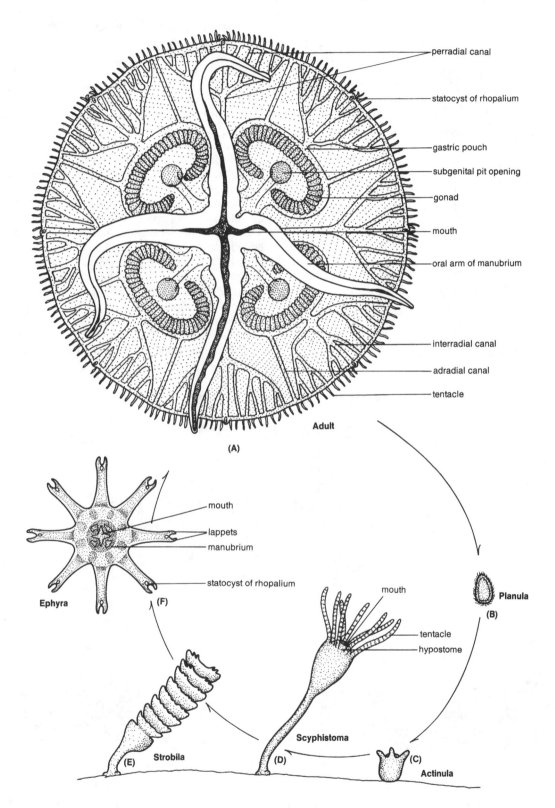

perradial canal

statocyst of rhopalium

gastric pouch

subgenital pit opening

gonad

mouth

oral arm of manubrium

interradial canal

adradial canal

tentacle

Adult

(A)

mouth

lappets

manubrium

statocyst of rhopalium

Ephyra

(F)

mouth

tentacle

hypostome

Scyphistoma

(D)

Planula

(B)

(C)

Actinula

(E)   **Strobila**

Figure 3.15 *Aurelia* life cycle.

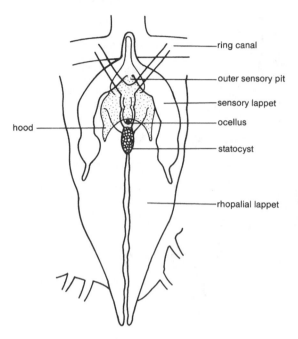

Figure 3.16 Rhopalium of *Aurelia*.

■ Observational Procedure: *Cassiopeia*

1. Obtain a specimen of *Cassiopeia* in a small dish of water and observe it with the dissection microscope. Note the flattened bell and lack of tentacles (Fig. 3.17). Instead of tentacles around the bell's margin as seen in *Aurelia*, *Cassiopeia* has a fringe of numerous, small **scallops**. Regularly spaced between certain scallops are 16 rhopalia. The rhopalia play an important role in controlling pulsations of the bell and in regenerating missing areas of the bell.

2. Excise a rhopalium with a sharp razor blade or needle. Place the rhopalium on a glass slide with a drop of water, cover with a cover slip, and examine with the compound microscope. A small amount of methylene blue added to the specimen may enhance your observation. Compare the rhopalium with that of *Aurelia*. Is a hood present? Unlike *Aurelia*, *Cassiopeia* lacks an eyespot.

3. Note that *Cassiopeia* lacks the distinct, central manubrium with a mouth and four separate oral arms observed in *Aurelia*. The four oral lobes are expanded and bifurcated in the adult to form eight, thick, gelati-

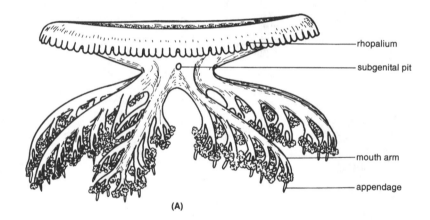

(A)

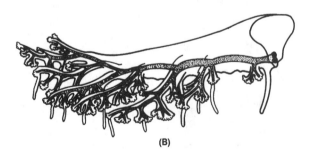

(B)

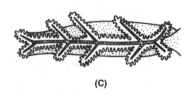

(C)

Figure 3.17 *Cassiopeia*. (A) Lateral view of adult *Cassiopeia* with oral surface down. (B) Lateral view of *Cassiopeia's* mouth arm showing the gastrovascular canals. (C) A mouth of *Cassiopeia* closed.

nous frilly arms. The original arm groove closes over, forming a branched **brachial canal** that opens to the oral surface by minute pores called secondary or brachial mouths. Using an illuminator, shine light through the specimen. This should permit the brachial canals to be seen. Numerous elongated appendages bearing many cnidocytes with nematocysts and mucus cells extend from the arms (Fig. 3.17). These are used in capturing food.

4. Cut off a small piece of the frilly arm at its distal end, make a wet mount, and observe under high power of the compound microscope. A small drop of methylene blue added to the specimen before the cover slip is affixed may aid your observation. You should see numerous, undischarged nematocysts.

5. Follow the brachial canals to the central stomach where extracellular digestion occurs. Numerous radial canals extend from the stomach; the canals form an anastomosing network of channels throughout the bell. A distinct ring canal, as observed in *Aurelia*, is lacking.

Examine the flattened aboral surface of the bell to see the numerous radial canals.

6. Lift up the bell and locate the four openings, called **subgenital pits**, located in the stalklike area between the arms and flattened bell. The pits lead to a cross-shaped space, the **subgenital porticus**, which lies below the stomach.

7. With a sharp razor blade or scalpel, cut the flattened bell from the stalk to better observe the four gonads. Note their curved shapes. The gonads lie on the edge of the central stomach above the subgenital pits. The pits do not connect to the gonads. The gametes are discharged into the porticus.

Cassiopeia is dioecious. The planula larva settles down and develops into a scyphistoma. One ephyra at a time buds from the scyphistoma. *Cassiopeia* lacks a strobila. The ephyra settles on the sea floor when it reaches about 2 cm in diameter, and there it develops into the adult. *Cassiopeia* also buds off small planulae from the scyphistoma. ■

C. Class Anthozoa

Over 6,000 species, including of sea anemones and corals, belong to the largest cnidarian class, Anthozoa [an-tho-ZO-a; G., *anthos*, flower + G., *zoon*, animal]. The name flower animal is appropriately assigned to these animals because many are highly colored and resemble flowering plants. These marine invertebrates occur in both deep and shallow waters of tropical and temperate seas but reach their greatest abundance in tropical areas.

All adult anthozoans are sessile polyps; there is no medusa. Anthozoan polyps may be elongated and thin-walled or short, rather squatty, and thick-walled. The polyps occur as solitary individuals (e.g., sea anemones) or as complex colonies with several hundred interconnecting polyps (e.g., corals).

The anthozoan polyp differs in a number of important features from hydrozoans and scyphozoan polyps. Instead of having a hypostome, the oral surface of an anthozoan polyp is typically expanded and flattened as an **oral disc** bearing an ovoid mouth and hollow tentacles studded with cnidocytes (Fig. 3.18). Radial muscles in the disc permit opening of the mouth during feeding. The number of tentacles varies from six to more than 100, and the cavities of the tentacles communicate with the gastrovascular cavity.

Anthozoans are the only cnidarians with the gastrovascular cavity divided by longitudinal (oral-aboral) **septa** or **mesenteries** and in having a pharynx (stomodaeum,

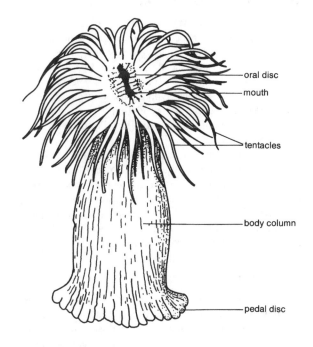

Figure 3.18 External view of sea anemone.

LIVERPOOL JOHN MOORES UNIVERSITY
LEARNING SERVICES

actinopharynx, esophagus, gullet) (Figs. 3.19 and 3.20). The septa are composed largely of mesoglea bounded by a layer of gastrodermis on either side. The septa that extend from the body wall to the pharynx are called **primary** (perfect or complete) septa (Fig. 3.19). In zoantharian anthozoans, there are **incomplete** (imperfect) septa that do not meet the pharynx and according to their lengths, they are called secondary, tertiary, and quaternary septa. The free ends of the septa near and below the pharynx form coiled **septal filaments** (mesenteric filaments) containing enzyme-producing cells and cnidocytes (Fig. 3.19).

The pharynx is an ectodermally lined, ciliated tube chiefly for ingestion of food. The pharynx connects the mouth and gastrovascular cavity, and typically bears one or more **siphonoglyphs** (sulcuses)—ciliated grooves or gutters for moving water through the mouth and into the gastrovascular cavity (Fig. 3.20). Coral polyps lack the siphonoglyphs.

Two subclasses, Zoantharia (or Hexacorallia) and Alcyonaria (or Octocorallia), comprise class Anthozoa. Zoantharians (sea anemones, stony corals) have their structures arranged in sixes (hexamerous) or multiples thereof. The several unbranched hollow tentacles are arranged in one or more circles around the mouth on the oral disc. Alyconarians (soft and horny corals) are octamerous with eight, hollow, pinnate tentacles arranged around the outer margin of the oral disc. Selected examples from the subclasses will be studied.

Subclass Zoantharia

Most species of class Anthozoa are zoantharians (hexacorallians). The 3,500 species include sea anemones and hard corals. The hexamerous symmetry and paired septa partitioning the gastrovascular cavity are two important distinguishing features of zoantharians. The pharynx bears at least one siphonoglyph in anemones, but in hard corals the siphonoglyph is absent. Tentacles of zoantharian polyps are usually simple and not like those of alyconarian anthozoans.

All septa in zoantharians are coupled, being arranged in a bilateral or biradial manner. Also, most zoantharians's septa occur close together in twos and are said to be paired. The gastrovascular cavity between pairs is the **exocoel** and the cavity between members of a septal pair is the **endocoel** (Fig. 3.19). The exocoels and endocoels communicate with the cavities of the tentacles.

Metridium

Metridium is a common sea anemone found along the Atlantic and Pacific coasts. It belongs to the order Actinaria which contains all typical sea anemones. *Metridium* will be studied in detail to illustrate the basic anatomy of an anthozoan polyp.

■ Observational Procedure: *Metridium*

1. Place a specimen of *Metridium* in a dissection pan and observe its general morphology (Fig. 3.18). Make a list of the similarities this specimen has to a hydra. In making this comparison consider size, body form, and internal structure.

2. Locate the **oral disc, body column**, and **base** (Fig. 3.18). The cylindrical-shaped body column can be divided into an upper, thin-walled **capitulum** and a lower, thick-walled, **scapus** (Fig. 3.20).

3. Between the capitulum and scapus is a groove, the **fosse**, bordered by a prominent fold of the body wall called the **collar** (parapet).

4. Note how the closed, aboral base is expanded into the **pedal disc** by which the organism attaches to a substrate. Although *Metridium* is sessile, it can detach and move to another site. The oral end contains an ovoid mouth in the center of the oral disc.

5. Locate the **peristome**, an area devoid of tentacles immediately surrounding the mouth.

6. Examine the numerous hollow tentacles bearing cnidocytes that border the peristome and cover the oral disc. Are the tentacles arranged in any definite pattern? Are all tentacles on the oral disc the same length? The tentacles contain longitudinal and circular muscle fibers and can be expanded and shortened in the living animal.

7. Find the mouth and the ciliated pharynx (Fig. 3.20). At one or both ends of the mouth and extending down into the pharynx locate the ciliated groove or gutter called the siphonoglyph that moves water into the gastrovascular cavity. How does the position and number of siphonoglyphs in the mouth superimpose a bilateral symmetry onto this radially symmetrical animal?

8. Using a sharp scalpel, cut longitudinally a preserved specimen cleanly and evenly into two equal halves beginning at the oral end and extending through the pedal disc. With another specimen, make two, clean cross sections through the body column. One section should transect the pharynx and the other section should be made below the pharynx and near the pedal disc. Locate the pharynx, siphonoglyph, and gastrovascular cavity (Figs. 3.19 and 3.20).

9. Study the location and arrangement of the paired mesenteries (or septa) in both longitudinal and cross sections. The mesenteries are paired and occur in sets or cycles. Those mesenteries that extend from the body wall to the pharynx are called primary or complete; those that extend from the body but do not meet the pharynx are incomplete (Fig. 3.19).

10. Locate the cycles of septa. The first cycle consists of six pairs of primary septa. The second cycle consists

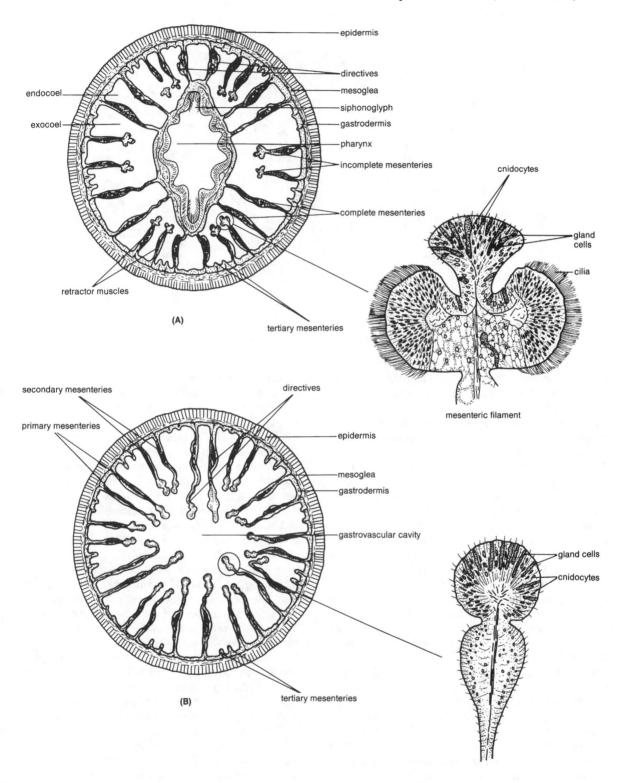

Figure 3.19 *Sea anemone.* (A) Cross section through pharynx showing two siphonoglyphs. Mesenteric filament enlarged. (B) Cross section below the pharynx. Mesenteric filament enlarged.

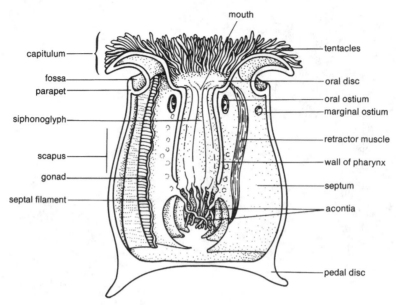

Figure 3.20 Oral-aboral section of *Metridium* showing internal structures.

of six pairs of secondary septa. The third cycle consists of 12 pairs of tertiary septa. Is there a fourth cycle of quaternary septa?

11. If you have difficulty seeing the septa cycles in the preserved specimen, they will be observed when prepared slides are studied.

12. In the longitudinally halved specimen, locate a longitudinal **retractor muscle** on one face of a septum (Fig. 3.20). This muscle occurs on the side of each paired septum facing the endocoel, except for the septa attached at the ends of the siphonoglyphs in the pharynx. On these paired, complete septa, called **directives**, the retractor muscles face the exocoels (Fig. 3.19). The retractor muscles retract the oral disc. (**NB**: The spacial relationship of septa, the muscles, and their spaces can be seen best in a prepared slide of a cross section of *Metridium*: see steps 15–21, below). Depending on how you made the longitudinal section, one to two ostia may be seen in the septa near the pharynx and body wall at the distal end of the animal. What function might these septal perforations serve?

13. At the free end of each incomplete septum locate the longitudinal cord or ridge called the **septal** or **mesenteric filament** (Fig. 3.19). The filament is trilobed in the region of the pharynx; those near the base are unilobed. The detail structure of a filament will be seen in the prepared cross section.

14. Locate the many, threadlike structures called **acontia** (Fig. 3.20). These might be more easily observed in the longitudinal section. Acontia are continuations of the

middle lobe of the lower septal mesenteries. They aid in defense and capture of prey, bear cnidocytes and enzyme producing gland cells, and can be protruded through the mouth and through minute openings, **cinclides**, located in the lower body column. When the column contracts, water is expelled through the cinclides.

Metridium is dioecious. Careful dissection and observation may reveal the gonads, which are thickened bands located behind a mesenteric filament (Fig. 3.20).

15. Examine a prepared slide showing a cross section of *Metridium* through the pharynx under low power of a compound microscope (Fig. 3.19). Locate the outer epidermis of columnar cells, the inner gastrodermis lining the gastrovascular cavity, and the middle mesoglea containing ameboid cells and muscle fibers in the body wall. The inner wall of the pharynx is folded and lined with epidermis, containing cnidocytes and gland cells. The outer covering of the pharynx is gastrodermis and the middle area is mesoglea.

16. Locate the siphonoglyph. Do you see more than one siphonoglyph? Observe the cilia of the pharynx and siphonoglyph under high power.

17. Study the arrangement of the mesenteries. Note that each mesentery consists of an inner layer of mesoglea surrounded by gastrodermis.

18. Study the structural relationship of a mesentery to the body wall from which it arises. How many pairs of primary and secondary mesenteries are present? Do you see any tertiary mesenteries?

19. Locate the retractor muscles on the secondary and primary septa. Note how the muscles face the endocoels. Do the muscles on the directives face the same way as those on the remaining septa?

20. Locate the clover-shaped mesenteric filament. The central lobe is nonciliated and contains gland cells that produce enzymes used in digestion and numerous cnidocytes. The two lateral lobes are ciliated and lack cnidocytes. What function might the cilia serve?

21. Study a longitudinal section on a prepared slide and identify as many of the above-mentioned structures as possible. ■

Stony Corals

True or stony corals are zoantharian anthozoans of the order Scleractinia (Madreporaria). There are two groups of hard corals: **hermatypic** and **ahermatypic**. Hermatypic corals are colonial reef builders, found in shallow tropical seas. Ahermatypic corals are noncolonial, nonreef builders, and may occur in rather deep and cold waters of higher latitudes. Most coral species are hermatypic with polyps 1–3 mm in diameter. The limy exoskeleton is secreted throughout the life of the polyp. Polyps of colonial corals are interconnected by a lateral fold of the body wall extending from the column and running along the surface of the colony. This results in the body tissues (epidermis, mesoglea, and gastrodermis) and the gastrovascular cavities of the polyps being continuous. The mesoglea is thin, but cellular.

Growth patterns and arrangement of the polyps in a colony are varied, resulting in different skeletal configurations. Coral deposits may be upright, branched, round, or flattened masses. Selected examples of common species are briefly described below. Your instructor may provide additional examples to observe.

■ Observational Procedure: Stony Corals

1. Examine the specimens available for study noting differences in density, growth form (branching or globular), and surface patterns. For example, do the specimens that branch all have the same branching patterns? How many branches occur before the patterns stops?

2. Often other sessile organisms utilize the coraline surface as a place of attachment. What evidence do you see that other organisms have attached to the specimens? What types of organisms have colonized the surface? Now reverse the question. Have any of these corals established themselves on another organism? What do these simple observations tell us about the biology of corals?

3. Each coral polyp is a miniature sea anemone fixed in a cup of calcium carbonate (Fig. 3.21) into which the living polyp can retract. In any specimen, what sort of variation is seen among the size of the cups? Are all of the cups of a uniform shape? How do these shapes differ?

4. The limy encasement is secreted by the epidermis of the polyp's base and lower area of the column. The nonliving exoskeleton of a single polyp is called a **corallite (calyx)**; that of the colony is the **corallum**. The bottom of the corallite is the basal plate (**tabula**) and that part of the cup containing the lower part of the polyp is the **theca**. Using a dissection microscope or a hand lens identify these structures in the specimens available for your study.

5. Using a hand lens examine the cut surfaces of several specimens. Do you see any evidence that the

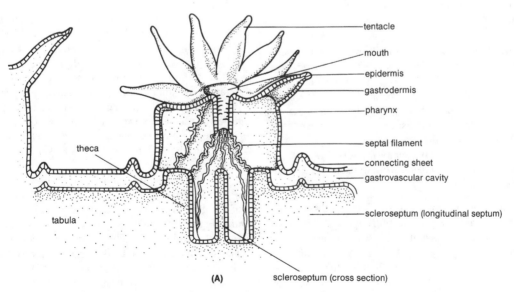

(A)

Figure 3.21 (A) Oral-aboral section of a hard coral polyp and its calcareous cup.

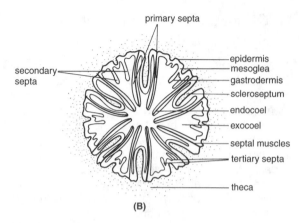

Figure 3.21 (B) Cross section of a coral polyp below the pharynx.

coral polyps are connected to one another through canals that perforate the corallum? If these cavities are not canals connecting one polyp to another, what are they?

6. Unlike sea anemones, coral polyps lack the pedal disc, siphonoglyphs, and septal perforations. The hollow tentacles occur in sets of six and the septa are usually hexamerous too. Six complete mesenteric septa typically alternate with six incomplete septa. Tertiary septa may occur. Locate the skeletal septa or ridges, called **sclerosepta**, present in each theca. These represent radial folds found between each complete septa, that are secreted by the underside of the polyp; this results in the lower part of the gastrovascular cavity divided by folds of tissue. Examine the specimens and find complete, incomplete, and tertiary septa. ■

Meandrina: **Brain Coral.** Note how the polyps are arranged close together in rows but that the rows are well separated. The resulting coral has ridges and valleys that appear somewhat like a human brain. The polyps of *Meandrina* are produced by intratentacular budding; however, the oral discs and columns never separate after new mouths are formed. Therefore, the polyps of a new row have many mouths, but share a common oral disc. *Meandrina sinuosa* occurs in Florida and the West Indies.

Astrangla: **Eyed or Star Coral.** This North Atlantic coastal coral forms small, encrusting colonies. The skeleton shows a pitted appearance because the polyps are well separated. The polyps in *A. danae* cannot withdraw into the cups. This species occurs from North Carolina to Massachusetts. Can you suggest where active budding may have been occurring in the specimen?

Fungia: **Mushroom Coral.** This solitary, rather deep coral has a mushroom-shaped skeleton. It is a common species of the Great Barrier Reef; *F. elegans* occurs in the Gulf of California. Individuals of *Fungia* can slide themselves over the substratum without the aid of currents. Is there a true theca in *Fungia*? Using a dissection microscope determine the arrangement of sclerosepta. Is the arrangement in *Fungia* the same as in the other corals examined?

Acropora: **Staghorn, Elkhorn, or Antler Coral.** Corals of this genus are mostly branched colonies with the corallum porous and of loose construction. This feature may be seen when a broken edge of a branch is examined on end. *Acropora* is a reef builder.

Porites: These corals are reef builders forming branched porous colonies. The polyps are close together and the calcareous cups are rather shallow. *P. porites* occurs in Florida and the West Indies.

Oculina: This genus forms branched, rather compact colonies; polyps are widely separated and spirally arranged. *Oculina diffusia* occurs from North Carolina to Florida.

Subclass Alcyonaria

Organ-pipe coral (*Tubipora*), sea fans (*Gorgonia*), sea pansies (*Renilla*), sea pens (*Stylatula*), sea whips (*Leptogorgia*), and their allies are alcyonarian cnidarians. All 3,000 of the known living species of alcyonarians are colonial and most abundant in warm tropical areas of the Indo-Pacific.

Unlike polyps of zoantharians, each alcyonarian polyp has eight, unpaired mesentaries that connect to the body wall and a relatively short pharynx (Fig. 3.22). Furthermore, each polyp bears eight pinnate tentacles that are arranged symmetrically. The number of tentacles and septa and their arrangement give alyconarians an octamerous radial symmetry. The feathery appearance of the tentacles is due to many finger-like projecting **pinnules** that arise from the main part of the tentacle. The tentacles are hollow; their cavities are extensions of the gastrovascular system. Cnidocytes and cilia on the tentacles capture food.

The gastrovascular cavities of the polyps are connected by gastrodermal tubes—branched continuations of the gastrovascular walls of the polyps—called **solenia** (Fig. 3.22). New polyps arise as buds or sprouts from the solenia. Arrangement of the solenia varies. The solenia are embedded in a thick mesoglea called **coenenchyme** and the entire outer surface of the colony is covered with epidermis. The thick matrix of coenenchyme also contains ameboid cells, skeletal ele-

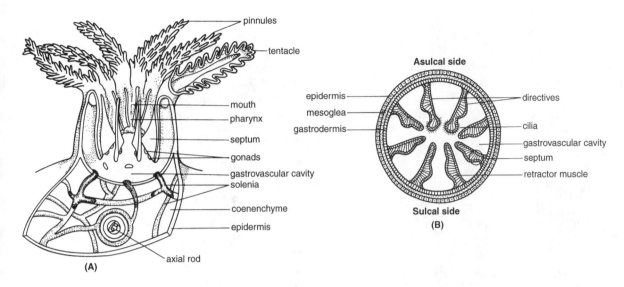

Figure 3.22 *Alcyonarian polyp.* (A) Aboral-oral section. (B) Cross section of the polyp below the pharynx.

ments (e.g., calcareous spicules and horny protein), and most of the body of the polyp. Only the oral ends of the polyps (**anthocodia**) project from the surface of the coenenchyme.

There is usually one siphonoglyph (or sulcus) located in the pharynx of the polyp. This ciliated groove, like that of other anthozoans, drives water into the gastrovascular cavity. The side of the polyp bearing the siphonoglyph is the sulcal side, and the opposite side the asulcal side.

Each septum on its sulcal side (e.g., toward the siphonoglyph) bears a strong, longitudinal retractor muscle (Fig. 3.22). The free edge of each septum below the pharynx bears a septal (or mesenteric) filament.

■ Observational Procedure: Alcyonaria

The orders of Alcyonaria differ in the nature of the skeleton and mode of colony formation. Following you will study a few examples.

Tubipora: Obtain the specimen of the organ-pipe coral (*Tubipora*). The order to which this genus belongs (Stolonifera), is characterized by the lack of the coenenchymal mass (Fig. 3.23). The polyps arise singly from a mat of **stolons**. Observe that the colony consists of upright, parallel polyps supported by hard skeletal tubes of fused spicules giving an impression of the pipes of an organ. Using a hand lens determine whether the tubes are solid. (NB: Back lighting will aid you in this observation.) The erect tubes are united at

spaced levels by transverse platforms. Iron salts in the skeleton impart a dull red color to the organism.

Gorgonia: The sea fan (*Gorgonia*) is a horny coral of the order Gorgonacea, which also includes the sea whips and sea feathers (Figs. 3.23). These highly colored, graceful colonies occur in tropical and subtropical waters, especially in the Indo-Pacific Ocean and in the subtropical Atlantic around Bermuda, the Bahamas, and West Indies. Because of their sessile nature and firm upright position, sponges, hydroids, and other invertebrates become attached to the sea fan. The most characteristic feature of these anthozoans is the horny, protein material called **gorgonin**, located in their skeletons. The polyps secrete the gorgonin.

1. Examine a sea fan closely (Fig. 3.23C). Note the short main trunk to which the colony was attached to the substratum by a holdfast. Extending from the trunk is a central axial rod of gorgonin. Emerging from the axial skeleton are numerous small branches united by cross connections into a lattice network. The many branches of the colony are in one plane, thus resulting in the fan-shaped pattern. A thin layer of coenenchyme covers the colony.

2. Observe the branches of the sea fan using the dissection microscope. Locate the small polyps (**anthocodia**) scattered over the stems and branches. If you are using a dried specimen, you will see openings where the once living polyps occurred. Can you see polyps on the main stem? Do polyps occur on both surfaces of the sea fan?

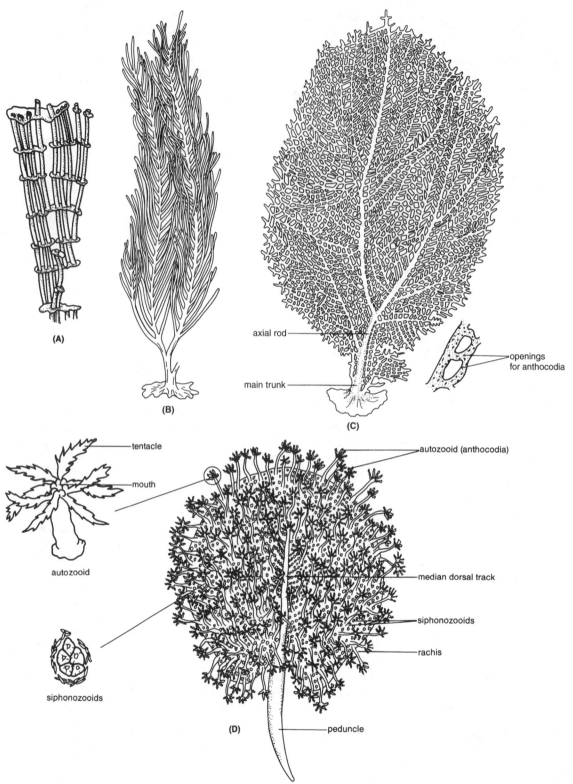

Figure 3.23 Selected examples of subclass Alcyonaria. (A) Piece of *Tubipora* (organ-pipe coral) skeleton. (B) Sea whip. (C) Sea fan (*Gorgonia*); magnified part of fan showing holes for polyps. (D) Sea pansy (*Renilla*), dorsal view. (From Hyman).

Renilla: Sea pansies (*Renilla*), of the order Pennatulacea, occur on soft bottoms of near-shore waters of the Atlantic, Gulf, and southern California coasts. The colony can uncover itself if buried by moving sediment.

1. Obtain a preserved or living *Renilla* and observe the specimen under the dissection microscope. The primary axial polyp (**rachis**) is leaflike or somewhat heart-shaped, and is supported by a short, proximal stalk (**peduncle**) to which the colony was anchored to a substrate by peristaltic contractions of the peduncle (Fig. 3.23D). If the organism is dislodged, it can re-anchor itself. Note the bilateral symmetry of the colony. The symmetry is established by a stem (or track) extending from the peduncle to about the middle of the rachis.

2. Do polyps occur on the peduncle? Are there polyps on the ventral surface? Many secondary polyps are embedded in the fleshy upper surface of the rachis. The secondary polyps are dimorphic and called **autozooids** and **siphonozooids**. Autozooids are feeding polyps, containing a mouth and six, pinnate tentacles. Siphonozooids are tiny, nonfeeding, warty protuberances arranged in clusters (Fig. 3.23D). Tentacles of siphonozooids are rudimentary. These polyps move water through the colony. The secondary polyps are formed by budding. Sexual reproduction, as in other anthozoans, results in a free-swimming planula larva.

Other Alcyonarians. Examine specimens of sea whips (*Leptogorgia*, order Gorgonacea), sea pens (*Stylatula*, order Pennatulacea), or other examples of alcyonarians. Study these with the dissection microscope and compare their structures with members previously studied. Note the branching patterns and locations of the polyps.

Slide of a Soft Coral. Using a dissection microscope or low power only of a compound microscope examine a prepared slide of an alcyonarian branch showing a few polyps. It is a good idea to examine several slides to get a good appreciation of the form of these small polyps.

1. Count the number of tentacles on the polyps and note the short fleshy pinnules that project at right angles from the tentacles.

2. On one or more of the polyps, you should be able to find the following: mouth, pharynx, and gastrovascular cavity, with its intercommunicating canals, the solenia.

3. The most prominent internal feature of the branch is the axial rod. Locate this dark structure and check it for birefringence. What does that tell you about its biomolecular structure? ■

D. Fossil Cnidarians

Cnidarians represent an ancient group of animals dating from the **Cambrian Period** (Fig. 3.1). Because of the soft-bodied nature of most individuals, the fossil record of cnidarians is scant. However, individuals with chitinous perisarcs and calcareous skeletons have been preserved as fossils. Extinct individuals of Hydrozoa, Scyphozoa, and Anthozoa have been found as fossils, but the largest number of extinct cnidarians is fossil corals. The extinct **rugose** (Order Rugosa or Tetracoralla) and **tabulate** (Subclass Tabulata) corals were important reef-builders throughout most of the **Paleozoic Era** (Fig. 3.24). Both groups died out in the **Permian Period**. The rugose corals were mostly solitary, cone-shaped corals with an outer skeleton having a wrinkled appearance as indicated by the name rugose (Fig. 3.24). The tabulate corals were colonial anthozoans with calcareous skeletal tubes containing tabulae, or horizontal platforms, on which the polyps rested.

These extinct corals were replaced by scleractinian and hydrozoan corals. Many well-known living genera such as *Acropora, Fungia, Favia,* and *Orbicella* have a long fossil history.

■ **Observational Procedure: Fossil Cnidarians**

Examine fossil specimens provided and note the similarities and differences to the present-day forms. What evidence informs you that these fossils were corals? Observe the presence of theca with sclerosepta in the specimen. In rugose corals note the wrinkled appearance of the outer skeleton. Locate the septal furrows, growth lines, septa, and fossia. In tabulate corals look for the presence of tabulae and mural pores. Your instructor will give some specific instructions to guide you further in your observations. ■

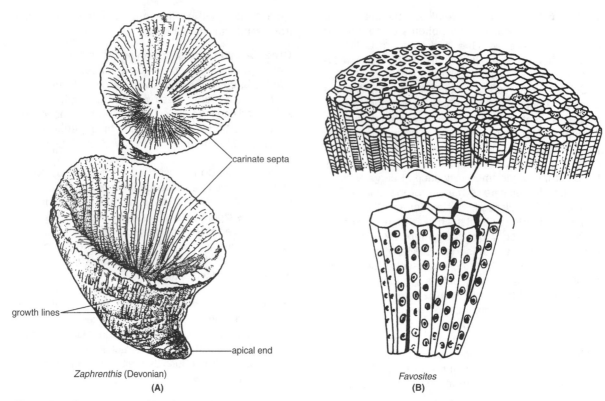

Figure 3.24 Two examples of fossil corals. (A) Rugose coral morphology. (B) Tabulate coral morphology.

Supplemental Readings

Bayer, F. M., and H .B. Owre. 1968. The Free-Living Lower Invertebrates. Macmillan Co., New York.

Brown, B. E., and J. C. Ogden. 1993. Coral bleaching. Sci. Am. 268(1): 64–70.

Burnett, A. L. (eds.). 1973. Biology of Hydra. Academic Press, New York.

Carre, C., and C. Carre. 1980. Triggering and control of cnidocyst discharge. Mar. Behav. Physiol. 7:109–117.

Dodson, S. I., and S. D. Cooper. 1983. Trophic relationships of the freshwater jellyfish *Craspedacusta sowerbyi* Lankester 1880. Limnol. Oceanogr. 28:345–351.

Fabricius, K. E., Y. Benayahu, and A. Genin. 1995. Herbivory in asymbiotic soft corals. Science 268: 90–92.

Faulkner, D., and R. Chesher. 1979. Living Corals. Clarkson N. Potter, New York.

Gierer, A. 1974. Hydra as a model for the development of biological form. Sci. Am. 231:44–54.

Goreau, T. F., and N. I. Goreau. 1979. Corals and coral reefs. Sci. Am. 241(2):124–136.

Grigg, R. W. 1972. Orientation and growth form of the sea fans. Limnol. Oceanogr. 17:185–192.

Hand, C. 1959. On the origin and phylogeny of the coelenterates. Syst. Zool. 8:191–202.

Harrison, F. W., and J. A. Westfall (eds.). 1991. Microscopic Anatomy of Invertebrates. Vol. 2: Placozoa, Porifera, Cnidaria, and Ctenophora. Wiley-Liss, New York.

Hessinger, D. A., and H. M. Lenhoff (eds.). 1988. The Biology of Nematocysts. Academic Press, New York.

Hughes, T. P., and J. B. C. Jackson. 1980. Do corals lie about their age? Some demographic consequences of partial mortality. Fission and fusion. Science 209:713–715.

Kaplan, E. H. 1982. A Field Guide to Coral Reefs of the Caribbean and Florida. Peterson Field Guide Series. Houghton Mifflin Co., Boston, MA.

Kramp, P. L. 1961. Synopsis of the medusae of the world. J. Mar. Biol. Assoc. U.K. 40:1–469.

Lane, C. E. 1960. The portuguese man-of-war. Sci. Am. 202:158–168.

Lang, J. 1973. Interspecific aggression by scleractinian corals: Why the race is not only to the swift. Bull. Mar. Sci. 23:260–279.

Lenhoff, H. M. 1983. Hydra Research Methods. Plenum Press, New York.

Lenhoff, H. M., L. Muscatine, and L.V. Davis (eds.). 1971. Experimental Coelenterate Biology. University of Hawaii Press, Honolulu, HI.

Lubbock, R. 1979. Chemical recognition and nematocyte excitation in sea anemone. J. Exp. Biol. 83:283–292.

Mackie, G. O. (ed.). 1976. Coelenterate Ecology and Behavior. Plenum Press, New York.

Mackie, G. O., P. A. V. Anderson, and C.L. Singla. 1984. Apparent absence of gap junctions in two classes of Cnidaria. Bio. Bull. 167:120–123.

Mackie, G. O., P. P. Pugh, and J. E. Purcell. 1987. Siphonophore biology. Adv. Mar. Biol. 24:97–262.

Muscatine, L., and H. M. Lenhoff (eds.). 1974. Coelenterate Biology: Reviews and New Perspectives. Academic Press, New York.

Newell, N. D. 1972. The evolution of reefs. Sci. Am. 226(6):54–65.

Oliver, W. A. 1980. The relationship of the scleractinian corals to the rugose corals. Paleobiology 6:146–160.

Pool, R. 1995. Coral chemistry leads to human bone repair. Science 267: 1772.

Rees, W. J. (ed.). 1966. The Cnidaria and Their Evolution. Academic Press, New York.

Rosen, B. R. 1982. Darwin, coral reefs, and global geology. BioScience 32:519–525.

Schlichter, D. 1982. Nutritional strategies of Cnidarians: The absorption, translocation, and utilization of dissolved nutrients by *Heteroxenia fuscescens*. Am. Zool. 22:659–669.

Sebens, K. P. 1979. The energetics of asexual reproduction and colony formation in benthic marine invertebrates. Am. Zool. 19:683–697.

Sebens, K. P. 1980. The regulation of asexual reproduction and indeterminate body size in the sea anemone, *Anthopleura elegantissima*. Biol. Bull. 158:370–381.

Sebens, K. P. 1984. Agonistic behavior in the intertidal sea anemone *Anthopleura xanthogrammica*. Biol. Bull. 166:547–472.

Shick, J. M. 1991. A Functional Biology of Sea Anemones. Chapman and Hall, London.

Slobodkin, L. B., and P. E. Bossert. 1991. The freshwater Cnidaria—or Coelenterates. *In*: J.H. Thorp and A.P. Covich (eds.). Ecology and Classification of North American Freshwater Invertebrates. Academic Press, New York, pp. 125–143.

Smith, F. G. W. 1971. A Handbook of the Common Atlantic Reef and Shallow-water Corals. University of Miami Press, Coral Gables, FL.

Stoddart, D. E. 1969. Ecology and morphology of recent coral reefs. Biol. Rev. 44:433–498.

Tardent P., and R. Tardent (eds.). 1980. Development and cellular biology of coelenterates. Elsievier/North Holland, Amsterdam.

Veron, J. E. N. 1995. Corals in Space and Time. Cornell University Press, Ithaca, N.Y.

Wood, E. M. 1983. Corals of the World. T.F.H. Publishers. Neptune City, NJ.

Wyttenbach, C.R. (ed.). 1974. The developmental biology of the Cnidaria. Am. Zool. 14:440–866.

Yonge, C. M. 1963. The biology of coral reefs. Adv. Mar. Biol. 1:209 –261.

Yonge, C. M. 1968. Living corals. Proc. Royal Soc. Lond. Ser. B. 169:329–344.

■

Phylum Ctenophora

Sea walnuts, comb jellies, and sea gooseberries are common names often applied to members of the phylum Ctenophora [te-NOF-or-a; G., *ctenos*, comb + *phora*, to bear]. The transverse, ciliary comb-shaped plates or **ctenes** located on eight bands called **comb rows** are the basis for the phylum name (Fig. 4.1A). The ctenes are used in locomotion.

There are approximately 100 carnivorous species in this marine group of **biradial** symmetrical invertebrates. Their biradial symmetry is established by (1) the bilateral symmetrical digestive system, which is composed of paired canals arising from a central stomach; (2) the radially arranged comb rows; and (3) the bilaterally arranged tentacles when present. Ctenophores are worldwide in distribution, and occur at all depths in the open sea as well as in coastal waters.

Ctenophora and Cnidaria (Exercise 3) are the two major phyla that constitute the Radiata because their members possess primary radial symmetry. Although both phyla have several similarities (e.g., body wall construction and gastrovascular digestive system), there are important differences. Unlike cnidarians, ctenophores are not colonial, they do not exhibit polymorphism, and they lack nematocysts. Instead of nematocysts, tentaculate ctenophores have peculiar cells called **collocytes** (G., *collod*, glue) that secrete a sticky substance for catching and holding prey such as copepods and other small organisms (Fig. 4.1B).

Ctenophores are monoecious except for the lobate ctenophore, *Ocyropsis*. The testes and ovaries develop as two bands located within each of the eight meridional canals located beneath the comb rows. Fertilization occurs in the seawater; most forms have a free-swimming **cydippid larva** that resembles an adult ctenophore of the order Cydippida.

Classification

The phylum is divided into two classes based primarily on the presence or absence of tentacles.

1. **Class Tentaculata.** Tentacles present. Order Cydippida: ovoid body with two branched tentacles that retract into pouches or sheaths; examples are *Pleurobrachia* and *Mertensia*. Order Lobata: compressed body in tentacular plane forming two oral lobes on either side of that plane. Tentacles are small and not located in pouches; examples are *Mnemiopsis*, *Bolinopsis*, and *Ocyropsis*. Order Cestida: belt-shaped body and compressed in tentacular plane; tentacles in pouches, but reduced and near the mouth; examples are *Cestum* and *Velamen*. Order Platyctenea: body flattened along an oral-aboral axis and adapted for creeping movements; examples are *Coeloplana* and *Ctenoplana*.

2. **Class Nuda.** Tentacles absent. Order Beroida: body cylindrical and compressed; mouth and pharynx enlarged; example is *Beroë*.

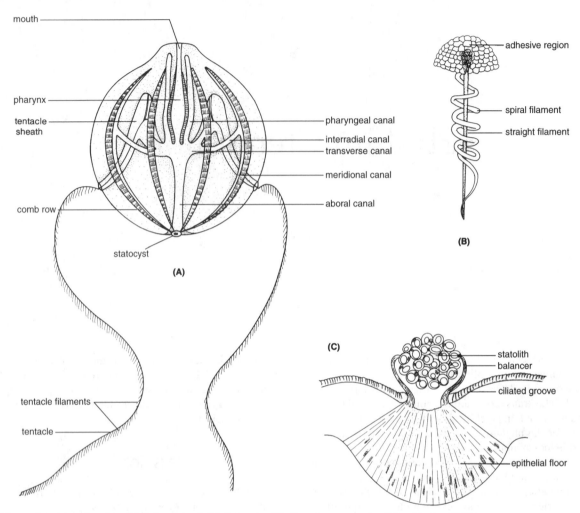

Figure 4.1. *Pleurobrachia.* (A) Entire animal. (B) Collocyte. (C) Aboral sense organ. (After Shelton 1982).

Pleurobrachia

Pleurobrachia, often called the sea gooseberry, is a tentaculate ctenophore. Individuals of the genus are commonly found in coastal waters of the Atlantic Ocean.

■ Observational Procedure: *Pleurobrachia*

1. Place a preserved *Pleurobrachia* in a small dish filled with water. Using a dissecting microscope, observe the ctenophore's ovoid shape (Fig. 4.1A). The mouth is located at the slightly narrower (oral) end. It leads to the pharynx, stomach, and aboral canal that terminates in aboral anal pores. Arising off either side of the stomach are **transverse canals** that extend as **interradial canals** connecting the eight **meridional canals**. Two **pharyngeal canals** run parallel to the

pharynx (Fig. 4.1A). This complex of canals constitutes the gastrovascular digestive system.

2. Locate the eight radially arranged comb rows located above the meridional canals. These orally-aborally arranged rows are often whitish in color. Examine closely one comb row and locate the ctenes, each composed of fused cilia arranged like teeth of a comb. Find the openings of the two **tentacular sheaths** which are located about one third of the way to the aboral end of the animal (Fig. 4.1A). Preserved specimens usually have their tentacles retracted into the sheaths; however, in life the extended tentacles (up to 15 cm long) with many lateral filaments function as a fishing line to ensnare prey.

3. At the aboral end is a sense organ called a **statocyst**, containing a statolith that helps coordinate beating of the cilia on the comb rows (Fig. 4.1C). With a razor blade or sharp scalpel, cut away the aboral end and place it in a shallow dish filled with water. Examine the statocyst under high power of a dissection microscope. Can you see the four ciliary tufts that support the calcareous statocyst in its dome-shaped covering? **4.** Cut away the outer wall of an aboral quarter of the body. Locate the pharynx, pharyngeal canals, and meridional canals of the digestive system. Locate the tentacular sheath. If the tentacles are retracted into the sheath, cut an entire sheath away and carefully lease the tentacle from its sheath. ■

Beroë

Beroë, the thimble jelly, is larger (5 × 10 cm) than *Pleurobrachia*. *Beroë* lacks tentacles and is pinkish in color.

■ Observational Procedure: *Beroë*

Obtain a specimen of *Beroë* in a small dish with water. Note its bell-shaped and laterally compressed body (Fig. 4.2). Observe the enlarged mouth. Locate the eight comb rows and highly branched gastrovascular cavity. ■

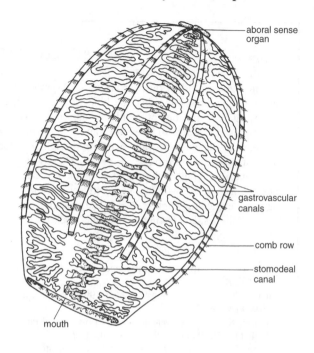

Figure 4.2. *Beroë*.

aboral sense organ

gastrovascular canals

comb row

stomodeal canal

mouth

Supplemental Readings

Franc, J. M. 1978. Organization and function of ctenophore colloblasts: an ultrastructural study. Biol. Bull. 155:527–541.

Greve, W. 1978. Ctenophores. Fiches Identif. Zooplancton 146.

Harbison, G. R. 1985. On the classification and evolution of the Ctenophora. *In*: Conway Morris, S., J.D. George, R. Gibson, and H.M. Platt (eds.). The Origins and Relationships of Lower Invertebrates. Systematics Association Special Volume 28. Clarendon Press, Oxford, pp. 78–100.

Harbison, G. R., L.P. Madin, and N.R. Swanberg. 1978. On the natural history and distribution of oceanic ctenophores. Deep-Sea Res. 25:233–256.

Horridge, G. A. 1974. Recent studies on the Ctenophora. *In*: L. Muscatine and H.M. Lenhoff (eds.). Coelenterate Biology. Academic Press, New York, pp. 439–458.

Pianka, H. D. 1974. Ctenophora. *In*: A. C. Giese and J. S. Pearse (eds.). Reproduction of Marine Invertebrates, Vol. 1: Academic Press, New York, pp. 201–265.

Reeve, M. R., and M .A. Walter. 1978. Nutritional ecology of ctenophores—a review of recent research. Adv. Mar. Biol. 15:249–287.

Reeve, M. R., M .A. Walter, and T. Ikeda. 1978. Laboratory studies of ingestion and food utilization in lobate and tentaculate ctenophores. Limnol. Oceanogr. 23:740–751.

Stanlaw, K. A., M. R. Reeve, and M. A. Walter. 1981. Growth, food, and vulnerability to damage of the ctenophore *Mnemiopsis mccradyi* in its early life history stages. Limnol. Oceanogr. 26:224–234.

Tamm, S. L. 1980. Cilia and ctenophores. Oceanus 23:50–59.

Tamm, S. L. 1982. Ctenophora. *In*: G.A.B. Shelton (ed.). Electrical Conduction and Behaviour in 'Simple' Invertebrates. Oxford University Press, Oxford, pp. 266–358.

Tamm, S. L. 1983. Motility and mechanosensitivity of macrocilia in the ctenophore *Beroë*. Nature 305:430–433.

ACOELOMATE BILATERIA

The metazoans studied thus far are either asymmetrical, radially symmetrical, or biradially symmetrical. The remaining phyla to be studied are bilaterally symmetrical, although members of several phyla have undergone additional reorganization of their body plan so that they exhibit secondary radial symmetry. Bilaterally symmetrical organisms (**Bilateria**) show three levels in the organization of their perivisceral or body cavity (i.e., the space between the gut and body wall): **acoelomates** [G., *a*, without + G., *coelo*, hollow], **pseudocoelomates** [G., *pseudo*, false], and **eucoelomates**. [G., *eu*, good]. These groupings generally are not given taxonomic status, and there is some question whether the pseudocoelomate condition actually exists. In the acoelomates, derivatives of the mesoderm completely fill the space between the body wall and gut, leaving the animal without a coelom (but compare Turbeville and Ruppert 1985). The acoelomates include four wormy phyla: Gastrotricha, Gnathostomulida, Platyhelminthes, and Nemertea (Rhynchocoela).

Ehlers, U. 1994. Absence of a pseudocoel or pseudocoelom in *Anophostoma vivipara* (Nematodes). Microfauna Marina 9: 345–350.

Harrison, F. W., and B. J. Bogitsh. (eds.). 1991. Platyhelminthes and Nemertina, Vol. 3. Microscopic Anatomy of Invertebrates. Wiley-Liss, New York.

Lorenzen, S. 1985. Phylogenetic aspects of pseudocoelomate evolution. *In*: S. Conway Morris, J. D. George, R. Gibson, and H. M. Platt (eds.). Clarendon Press, Oxford, pp. 210–223.

Nielsen, C. 1995. Animal Evolution. Oxford University Press, Oxford.

Turbeville, J. M., K. G. Field, and R. A. Raff. 1992. Phylogenetic position of phylum Nemertini, inferred from 18S rRNA sequences: molecular data as a test of morphological character homology. Mol. Biol. Evol. 9:235–249.

Turbeville, J. M., and E. E. Ruppert. 1985. Comparative ultrastructure and the evolution of nemertines. Am. Zool. 25:53–71.

Sterrer, W., M. Mainitz, and R. M. Rieger. 1985. Gnathostomulida: enigmatic as ever. *In*: S. Conway Morris, The Origins and Relationships of Lower Invertebrates. J. D. George, R. Gibson, and H. M. Platt. (eds.). Clarendon Press, Oxford, pp. 181–199.

Wallace, R. L., C. Ricci, and G. Melone. 1996. A cladistic analysis of pseudocoelomate (aschelminth) morphology. Invertebrate Biology 115 (2) 104–112.

■

Phylum Platyhelminthes

Platyhelminthes [PLAT-e-hel-MIN-thez; G., *platy*, flat + G., *helminth*, worm] are acoelomate, nonsegmented, dorsoventrally flattened worms. There are more than 15,000 species of flatworms which range in size from microscopic to an incredible 30 m. Most species in this phylum are entirely endo- or ectoparasitic.

There is no coelom in flatworms. The internal organs are embedded within a compact mass of **parenchyma**, a tissue of mesodermal origin, functioning in secretion, storage, and regeneration. The repetition of body parts in the tapeworms is not considered to be true segmentation as found in Phylum Annelida (Exercise 13). Usually a digestive system, which is incomplete (i.e., no anus), is present only in the turbellarians and flukes. Tapeworms lack a digestive system.

Classification

Older classification schemes subdivided Platyhelminthes into three classes: Turbellaria, Trematoda, and Cestoda. However, recent work recognizes that Trematoda is more complex than originally thought and it has been subdivided into two classes, the Monogenea and a more restrictively defined Trematoda (Subclasses Digenea and Aspidobothria). Some schemes separate each of these two subclasses into their own class. Ehlers (1985) and Brooks et al. (1985) offer new systems of classification constructed on the basis of phylogenetic systematics. These systems lack an easily defined class hierarchy (see also Nielsen 1995). For instructional purposes, we will use the four-class scheme.

1. **Class Turbellaria**. Relatively small flatworms with ciliated epidermis richly supplied with glands. Distinct rod-shaped structures called rhabdoids (including rhabdites) are among the glandular secretions. All possess a blind gut with oral opening usually located midventrally. Most are free-living, but some are commensal or parasitic. Twelve orders are recognized. Examples: *Dugesia* (the common pond planarian), *Bdelloura* (a commensal on the gills of horseshoe crabs), *Stenostomum* (a common, microscopic freshwater turbellarian), and *Bipalium* (a terrestrial turbellarian in gardens and greenhouses in the southern United States).

2. **Class Trematoda**. Moderate-sized, leaf- to wormlike parasitic flatworms, characterized by having one or more prominent, ventral attachment structures. Life cycles may be simple or complex, as endo- or ectoparasites of both invertebrate and vertebrate hosts. Examples: *Aspidogaster conchicola* (freshwater clam fluke), *Clonorchis* (= *Opisthorchis*) *sinensis* and *Fasciola hepatica* (human and sheep liver flukes, respectively), and *Schistosoma* spp. and *Trichobilharzia* spp. (human and bird blood flukes, respectively).

3. **Class Monogenea**. Small ectoparasites of vertebrates, especially fish. A characteristic attachment organ, the opisthaptor, at the posterior end. Examples: *Dactylogyrus* (a pest of hatchery fish), *Gyrodactylus* (gills of trout, bluegills, and goldfish), and *Polystoma nearcticum* (found in the urinary bladders of toads).

4. **Class Cestoda**. Elongate, obligate, endoparasitic flatworms lacking a gut. Most possess an anterior knoblike holdfast called the "head" or scolex. Subclasses Cestodaria includes tapeworms of fish and turtles having a leaflike body shape. Example: *Gyrocotyle parvispinosa* (a parasite of fish). Subclass Eucestoda (true tapeworms) parasitize all vertebrate classes. These forms possess an elongate body or strobila consisting of many repeated units called proglottids. Examples: *Dipylidium caninum* (dog tapeworm), *Moniezia expansa* (sheep tapeworm), *Taenia solium* (pig tapeworm), and *Taeniarhynchus saginatus* (beef tapeworm).

A. Class Turbellaria

Turbellaria [ter-be-LAR-e-a; L., *turbella*, a little crowd] is a moderate-sized class of flatworms comprising about 3,000 species. Many of them have an oval to elongate, dorsoventrally flattened shape and range in size from microscopic to over 50 cm long; most are less than 5 mm long. Turbellarians are predominantly free-living, although many commensal and parasitic species are known. Members of this class are found in freshwater, marine, and terrestrial habitats. All have a simple life history.

Turbellarians have a single-layered epidermis which is cellular or **syncytial**. In most species the epidermis is completely ciliated all around and amply provided with glands; some have ciliation restricted to special areas of the epidermis such as a creeping sole. One prominant secretory product in the epidermis is the **rhabdoid**. There are several different types of rhabdoids, but one common type is the **rhabdite**.

A mouth opens into the gut, in most species through a simple or complex pharynx that in some forms may be protruded through the mouth to feed. The location of the mouth is highly varied within the class. The pharynx in turn opens into a blind gut. In some lower turbellarians a true digestive cavity is lacking. Other forms have a simple sac or complex intestine with many diverticula. Digestion begins extracellularly, and is completed intracellularly.

■ Observational Procedure: Turbellaria

Live Specimens

1. If time and climatic conditions permit, field collections of living planarians will be made. When making collections in freshwater lakes, streams, ponds, or springs, examine the undersurface of submerged objects such as stones, old leaves, and wood. If present, planarians will be seen attached to substrates sometimes in clusters (hence the use of the Latin term *turbella*).

2. Carefully remove several animals from the holding jar and place them in a shallow dish containing some pond water. Observe the specimens under a dissection microscope and note their general manner of locomotion and any behaviors they may exhibit when obstacles are encountered. Using a probe, turn a specimen upside down and watch how it rights itself.

3. Place a specimen in a depression slide filled with pond water, put a cover glass over the specimen, and observe it with low power of a compound microscope. As the animal moves about, attempt to examine its dorsal and ventral surfaces. The ventral surface is ciliated and ciliary action may be observed as the animal twists and turns.

4. To aid in this observation add a small amount of carmine or carbon black powder to the depression and observe the movement of the particles brought about by the action of the cilia.

5. If time permits, your instructor may provide instructions for observing planarian feeding habits or other behaviors, or directions for exercises in planarian regeneration. The study of other turbellarians (e.g., *Bipalium, Stenostomum*) also may be undertaken.

Whole Preserved Specimens

1. Obtain a prepared slide of a whole, stained specimen of the freshwater planarian *Dugesia*. Locate the anterior and posterior ends. The earlike structures at the anterior end are chemosensory lobes called **auricles**. Near the middle of the worm is the muscular pharynx; this protrusible structure is housed in a pharyngeal cavity. In an animal that has been fed an opaque material such as carbon black, trace the outline of the intestine (Fig. 5.1). How many main branches of the gut are present? Note that each main branch is branched into numerous diverticula.

2. Locate the pair of **eyespots** at the anterior end. These are special photoreceptive sensory organs and are of the inverse pigment-cup type. The shape and position of the pigment cup in relation to the photoreceptive cells gives the planarian an interesting appearance. Observe a cross section of a planarian at the level of the eyespots. Note the shape of the pigment cups. Which way do they face? What function does the black pigment serve?

3. Compare a whole-mount slide of *Bdelloura* to that of *Dugesia* (Fig. 5.2). *Bdelloura* is a marine turbellarian that is commensal on gills of the horseshoe crab,

POSTERIOR

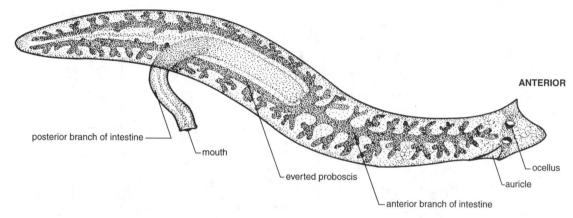

Figure 5.1. Dorsal view of the freshwater turbellarian *Dugesia*.

Limulus. At the anterior end of *Bdelloura*, note the paired eyespots, but absence of auricles. Under each eye locate a ganglion and trace the paired ventral nerve cords which leave these ganglia and run the length of the animal. In the middle of *Bdelloura* is the large pharynx and just posterior to it a copulatory organ.

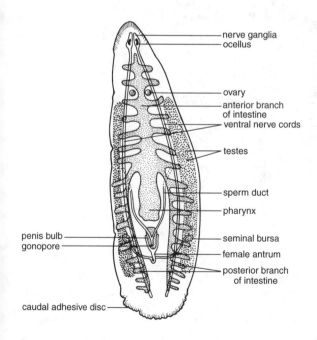

Figure 5.2. Dorsal view of the marine commensal turbellarian *Bdelloura*.

Cross Sections

Examine several cross sections of *Dugesia* and make the following observations.

1. Epidermis. Observe the epidermal surfaces and locate the rhabdoids (Fig. 5.3). Do they have a uniform distribution? Examine the parenchyma for the presence of rhabdoids. Where are these structures concentrated? Where are they absent? Observe the lateral edges of the section. Are rhabdoids present there? In this area locate the special eosinophilic glandular cells, which produce a secretion enabling the worm to adhere to surfaces.

The next observations will require the use of oil immersion techniques. If you are not familiar with these procedures ask your instructor for assistance. (NB: This is a difficult investigation; one that requires a very good slide.) Using oil immersion techniques, examine the dorsal and ventral surfaces of the cross section for the presence of cilia. Are cilia uniformly distributed on the dorsal and ventral surfaces? If not where are cilia located?

2. Muscular system. Just under the epidermal basement membrane, locate the muscle layers. These are made up of circular, longitudinal, and diagonal muscles (Fig. 5.3).

3. Parenchyma. Examine the parenchymal tissue and note that this tissue completely fills the body with a loose cellular matrix (Fig. 5.4). However, there are cavities present in the body of the worm. What do they represent? Passing through the parenchyma locate numerous dorsoventral muscle fibers. How do they function in locomotion?

4. Gut. Locate the gut in a cross section. How many regions of the gut do you see in a single cross section of the planarian worm? Relate these numerous sections of the gut to what you saw in the whole worm. Note the large vacuolated phagocytic cells of the gastrodermis.

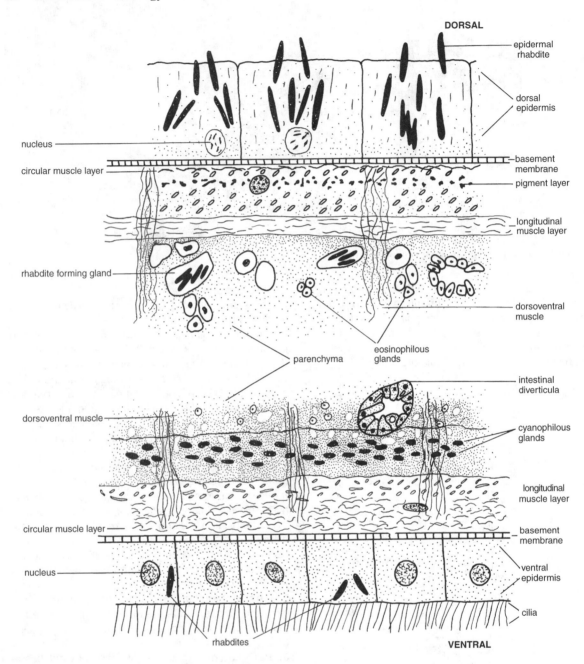

Figure 5.3. Cross sections of the dorsal and ventral body wall of *Dugesia*.

Examine the gastrodermis of your preparation carefully. In some, the gregarian parasite (*Lankesteria planariae*) may be present. This parasite may be distinguished from cells of the planarian gastrodermis by its distinct nucleation and by the larger, pear-shape of the cell.

5. Pharynx. This protrusible organ is capable of being projected through the mouth. Carefully examine cross and longitudinal sections of the pharynx. Observe how it is located in the pharyngeal cavity. What is the rela-

tive size of the pharynx in comparison to the rest of the body? Find the outer ciliated layer and muscle layers of the pharynx. Compare the relative size of the muscles comprising the pharynx to that of the rest of the body. What does this comparison tell you about the feeding biology and the locomotion of this flatworm? Using a longitudinal section through the level of the pharynx, determine the anatomical relationship between the pharynx and the mouth. ■

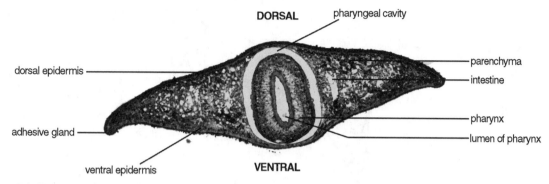

DORSAL
pharyngeal cavity
dorsal epidermis
parenchyma
intestine
pharynx
lumen of pharynx
adhesive gland
ventral epidermis
VENTRAL

Figure 5.4. Photomicrograph of a cross section of *Dugesia* at the level of the pharynx.

B. Class Trematoda

Trematoda [trem-a-TO-da; G., *trema*, a hole] is a large class containing about 11,000 species of parasitic flatworms. Trematodes, or **flukes** as they are commonly called, are similar in size and shape to the turbellarians. Most are dorsoventrally flattened; however, some have a cylindrical body. As a group they range in size from microscopic to several meters in length. Flukes have enormous medical, veterinary, and economic importances. Together with the class Cestoda (tapeworms) and phylum Nematoda (unsegmented roundworms, Exercise 10), flukes are the subject of the field of **helminthology** [G., *helminth*, worm + G., *ology*, study].

Trematodes possess one or more ventral suckers, usually an oral one surrounding the mouth and another located within the anterior one third of the body. These may be armed with many spines or hooks. The fluke body wall differs from that of turbellarians. Although once believed to be nonliving, the **tegument** is a complex, living syncytium which may be ornamented with minute spines.

Trematodes may have a simple life cycle involving a single host (subclass Aspidogastrea) or a very complex one involving several larval stages and more than one host (subclass Digenea). In the Digenea, development may proceed through as many as five different larval forms: **miracidium, sporocyst, redia, cercaria,** and **metacercaria.** Some of these are small free-swimming stages which are involved in the location of the next host. Others are encysted within a host or on a substrate.

It is not possible to cover this diverse class completely in an introductory laboratory; therefore, exercises for only a few selected species are provided here. At this point, it will be profitable if students review the life history of the digenetic fluke, *Fasciola hepatica* (Fig. 5.5).

■ Observational Procedure: Trematoda

Adults

Fasciola hepatica (Sheep Liver Fluke). *Fasciola hepatica* is a large (ca. 30 mm), leaf-shaped fluke often found in the livers of deer, goats, cattle, and sheep.

1. Observe a slide of a stained specimen and identify the anterior and posterior ends (Fig. 5.6). The body tapers abruptly at the anterior end, giving the effect of a pointed head with shoulders. This region is called the **oral cone.**

2. Locate the **oral sucker** which surrounds the mouth. The mouth leads to the pharynx which connects to a short esophagus and then to a biramus intestine (intestinal ceca). Trace the arms of the intestine that run the length of the fluke. Does the intestine branch into small diverticula as in *Dugesia*?

3. Just posterior to the pharynx locate the large ventral sucker called the **acetabulum**, and in front of it the common genital pore. The genital pore receives both male and female gametes. Locate the highly branched testes that lie in tandum in the central region. Bilaterally symmetrical organisms usually have paired organs on opposite sides of the body. How might this tandum arrangement of the testes affect body shape? Vasa efferentia (**vas deferens**) transport sperm from the testes to the seminal vesicle. The seminal vesicle and **cirrus sac** may be seen to one side of the genital pore.

4. Locate the tubular **uterus** that courses posteriorly from the genital pore. It may contain many eggs. (NB: Although called eggs, these are really developing

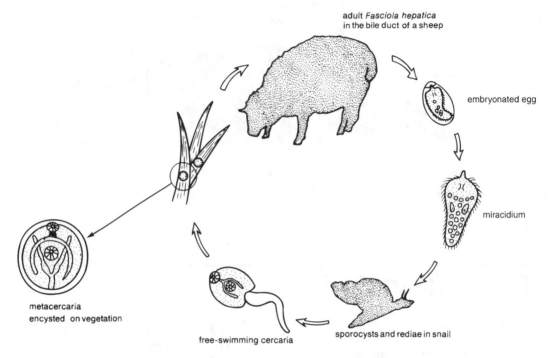

Figure 5.5. Life cycle of the digenetic trematode *Fasciola hepatica*.

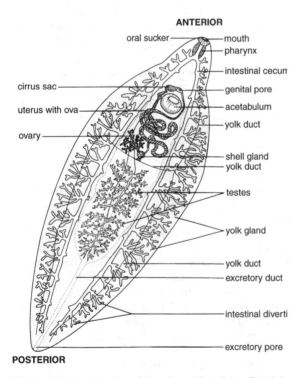

Figure 5.6. Ventral view of the sheep liver fluke *Fasciola hepatica*.

embryos still encased in the egg shell.) The uterus terminates at the shell gland that receives yolk from the large lateral **vitellaria** (yolk glands). The vitellaria may obscure the intestinal ceca.

5. On the right side of the fluke find the ovary that also terminates at the shell gland. It is branched, but darker than the testes and lies anterior to them.

6. At the posterior end of the animal, locate the excretory pore that drains the excretory vesicle. Is this structure an anus? How can you demonstrate that it is not?

7. Examine a cross section of *Fasciola*. (A longitudinal section will work for these observations.) Determine the dorsal and ventral surfaces. Is the body surface ciliated? Are rhabdites present? Locate the numerous scale like spines that are buried in the tegument. What function might they serve in this parasite?

8. Below the tegument find the muscle layers and note the dorsoventral muscles passing through the parenchyma. In cross sections from the anterior region of the worm, locate uterus containing shelled eggs and laterally find the intestinal ceca and vitellaria.

9. In cross sections from the central region of the worm, locate the testis. Laterally find the intestinal diverticula and vitellaria.

Fasciolopsis buski (Human Intestinal Fluke). *Fasciolopsis buski* is an important fasciolid which parasitizes people living in Southeast Asia. This parasite has a life history similar to that of *Fasciola* except that it enhabits the intestine. There are a few other notable differences in the body plan of these species. If slides are available, compare the general anatomy of *Fasciolopsis buski* to *Fasciola hepatica*.

1. Note that *Fasciolopsis buski* is much larger (up to 75 mm long), lacks the oral cone, and its oral sucker and acetabulum is larger than that of *Fasciola*. One way to visualize the relative proportions of the oral sucker and the acetabulum in these two species is to examine their musculature for birefringence. How extensive are these these structures? Considering the differences in habitat of the two species, what might be the significance of a larger (stronger) acetabulum? Using a dissection microscope, re-examine the pharynx of *Dugesia* and compare the relative size of the pharynx of that turbellarian to that of the two flukes you have just studied. Why might the size of the pharynx in the planarian be disproportionately longer?

2. Compare and contrast the intestine of *Fasciolopsis* and *Fasciola*. How are they similar? How is the intestine of *Fasciolopsis* simpler? What might this indicate about the foods of these two parasites?

3. The reproductive systems of the two species are essentially the same. Testes lie in tandum in the central region of the body, a smaller ovary lies ahead of the testes next to the shell gland, and large vitellaria are found along the lateral margins of the fluke.

Clonorchis sinensis (Chinese Liver Fluke). The opisthorchiid, *Clonorchis* (= *Opisthorchis*) *sinensis*, like *Fasciola*, lives in the liver of its host (Fig. 5.7). However, it is a smaller fluke (ca. 25 mm long) that lacks tegmental spines. If slides are available, compare this fluke to the fasciolids previously studied.

Note that in this case the oral sucker is slightly larger than the acetabulum. Is the intestine similar to *Fasciola* or *Fasciolopsis*? Two large, branched testes lie in tandum near the posterior end, and just anterior to them, lies a small three-lobed ovary. From the ovary the uterus courses anteriorly to the genital pore which lies in front of the acetabulum. To either side of the uterus are the moderately sized vitellaria.

Schistosome Flukes. Another important group of flukes are the **schistosomes** (blood flukes) including those which afflict humans: *Schistosoma mansoni, S. japonicum,* and *S. haematobium* (Fig. 5.8). The first two are found in veins of the intestine and the latter in veins of the urinary bladder, Schistosomiasis is endemic in Africa, in parts of the Middle East, Southeast Asia, and Japan, and in eastern and northern South America. The following observations may be done using any of the three species noted above.

1. Observe prepared slides of *Schistosoma* (male and female). Identify the anterior and posterior ends. How does the shape of *Schistosoma* differ from the other flukes studies so far? There is a distinct size difference between the males and females; which is larger?

2. Find the oral and ventral suckers of the specimens. Are there differences in the sizes of the suckers?

3. Locate the **gynecophoral canal** (groove) on the male's ventral surface. Does the female have this groove? What is the significance of this structure? If available, observe a slide that shows the sexes in copula.

4. Observe a prepared slide of *Schistosoma* eggs and note the lateral **spine**. The eggs are responsible for some of the very serious pathologic aspects of *schistosomiasis*. After being liberated from the female the eggs pass through the intestinal wall where they do considerable damage. Do embryos possess any locomotory apparatus that would permit them to move through the muscle wall of the small intestine? If not, how are they capable of accomplishing this feat?

5. Eggs also may be swept into the liver through the hepatic portal system where they become lodged. Observe slides of *Schistosoma* eggs in the wall of the intestine and in the liver. Note any evidence of tissue damage done to the host's organs in the region immediately surrounding the eggs. In the liver, inflammation at the site of each egg will eventually lead to a host reaction called a **granuloma** or **pseudotubercle**.

6. Under low power, scan your slide for pseudotubercles. How many do you see in your field of view? Measure the size of the egg and compare it to the thickness of the layer of granulomatous cells produced by the host. Make a judgment of how much of the host's tissue is affected by the parasite.

7. If material is available, examine the other *Schistosoma* species. How are they similar; how are they different? How do the schistosomes differ from the other flukes you have examined? Constructing a table may help you organize this comparison.

General Study of Fluke Larval Stages
Obtain prepared slides of the five larval stages from species such as *Fasciola hepatica, Schistosoma* species, or any other digenetic trematode (Fig. 5.9). The basic sequence of stages is miracidium, sporocyst, redia, cercaria, and metacercaria (Fig. 5.5). However, several important modifications may be found in different species. Sporocyst and redia stages may have more than one generation (mother and daughter) or may be absent; the metacercaria stage may be absent also. Depending on the quality of the preparation, the internal structures of the larval stages shown in Fig. 5.9 are often not visible in stained preserved specimens. Sometimes phase contrast lighting

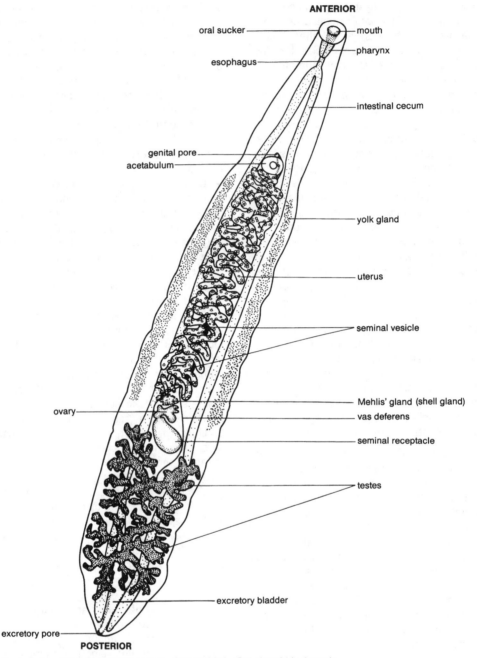

Figure 5.7. Ventral view of the Chinese liver fluke *Clonorchis* (= *Opisthorchis*) *sinensis.*

is helpful in identifying these organs. Whenever possible, you should first study whole mounts of the larval stages before examining sections.

1. Miracidium. The miracidium is a small, free-swimming, ciliated larva that resembles a protozoan. Observe a slide of miracidia. What evidence suggests that this organism is a metazoan and not a protozoan? Attempt to locate as many of the following features as

possible: ciliated epidermis, apical papilla, apical eyespot, developing sporocyst. Sometimes the best that one can do is to determine the anterior and the posterior ends; it all depends upon the quality of the slide.

2. Sporocyst. This larval generation absorbs nutrients directly from host tissues and reproduces another embryonic stage (daughter sporocysts, redia, or cercaria). Look for evidence of these stages within the

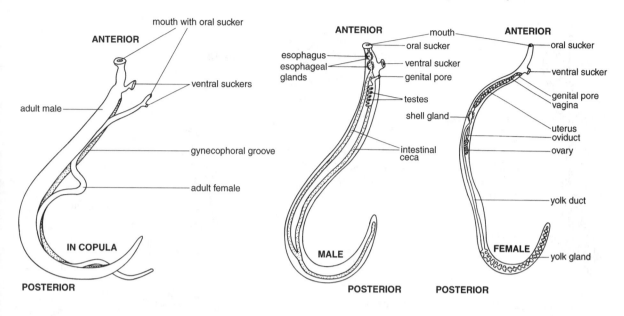

Figure 5.8. Generalized anatomy and morphology of male and female *Schistosoma mansoni*.

sporocyst. Sections of tissue parasitized by sporocysts also may be examined. Can you distinguish host tissue from that of the parasite? How can you tell the difference? Are daughter stages evident within these sections?

3. Redia. Rediae leave the sporocyst and relocate in the host's digestive gland or gonad. Note the elongate shape of this stage and the presence of any embryos (daughter rediae or cercariae). At the anterior end locate the mouth and muscular pharynx that leads to a small saccate gut. Towards the posterior locate the two locomotory processes. Did the sporocyst stage possess locomotory processes? Why might redia possess such appendages while sporocysts do not?

4. Cercaria. This stage leaves the intermediate host and either locates its definitive host directly or encysts in another intermediate host, becoming a metacercaria. Note the tail used for locomotion. The nature of the tail is important in distinguishing the types of cercaria. Is the tail forked? Attempt to locate the oral and midventral (acetabulum) suckers, and the biramous intestine. Are cystogenous glands present in the cercariae? These secretory structures produce the metacercarial cyst. (Phase contrast lighting may help you in these observations.)

5. Metacercaria. In most trematodes this encysted stage precedes the adult. Encystment may take place on the surface of a plant, as in *Fasciola hepatica*, or in

a second intermediate host such as a fish, as in *Clonorchis sinensis*. In *Fasciola*, metacercariae are small, flattened, caplike structures. Scan slides of metacercariae of *Fasciola* and *Clonorchis*. In *Fasciola*, ignore the plant debris often associated with the preparation. Occasionally oral and ventral suckers, pharynx, and biramous intestine may be seen within the cyst. Unfortunately, sometimes the best you can do is to make out the cyst wall.

Observational Procedures for Live Larval Stages
Caution: All participants in this exercise are to wear rubber gloves. The cercariae isolated in the following exericse may be members of the genera of bird schistosomes. These larvae must be considered to be a biological hazard as they cause a rash upon penetrating your skin. This condition is known as cercarial dermatitis or swimmers itch. While not generally considered to be serious, there is an increased sensitization with successive encounters. Wash your hands and all instruments in soapy water upon completion of the observations. Spills should be cleaned up immediately; direct contact of the pond water with your skin should be attended to quickly.

Sporocysts and Rediae. Obtain a snail from your instructor and place it in a small glass dish containing a small amount of water. Using the blunt end of a metal dissection probe, crush the snail, working from the apex

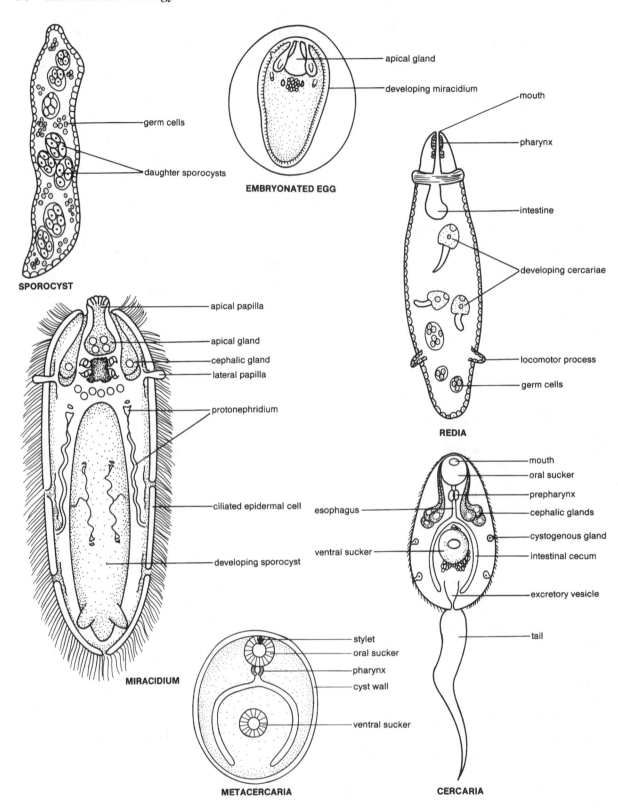

Figure 5.9. Fluke larval stages.

down the body. Do not completely pulverize the tissues as you may destroy the larval stages. Take up some of the material into a Pasteur pipette and place it into the well of a depression slide. This material will be examined for the larval stages, so the slurry should not be too thick. Dilute the preparation with water as necessary. Using low power of a compound microscope scan the slide for sporocysts and rediae (Fig. 5.9). Cercariae also may be present in these preparations, but they may not be mature. Once you have found these stages, repeat the observations you made on the prepared slides.

Live Cercariae

1. Place freshly collected snails or fingernail clams into separate vials (ca. 25–50 ml) filled with filtered pond water. (Do not use chlorinated tap water as the chlorine may be sufficiently concentrated to kill the mollusks. Tap water that has been aged for 2–3 days should be adequate.)

2. Examine each vial the next day for free-swimming cercariae. To do this, hold a vial in front of a piece of black paper and illuminate it from the side using a high-intensity lamp attached to a flexible arm. Try shining the lamp from different angles to achieve the best lighting. The cercariae will appear as tiny white specks moving about in the water. Use a Pasteur pipette to transfer all of the suspected cercariae to the well of a depression slide to examine them more closely.

3. Did you isolate cercariae or something else? How would you know whether the organisms are cercariae or not? If the organisms were cercariae, continue with the next two steps in the exercise.

4. How does this larval stage move? How fast do they move? To examine their internal anatomy, add a drop of a very dilution solution of neutral red or Nile blue to the depression slide. After a few minutes transfer the specimens to a regular microscope slide and add a cover slip, making sure that you do not crush the larvae. What structures can you see?

5. Were all of the mollusks examined infected with cercariae? About how many cercariae per mollusk were released? What does this indicate about the parasitic burden of these mollusks? Knowing where the snails were collected provides information on the trematode fauna of the habitat. Would you say that it is high or low?

Other Trematodes

Unfortunately, in recent years many biological supply houses have dropped a number of useful specimens from their product lines. However, the specimens available for study at your institution may include some of these interesting species. As time and materials permit, review the following specimens.

1. *Prosthogonimus*. This small digenetic fluke of the family Prosthogonimidae is a common parasite of birds, especially ducks, living in the Great Lakes region of North America. Adults of *Prosthogonimus* parasitize the oviduct of birds, and in doing so, occasionally becomes lodged in the bird's eggs. Examine a whole mount slide of *Prosthogonimus*. Note that the adult has a wide flattened body. Based on the habitat of this species, speculate on the significance of this morphology. Because prepared specimens usually have a transparent body you should easily be able to identify all of the structures discussed in the observational procedures for *Fasciola*.

2. *Metagonimus*. This intestinal parasite infects a variety of mammals, including humans, in countries of the Far East and surrounding the Baltic Sea. Infections develop from eating undercooked fish. Examine available slides of whole mounts; attempt to locate the structures seen in other flukes.

3. *Echinostoma*. This species is a widely distributed parasite of the ileum of birds and mammals, including humans. Infection of the definitive host occurrs by eating mollusks, fish, or amphibians infected with metacercariae. Examine available slides of whole mounts; attempt to locate the structures seen in other flukes.

4. *Paragonimus*. This small (5 × 10 mm) fluke usually lives in the lungs of humans and other mammals, encapsulated by connective tissue. However, they are also known to infect other organs. Infections begin by the consumption of undercooked crustaceans. Examine available slides of whole mounts; attempt to locate the structures seen in other flukes. ■

C. Class Monogenea

Monogenea [mon-o-GEN-e-a; G., *mono*, one + G., *gene*, descent] is a relatively small class of some 1,100 species that are all parasitic on freshwater and marine ectothermal vertebrates, especially fish. They tend to be ectoparasites, but some are found in the mouth, rectum, and urinary bladders. Monogeneans are similar in size and shape to the flukes, except for the characteristic posterior attachment organ, the **opisthaptor**. Anteriorly the **prohaptor**, an organ of feeding and attachment, resembles the oral sucker of flukes. Monogeneans are small, ranging in size from 0.03 to 30 mm in length. Unlike the flukes, there is but a single larval stage, the **oncomiracidium**. This free-swimming larva attaches to a host and then gradual develops into an adult. Few monogeneans are serious pathogens. However, under crowded conditions in fish hatcheries, massive infestations can result in serious erosion of tissues and lead to secondary infections which can cause massive die-offs.

■ Observation Procedure: Monogenea

Examine slides of *Gyrodactylus* (Fig. 5.10) or *Neobenedenia* (= *Benedenia*). Both are ectoparasites of fish. At the posterior end, locate the large ventral sucker (opisthaptor). How is this sucker armed? Does it possess centrally located hooks? Are there other smaller hooks (hooklets) along the margin of the sucker? Given that monogeneans are ectoparasites that feed on their host's tissues and body fluids, how does the opisthaptor provide for an effective means of attachment? At the anterior end, locate the forked adhesive organ (**head organs**), large muscular pharynx, and biramous intestine. In *Gyrodactylus*, the normal oncomiracidium is not present. However, a juvenile worm may be seen developing internally in this viviparous monogenean. (NB: You may need to view several specimens to see all of the internal structures. Depending on the quality of the preparation, sometimes the best you can do is to identify the anterior and posterior ends and see that the opisthaptor is a large sucker.) ■

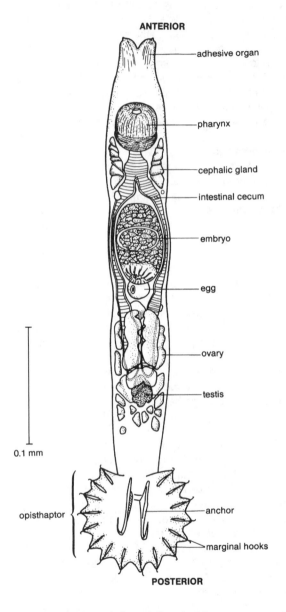

Figure 5.10. Ventral view of *Gyrodactylus*, a monogenetic flatworm.

D. Class Cestoda

Cestoda [ses-TO-da; G., *cest*, girdle + G., *oda*, resemblance] is a moderate-sized class comprising about 4,000 species of endoparasitic flatworms commonly known as **tapeworms**. Although they are dorsoventrally flattened, cestodes differ from other flatworms in a number of ways. Tapeworms completely lack a mouth and digestive tract. Most have a long, ribbon-like body that may be many meters in length; however, some cestodes are less than 3 mm long. The body is usually comprised of three regions: **scolex** (head), **neck**, and **strobila** (body). Adults are found lodged in the intestine of a vertebrate, where they utilize the nutrients of their host. Larval stages often cause considerable damage to their intermediate hosts. Humans infected with tapeworms, normally found in animals (a condition called **zoonosis**), may have serious complications (e.g., **hydatid cysts** of *Echinococcus granulosus*). Cestodes have enormous medical, veterinary, and economic importances.

The strobila is made up of many identical units called **proglottids**, except in the subclass Cestodaria whose members have a short undivided body. The strobila looks as if it were segmented, but tapeworms lack true segmentation. Production of proglottids is similar to strobilization of the ephyra from the strobila stage of the scyphozoan jellyfish, *Aurelia* (Exercise 3). Proglottids are formed sequentially in the neck region and are at first relatively undifferentiated, but they mature as they grow older. Therefore, proglottids are progressively older and more mature the further away from the anterior end they are located. Eggs of the very oldest proglottids have been fertilized and these proglottids with expanded uteri are said to be **gravid**. There are no septa (walls) separating proglottids and the tissues grade from one proglottid to the next. These characteristics are very different from the production of true segments in segmented phyla (e.g., Annelida, Exercise 13).

The scolex acts as an anchor and is it modified in different species to include various holdfast organs, with bulbous ends, spines, hooks, suckers, and several types of glands. The scolex also contains most of the animal's neural ganglia and sensory receptors. The tapeworm body wall is a living tissue (tegument) covered by numerous microscopic projections (**microtriches**). These microvilli-like structures are responsible for absorbing nutrients from the host's gut.

Tapeworms have an immense reproductive capacity, some being capable of releasing one million eggs per day! Tapeworms usually have one or more larval stages, each with its own intermediate host. However, there are many variations in the life cycle and several different types of intermediate stages are recognized.

It is not possible to cover this diverse class completely; therefore, we will study only a few species in some detail. Other species will be introduced, if they are available. At this point it will be profitable if students review the life history of a typical tapeworm (Fig. 5.11).

■ Observational Procedure: Cestoda

Scolices
Obtain whole mount slides of the anterior ends of several different tapeworms (e.g., *Diphyllobothrium latum, Echinococcus granulosus, Moniezia spp., Taenia solium,* and *Taeniarhynchus saginatus*) and examine the scolices of each (Fig. 5.12). What types of structural adaptations do scolices have to act as holdfast organs? Do they all have the same type of armaments? Are there any unusual structures? Compare the scolices of *D. latum* to *T. solium*. Are they the same? Does *D. latum* have a scolex comparable to the other species studied? Formulate a table that compared the scolices of the species you examined.

Strobila
Observe a whole specimen of a tapeworm.
1. As best you can, determine the length of the specimen. Is the worm of uniform width throughout? Identify the scolex, neck, and strobila of this specimen. What criteria did you use to make this differentiation? Note the individual proglottids of the strobila. How large (e.g., in width) is the scolex in comparison to the width of the neck? The width of the strobila? Some tapeworms can achieve remarkable lengths, but they do not continue to increase in size indefinitely. The length of the adult worm changes: some proglottids are shed, while others are added. How does this fact and the widths of the body inform us of the relative strength of the tissues composing the scolex, neck, and strobila?
2. Turn your attention back to the slides of tapeworm scolices and examine the neck regions (Fig. 5.12). This is the area where young proglottids are being formed. Do the proglottids begin at a specific point on the neck or do they gradually take form? Is there any differentiation of internal organs noticeable in the proglottids of the neck? Look for two strips of lighter tissue which extend posteriorly from the scolex. These are the **excretory canals**.
3. If available, examine a slide of a longitudinal section of *Taenia*. (Please note that cross sections will not do for these observations.) What structures can you see in this section (Fig. 5.13)? Focus your attention on the

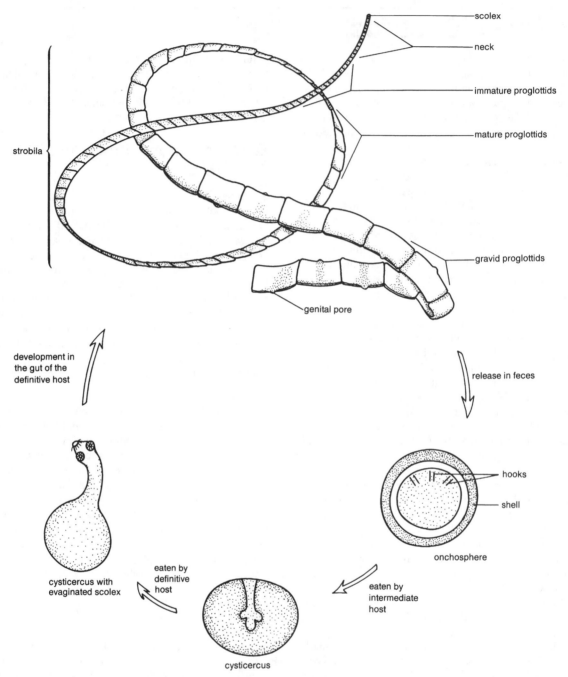

Figure 5.11. Life cycle of the pig tapeworm, *Taenia solium*.

region of the junction of two proglottids. What signifies the termination of one proglottid? Are there septa between the proglottids? Consider the significance of this fact in relationship to the term segmentation? Reconsider the strobila stage of the scyphistoma polyp of the cnidarian *Aurelia* (Exercise 3B). Compare and

contrast strobilization in the tapeworm and in cnidarians. How are they similar? How do they differ? (The concept of segmentation will come up again in later exercises, so it is important to begin to consider it now.)
4. Obtain a whole mount slide that includes several stages of development of proglottids of *Taenia* (or

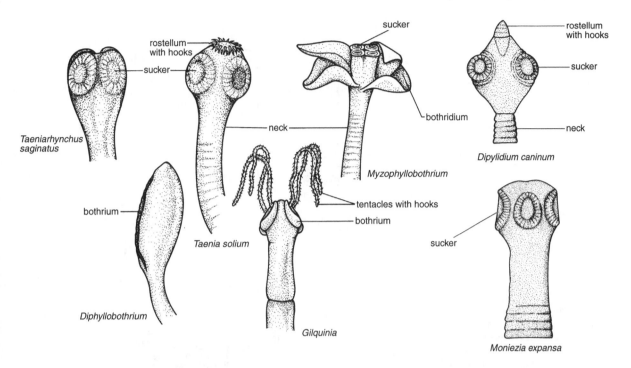

Figure 5.12. Representative tapeworm scolices.

another tapeworm), and with the aid of Fig. 5.14, identify the components of the male and female reproductive systems. In gravid proglottids, eggs fill the uterus and may obscure parts of the reproductive systems. Compare mature and gravid proglottids. How has the morphology of the uterus been changed by the storage of eggs?

5. Obtain a slide of a whole mount of the canine or hydatid tapeworm *E. granulosus*. Compare the anatomy of *Taenia* with that of *E. granulosus* (Fig. 5.15). Note that in *E. granulosus*, the entire worm is only a few millimeters long and contains only three proglottids. The scolex has a **rostellum** [L., little beak] with a double row of hooks and four suckers.

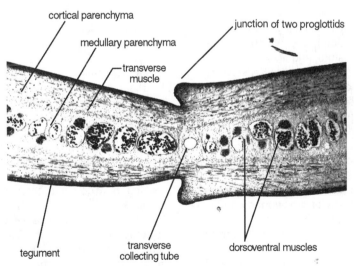

Figure 5.13. Photomicrograph of a longitudinal section of two proglottids of *Taenia*.

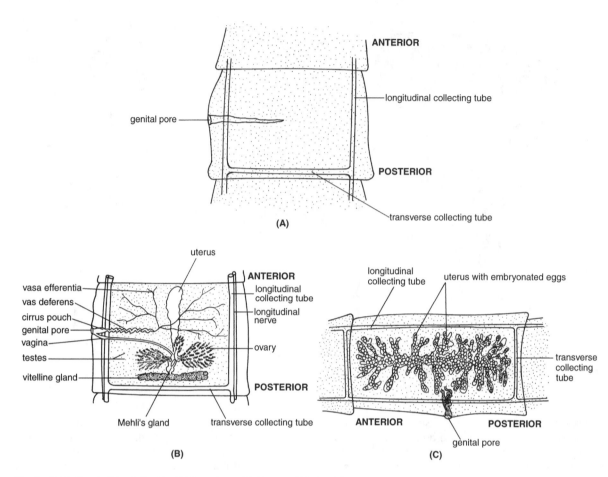

Figure 5.14. Proglottids of *Taenia*. (A) Immature. (B) Mature. (C) Gravid.

Immediately behind the scolex is a very short neck and then three proglottids. Determine which of these proglottids is immature, mature, and gravid.

6. Observe a prepared slide with cross sections of a sexually mature proglottid of *Taenia* (Fig. 5.16). On the surface, note the tegument and just beneath it the layers of circular and longitudinal muscle tissues. The tissue from this region inward is called the **cortex** region of the parenchyma. Look for the two parallel layers of transverse muscles which surround the reproductive systems and a less dense central parenchyma called the **medulla**. Unlike the parenchyma of turbellarians and flukes, the parenchyma of tapeworms contains few dorsoventral muscle fibers.

7. At the lateral margins of the proglottid, locate the large excretory ducts.

8. Examine a slide with serial sections of a single mature proglottid, and with aid of Fig. 5.14, determine where the approximate postion of each section would

be in the proglottid. Identify the following structures in the sections: testes, ovary, uterus, vitelline gland.

Larval Stages

1. Observe a slice of **measly pork** on display and note the pits scattered throughout the meat. These are the **bladderworm** or **cysticercus** stage of the pig tapeworm (*T. solium*) (Fig. 5.11). Do you think that a single cysticercus could go undetected during meat processing? What about on your dinner plate? What function does thorough cooking serve here? Besides proper cooking, how do societies avoid parasitization by this species?

2. Note that some cysticerci contain a bulbous structure on the inside. This is the invaginated scolex. Describe what it looks like. Examine a whole mount of a cysticercus from *Taenia* (or another species) and locate the scolex. How does the orientation of the hooks compare to that in the adult? Why

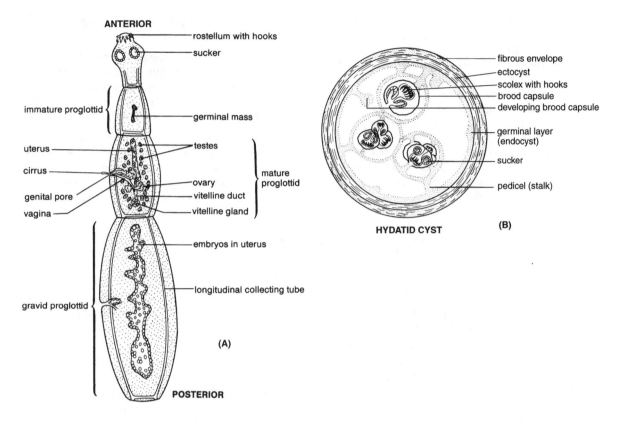

Figure 5.15. *Echinococcus granulosus.* (A) Adult. (B) Larval forms.

is there a difference? What process produces the reorientation (reorganization) of the hooks from the larval to the adult condition? Where and when does this happen?

3. If available, compare cysticerci from other tapeworms (e.g., *Hymenolepis*) to that of *Taenia.* How are they similar? How do they differ? Remember your conclusions to these questions in the following study of the hydatid tapeworm.

4. Re-examine the whole mount slide of *E. granulosus.* This species demonstrates asexual reproduction

in its larval stage. Once the **oncosphere** has implantated in the lungs or liver of an intermediate host, it develops into a structure called the **hydatid cyst** (Fig. 5.15). Eventually, inside the main portion of the cyst, many **brood capsules** are asexually produced. Within each brood capsule a dozen or more tiny protoscolices are formed. Individual protoscolices and whole brood capsules may be freed within the cyst and then are known collectively as **hydatid sand**.

5. Observe a cross section of a hydatid cyst and locate the structures indicated below (Fig. 5.15). The

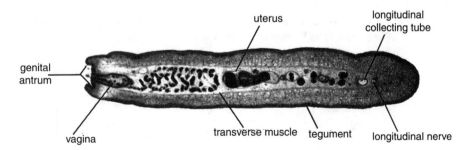

Figure 5.16. Cross section of a proglottid from *Taenia.*

outer part of the cyst is a thick, noncellular, fibrous envelope formed from host tissues. Just inside the envelope is a germinal layer from which brood capsules and independent protoscolices are budded. If the cyst walls of a hydatid cyst is ruptured by a predator during a meal, the hydatid sand (protoscolices) will be released all over the flesh of the prey. How might this event function to distribute tapeworms throughout a pack of wolves?

(NB: The zoonotic condition of having a hydatid cyst is very serious in humans. Fluid from the hydatid cyst will cause anaphylactic shock in its host if it is released due to trauma or during surgery. Consider the life cycle of this parasite and determine a scenario that would lead to human **hydatidosis**.)

Other Tapeworms and Live Specimens
Compare the general morphology of other species that are available for study to the species examined in this exercise. *If exercises with live worms are to be done, follow the instructions of your instructor, explicitly.*

Exercise Review
The Platyhelminthes is a complex phylum comprised of four unique classes. To reinforce your studies, we suggest that you construct a table that compares and contrasts the distinctive characteristics of each class. ■

Supplemental Readings*

Ansari, N. (ed.). 1973. Epidemiology and Control of Schistosomiasis (Bilharziasis). S. Karger AG, Basal. (TR)

Aral, H. P. (ed.). 1980. Biology of the Tapeworm *Hymenolepis diminuta*. Academic Press, New York. (C)

Boros, D. L. 1976. Schistosomiasis mansoni: a granulomatous disease of cell-mediated immune etiology. Ann. N.Y. Acad. Sci. 278: 36–46. (TR)

Bronsted, H. V. 1955. Planarian regeneration. Biol. Rev. 30: 65–126. (TU)

Brooks, D. R., R. T. O'Grady, and D. R. Glen. 1985. Phylogenetic analysis of the Digenea (Platyhelminthes: Cercomeria) with comments on their adaptative radiation. Can. J. Zool. 63: 411–443. (TR)

Brown, H. W., and F. A. Neva. 1983. Basic Clinical Parasitology. Appleton-Century-Crofts, Norwalk, CT. (TR, C)

Camargo, C. A., and W. H. Marshall. 1987. Radiological diagnosis of neurocysticercosis. Parasiol. Today 3: 30–31. (C)

Crellin, J. R., F. L. Andersen, P. M. Schantz, and S. J. Condie. 1982. Possible factors influencing distribution and prevalence of *Echinococcus granulosus* in Utah. Am. J. Epidemiol. 116: 463–474. (C)

Crezée, M. 1982. Turbellaria. *In*: S. Parker (ed.). Synopsis and Classification of Living Organisms, Vol. 1. McGraw-Hill, New York, pp. 718–740. (TU)

Ehlers, U. 1985. Comments on a new phylogenetic system of the Platyhelminthes. Hydrobiologia 132: 1–12. (G)

Erasmus, D. A. 1972. The Biology of Trematodes. Edward Arnold, London. (TR)

Gamble, W. G., M. Segal, P. M. Schantz, and R. L. Rausch. 1979. Alveolar hydatid disease in Minnesota. First human case acquired in the contiguous United States. JAMA 241: 904–907. (C)

Harrison, F. W., and B. J. Bogitsh (eds.). 1991. Microscopic Anatomy of Invertebrates, Platyhelminthes and Nemertina, Vol. 3. Wiley-Liss, New York. (G)

Jennings, J. B. 1957. Studies on feeding, digestion, and food storage in free-living flatworms. Bio. Bull. 112: 63–80. (TU)

Kenk, R. 1972. Freshwater Planarians (Turbellaria) of North America. Environmental Protection Agency, Washington, DC. (TU)

Kolasa, J. 1991. Flatworms: Turbellaria and Nemertea. *In*: J. H. Thorp and A. P. Covich (eds.). Ecology and Classification of North American Freshwater Invertebrates. Academic Press, New York, pp. 145–169. (TU)

Komiya, Y. 1966. *Clonorchis* and clonorchiasis. Adv. Parasitol. 4: 53–106. (TR)

Loker, E. S. 1983. A comparative study of the life-histories of mammalian schistosomes. Parasitology 87: 343–369. (TR)

Martin, G. 1978. A new function of rhabdites: mucus production for ciliary gliding. Zoomorphologie 91: 235–248. (TU)

*The supplemental literature listed here is coded to indicate the general topic covered by the paper: G = general Platyhelminthes; TU = Turbellaria; TR = Trematoda; C = Cestoda.

McManus, D. P., and J. D. Smyth. 1986. Hydatidosis: changing concepts in epidemiology and speciation. Parasitol. Today 2: 163–167. (C)

Montgomery, J. R., and S. J. Coward. 1974. On the minimal size of a planarian capable of regeneration. Trans. Am. Microsc. Soc. 93: 386–391. (TU)

Olsen, O. W. 1974. Animal Parasites: Their Life Cycles and Ecology, 3rd ed. University Park Press, Baltimore, MD. (TR, C)

Pantelouris, E. M. 1965. The Common Liver Fluke, *Fascicola hepatica* L. Pergamon Press, Oxford, UK. (TR)

Reinhard, E. G. 1957. Landmarks of parasitology. I. The discovery of the life cycle of the liver fluke. Exp. Parasitol. 6: 208–232. (TR)

Reynoldson, T. B., and A. D. Sefton. 1976. The food of *Planaria torva* (Müller) (Turbellaria-Tricladida), a laboratory and field study. Freshwater Biol. 6: 383–393. (TU)

Rieger, R. M. 1981. Morphology of the Turbellaria at the ultrastructural level. Hydrobiologia 84: 213–229. (TU)

Ruttenber, A. J., B. G. Weniger, F. Sorvillo, R. A. Murry, and S. L. Ford. 1984. Diphyllobothriasis associated with salmon consumption in Pacific coast states. Am. J. Trop. Med. Hyg. 33: 455–459. (C)

Schell, S. C. 1970. How to know the Trematodes. Brown, Dubuque, IA. (TR)

Schell, S. C. 1985. Trematodes of North America North of Mexico. University Press of Idaho, Moscow, ID. (TR)

Schmidt, G. D. 1970. How to Know the Tapeworms. Brown, Dubuque, IA. (C)

Schmidt, G. D. 1986. Handbook of Tapework Identification. CRC Press, Boca Raton, FL. (C)

Smith, J. S. Tyler, M. B. Thomas, and R. M. Rieger. 1982. The morphology of turbellarian rhabdites: phylogenetic implications. Trans. Am. Microsc. Soc. 101: 209–228. (TU)

Spiegelman, M., and P. Dudley. 1973. Morphological stages of regeneration in the planarian, *Dugesia tigrina*: a light and electron microscope study. J. Morph. 139: 155–183. (TU)

Stoll, N. R. 1947. This wormy world. J. Parasitol. 72: 492–497. (TR)

Thompson, R. C. A. 1986. The Biology of *Echinococcus* and Hydatid Disease. George Allen and Unwin, London. (C)

Tyler, S. (ed.). 1991. Turbellarian Biology. Hydrobiologia 227: 1–398. (TU)

Waren, K. S. 1973. The pathogology of schistosome infections. Helminth. Abstr. Ser. A 42: 591–633. (TR)

Warren, K. S., and V. A. Newill. 1968. Schistosomiasis. The Press of Case Western Reserve University, Cleveland, OH. (TR)

■

Phylum Nemertea (Rhynchocoela)

Nemertea [ne-MER-te-a; G., *nemerte*, unerring] is a phylum comprising about 900 species of nonsegmented, dorsoventrally flattened worms, lacking a body cavity. They are commonly known as ribbon worms because of their length and shape. Nemerteans (also nemertines) range in size from a few millimeters to an incredible 30 m in *Lineus longissimus*; the average length is <20 cm long. Most nemerteans are marine and benthic, but pelagic forms and freshwater and terrestrial species are known. Most nemerteans are carnivorous; a few are parasitic.

Nemerteans have several features also found in turbellarians, including a ciliated epidermis, protonephridia, and a bilateral nervous system. The space between the gut and the body wall also is filled with parenchyma, but this condition may not be acoelomate in the traditional sense. Turbeville and Ruppert (1985) suggest that both nemerteans and flatworms have evolved from coelomate taxa, perhaps the annelids. Nevertheless, nemerteans are unlike flatworms in three important characteristics. (1) They possess an eversible, tubular **proboscis** housed in a special, fluid-filled cavity (specialized coelom) called the **rhynchocoel**. For this reason the alternative phylum name **Rhynchocoela** [RING-ko-SE-la; G., *rhynchos*, snout + G., *coelo*, hollow] is often used. The proboscis is used in food capture, but has a different embryological origin from the mouth. Prey is seized by the proboscis which is everted from the rhynchocoel by hydrostatic pressure created by muscular contractions. The accuracy of the proboscis is said to be unerring, hence the phylum name, Nemertea. (2) There is a complete digestive system. (3) A closed vascular system, which may be homologous with the coelom of annelids, is present (Turbeville and Ruppert 1985).

Classification

The division of nemerteans into two classes is based on position of the mouth relative to the cerebral ganglia and on the nature of the proboscis.

1. **Anopla** [AN-o-pla; G., *anoplos*, unarmed]. Nemerteans with mouth below or posterior to the cerebral ganglia. Proboscis is unarmed. Examples: *Cerebratulus* and *Lineus*.

2. **Enopla** [EN-o-pla; G., *enoplos*, armed]. Nemerteans with mouth anterior to the cerebral ganglia, in most forms mouth and proboscis pore are united with a common opening. Proboscis is armed. Example: *Prostoma*.

■ Observational Procedure: *Cerebratulus*

1. Examine a nemertean such as *Cerebratulus* and identify the anterior and posterior ends (Fig. 6.1). Note that the body is long, somewhat dorsoventrally flattened, and nonsegmented.

2. Locate the proboscis pore at the very tip of the anterior end. Some specimens may have the proboscis partly extended. In *Cerebratulus* the large, slitlike, ventral mouth is near the anterior end.

3. On the lateral margins just anterior to the mouth are two thin, longitudinal grooves called **cephalic slits**. They probably function as chemoreceptors. At the very tip of the posterior end is the anus.

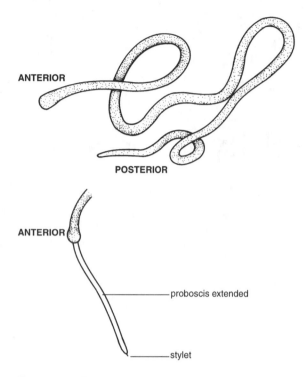

Figure 6.1.   External view of a nemertean worm.

4. Examine several cross sections of a nemertean. Note that the body is covered by an epidermis of ciliated columnar glandular cells (Fig. 6.2). Beneath the epidermis locate a layer of connective tissue called the **extracellular matrix** or **dermis**. It occasionally has extensions that go deeper into the body. Just under the dermis locate the circular and longitudinal muscles layers.

5. Note the absence of a coelom. The body is filled with circular and longitudinal muscles and parenchymal tissue. In the central part of the animal locate the proboscis and below it the gut with its vacuolate phagocytic cells. Note that the hollow proboscis is constructed of layers of circular and longitudinal muscles. How does this tubular structure function? Surrounding the proboscis is a cell-lined, fluid-filled cavity called the rhynchocoel.

6. Also present within the body of nemerteans are lateral blood vessels, lateral nerves, protonephridia, and gonads. Attempt to locate these structures in the cross sections available for study (Fig. 6.2).

7. Examine a slide of a helmet-shaped, **pilidium larva** [G., *pilidium*, small felt cap] (Fig. 6.3). Note that the larva is covered with cilia, and at the aboral pole is an apical sensory organ with a tuft of long cilia. Find the foregut (**stomodeum**) flanked on either side by the oral lobes. The foregut leads to the small blind midgut. The anus develops later by an ectodermal invagination when the larva undergoes metamorphosis. The large internal cavity is the blastocoel.

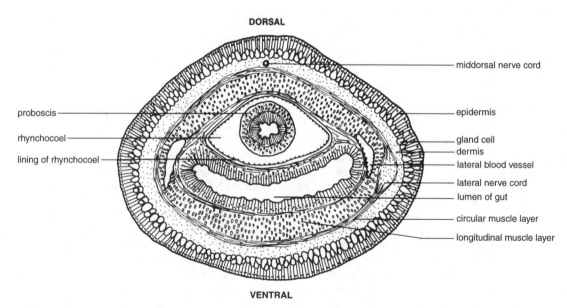

Figure 6.2.   Cross section of a nemertean worm.

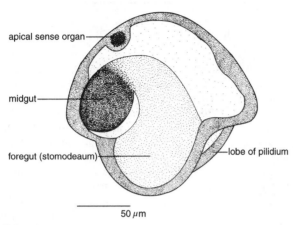

apical sense organ

midgut

foregut (stomodeaum)

lobe of pilidium

50 μm

Figure 6.3.   Side view of a pilidium larva.

Supplemental Readings

Gibson, R. 1972. Nemerteans. Hutchinson University Library, London.

Gibson, R. 1976. Freshwater nemerteans. Zool. J. Linn. Soc. 58: 177–218.

Gibson, R. 1985. The need for a standard approach to taxonomic descriptions of nemerteans. Am. Zool. 25: 5–14.

Gibson, R., and J. Moore. 1976. Freshwater nemerteans. Zool. Linn. Soc. 58: 117–218.

McDermott, J. J., and P. Roe. 1985. Food, feeding behavior and feeding ecology of nemerteans. Am. Zool. 25: 113–125.

Moore, J., and R. Gibson. 1985. The evolution and comparative physiology of terrestrial and freshwater nemerteans. Biol. Rev. 60: 267–312.

Norenburg, J. L. 1985. Structure of the nemertean integument with consideration of its ecological and phylogenetic significance. Am. Zool. 25: 37–51.

Riser, N. W. 1985. Epilogue: nemertinea, a successful phylum. Am. Zool. 25: 145–151.

Stricker, S. A., and R. A. Cloney 1981. The stylet apparatus of the nemertean *Paranemertes peregrina*: its ultrastructure and role in prey capture. Zoomorphology 97: 205–223.

Turbeville, J. M. 1991. Nemertean. *In*: F. W. Harrison and B. J. Bogitsh (eds.). Microscopic.

Turbeville, J. M. 1991. Nemertinea. *In*: F. W. Harrison and B. J. Bogitsh (eds.). Microscopic Anatomy of Invertebrates. Wiley-Liss, New York, pp. 285-328.

Turbeville, J. M., K. G. Field, and R.A. Raff. 1992. Phylogenic position of phylum Nemertini inferred from 18s rRNA sequences: molecular data as a test of merphological character homology. Mol. Biol. Evol. 9:235-249.

Turbeville, J. M., E. E. Ruppert. 1985. Comparative ultrastructure and the evolution of nemerteans. Am. Zool. 25:53-71.

PSEUDOCOELOMATE PHYLA

Traditionally a body cavity (perivisceral cavity) formed from a persistent blastocoel and lacking a mesodermal lining (i.e., a cellular peritoneum) is called a **pseudocoelom** (pseudocoel). Using this description, zoologists have designated six phyla as pseudocoelomates: Acanthocephala, Kinorhyncha, Loricifera, Nematoda, Nematomorpha, and Rotifera. To this list some zoologists add Phylum Priapulida. Because these phyla possess certain similar features, some of these phyla have been grouped within the **Aschelminthes** [AS-kel-MIN-theez; G., *ascus*, bladder + G., *helmins*, worm] or **Nemathelminthes** [Nee-mat-hel-MIN-theez; G., *nema*, thread]. This practice is more common in European publications, and is not generally followed in the United States.

Recent studies have concluded that there is no good reason to unite these taxa. Perhaps the most sigificant argument against their union is the fact that the pseudocoel is no longer considered to be an important feature; it may well be an artifact of nutritional status of the animal. However, in spite of this new information, there are pedagogical reasons to study these wormy phyla as a group. For example, most are less than 1 cm long and possess an outer, noncellular cuticle. Further, their bodies lack a well-formed head and gas exchange and circulatory systems. Other important features found in many of these phyla include a complete digestive tract with muscular pharynx, a protonephridial system, and eutely (consistancy in the total number of cells). From a teaching standpoint it makes sense to direct our attention to these common features.

Ehlers, U., W. Ahlrichs, C. Lemburg, and A. Schmidt-Rhaesa. 1996. Phylogenetic systematization of the Nemathelminthes (Aschelminthes). Verh. Dtsh. Zool. Ges. 89 (in press).

Garey, J. R., L. Y. Mackey, J. M. Brooks, B. Winnepenninckx, and T. Backeljau. 1995. Animal phylogeny: ribosomal RNA studies of Aschelminthes. J. Cell. Biochem., suppl. 19B: 345.

Garey, J. R., T. J. Near, M. R. Nonnemacher, and S. A. Nadler, 1996. Molecular evidence for Acanthocephala as a sub-taxon or Rotifera. J. Mol. Evol. (in press).

Kristensen, R. M. 1991. Kinorhyncha. *In*: Microscopic Anatomy of Invertebrates, Vol. 4. F. W. Harrison, and E. E. Ruppert (eds.) Wiley-Liss, New York, pp. 377–404.

Kristensen, R. M. 1991. Loricifera—a general biological and phylogenetic overview. Verh. Dtsch. Zool. Ges. 84: 231–246.

Kristensen, R. M. 1991. Loricifera. *In*: Microscopic Anatomy of Invertebrates, Vol. 4. F. W. Harrison and E. E. Ruppert (eds.). Wiley-Liss, New York, pp. 351–375.

Lorenzen, S. 1985. Phylogenetic aspects of pseudocoelomate evolution. *In*: The Origins and Relationships of Lower Invertebrates. S. Conway Morris, J. D. George, R. Gibson, and H. M. Platt. (eds.). Clarendon Press, Oxford, pp. 210–223.

Rieger, R. M. 1976. Monociliated epidermal cells in Gastrotricha: significance for concepts of early metazoan evolution. Z. Zool. Syst. Evolutionsforsch. 14: 198–226.

Rieger, R. M. and S. Tyler. 1995. Sister-group relationship of Gnathostomulida and Rotifer-Acanthocephala. Invertebr. Biol. 114: 186–189.

Ruppert, E. E. 1991. Introduction to the aschelminth phyla: a consideration of mesoderm, body cavities, and cuticle. *In*: Microscopic Anatomy of Invertebrates, Vol. 4. F. W. Harrison, and E. E. Ruppert (eds.). Wiley-Liss, New York, pp. 1–17.

Wallace, R. L., C. Ricci and G. Melone. 1996. A cladistic analysis of pseudocoelomate (aschelminth) morphology. Invertebr. Biol. 115: 104–112,

Winnepenninckx, B. T., L. Y. Backejau, J. M. Mackey, Brooks, R. De Wachter, S. Kumar, and J. R. Garey. 1995. 18S rRNA data indicate that aschelminthes are polyphyletic in origin and consist of at least three distinct clades. Mol. Biol. Evol. 12: 1132–1137.

■

Phylum Gastrotricha

Gastrotricha [GAS-tro-TRIK-a; G., *gastro*, stomach, + G., *trich*, hair] is a small phylum (ca. 450 species) of marine and freshwater wormlike, bilaterally symmetrical animals. Commonly, they are found gliding with their ventral surface in contact with other organisms or inanimate objects. Three characteristic features of gastrotrichs may be seen readily (Fig. 7.1): (1) the presence of a forked tail, especially in the freshwater species; (2) a modified cuticle surface with spine- or scale-like structures covering the dorsal side; (3) the patterned distribution of cilia which are more or less restricted to the ventral side as the phylum name suggests.

Classification

The phylum is divided into two orders.

Macrodasyida. Marine gastrotrichs with elongate, cylindrical body. Posterior end of various shapes. Adhesive tubes usually present along sides of the body and at anterior and posterior ends. Example: *Macrodasys*.

Chaetonotida. Marine and all freshwater gastrotrichs. Body bowling pin-shaped with a neck and ending in a forked tail. Adhesive tubes are found only at the posterior end. Examples: *Chaetonotus* and *Lepidodermella*.

■ Observational Procedure

1. Examine a culture of the chaetonotid gastrotrich, *Lepidodermella*, under a dissection microscope. You should be able to see the animals gliding about on the bottom of the culture dish. At this magnification, could they be mistaken for protozoans? Why? What features should one look for to tell the two groups apart.

2. Place a ring of methyl cellulose on a slide and add 1–2 drops of the gastrotrich culture into the center of the ring. Gently mix the methyl cellulose and culture fluid with a toothpick and then carefully apply a cover slip to the preparation, avoiding the introduction of air bubbles. Using low power, examine your slide for gastrotrichs. Describe how these animals move? At this magnification, could gastrotrichs be mistaken for protozoans? What distinctive characteristics do each possess? Be careful not to let the slide dry out during your examinations. This may be avoided by periodically adding a drop of culture fluid to the edge of the cover slip.

3. Switch to higher magnification to make additional observations. Note the distinct scaly appearance (Fig. 7.3). The body consists of a rounded, anterior head region, a slightly and gradually constricted neck, and an expanding elongate trunk with a forked tail. With the appropriate lighting, you may be able to see the anterior sensory bristles, mouth, pharynx, stomach-intestine, and reproductive system. However, sometimes it is necessary to stain your specimens to see their internal anatomy (see Step 4).

4. Make another preparation, again using a ring of methyl cellulose. However, this time apply a single drop of neutral red stain to the fluid before you place the cover slip on the slide. If the specimens do not stain

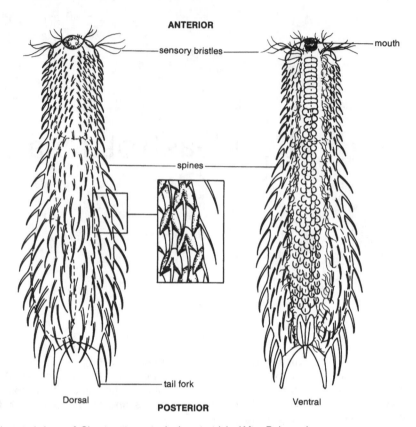

ANTERIOR

sensory bristles ——— mouth

spines

tail fork

Dorsal

POSTERIOR

Ventral

Figure 7.1. Dorsal and ventral views of *Chaetonotus*, a typical gastrotrich. (After Balsamo)

too darkly, you should be able to differentiate many of the internal structures noted above (Fig. 7.4).

5. Students should examine samples taken from field sites or from a laboratory fish tank for other specimens. Knowing how gastrotrichs move, where should you look to find specimens? Such searching may be tedious, but don't give up easily; a new and unusual specimen may be only one slide away.

One genus often present in such samples is *Chaetonotus* (Fig. 7.1). In *Chaetonotus* the dorsal side is covered with **spines** as well as **scales**. As the animal turns, you should be able to note the ventral cilia. Short sensory bristles may be seen at the anterior and lateral parts of the head region. Dense bodies located at the base of each tail fork are the adhesive glands. ■

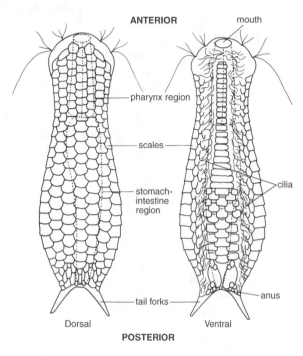

Figure 7.3. Dorsal and ventral views of *Lepidodermella*. (After Balsamo.)

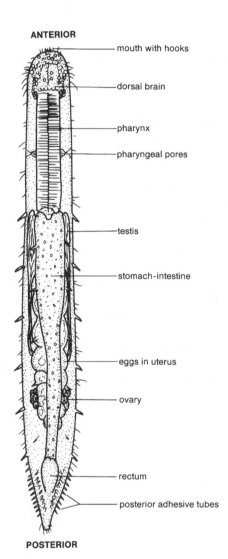

Figure 7.2. Dorsal view of a marine gastrotrich, *Macrodasys*. (From Hyman after Remane.)

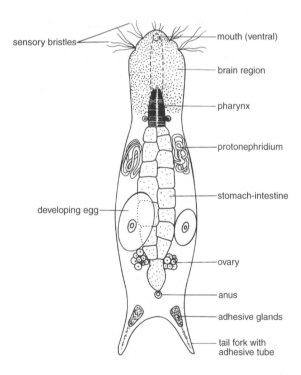

Figure 7.4. Dorsal internal view of a typical freshwater gastrotrich.

Supplemental Readings

Bennett, L. W. 1979. Experimental analysis of the trophic ecology of *Lepidodermella squammata* (Gastrotricha: Chaetonotida) in mixed culture. Trans. Am. Microsc. Soc. 98: 254–260.

Boaden, P. J. S. 1985. Why is a gastrotrich? *In*: S. Conway Morris, J. D. George, R. Gibson, and H. M. Platt (eds.). The Origin and Relationships of Lower Invertebrates. Clarendon Press, Oxford, pp. 248–260.

Brunson, R. B. 1950. An introduction to the taxonomy of the Gastrotricha with a study of eighteen species from Michigan. Trans. Am. Microsc. Soc. 69: 325–350.

Brunson, R. B. 1963. Aspects of the natural history and ecology of the Gastrotricha. *In*: E. Dougherty (ed.). The Lower Metazoa. University of California Press, Berkeley, CA, pp. 473–478.

Harrison F. W., and E. E. Ruppert (eds.). 1991. Microscopic Anatomy of Invertebrates, Aschelminthes, Vol. 4. Wiley-Liss, New York.

Hummon, W. 1966. Morphology, life history, and significance of the marine gastrotrich, *Chaetonotus testiculophorus* n.sp. Trans. Am. Microsc. Soc. 85(3): 450–457.

Hummon, W. 1982. Gastrotricha. *In*: S. P. Parker (ed.). Synopsis and Classification of Living Organisms, Vol. 1. McGraw-Hill, New York, pp. 857–863.

Hummon, M. R. 1984. Reproduction and sexual development in a freshwater gastrotrich. 1. Oogenesis of parthenogenic eggs (Gastrotricha). Zoomorphology 104: 33–41.

Hummon, M. R. 1984. Reproduction and sexual development in a freshwater gastrotrich. 2. Kinetics and fine structure of postparthenogenic sperm formation. Cell Tissue Res. 236: 619–628.

Hummon, M. R. 1984. Reproduction and sexual development in a freshwater gastrotrich. 3. Postparthenogenic development of primary oocytes and the X-body. Cell Tissue Res. 236: 629–636.

Rieger, G. E., and R. M. Rieger. 1977. Comparative fine structure study of the Gastrotrich cuticle and aspects of cuticle evolution within the aschelminthes. Z. Zool. Syst. Evolutionsforsch. 15: 81–124.

Rieger, R. M. 1976. Monociliated epidermal cells in Gastrotricha: significance for concepts of early metazoan evolution. Z. Zool. Syst. Evolutionsforsch. 14: 198–226.

Sacks, M. 1964. Life history of an aquatic gastrotrich. Trans. Am. Microsc. Soc. 83: 358–362.

Strayer, D. L., and W. D. Hummon. 1991. Gastrotricha. *In*: J. Thorp and A. Covich (eds.). Ecology and Classifications of North American Freshwater Invertebrates. Academic Press, New York, pp. 173–185.

Thane-Fenchel, A. 1970. Interstitial gastrotrichs in some south Florida beaches. Ophelia 7(2): 113–138.

■

Phylum Rotifera

Rotifera [ro-TIF-e-ra; L., *rota*, wheel + L., *ferre*, to bear] is a small phylum of some 2,000 species of bilaterally symmetrical, pseudocoelomates, having two distinguishing features: (1) a ciliated, apical region called the **corona**, used in locomotion and food gathering (Figs. 8.1 and 8.2), and (2) a muscular pharynx (**mastax**) equipped with complex jaws of seven pieces called **trophi** (Fig. 8.3). The name "wheel bearer" is an allusion to the corona which resembles a rotating wheel due to the metachronal beat of its cilia. In one group of rotifers (Collothecaceae) cilia are nearly lacking and long setae are used in prey-capture (Fig. 8.2).

Most rotifers are small (100–1,500 µm) and saccate to cylindrical in shape. Typically they are found in freshwater habitats where they are often very abundant, although several species are exclusively marine. One class (Bdelloidea) inhabits films of water found in soils or covering terrestrial plants such as mosses. Most rotifers are free moving, either by swimming or crawling, but many sessile species live attached to freshwater plants (Fig. 8.2). Nearly all rotifers are solitary, but about 25 species form colonies. Although very different in size and general biology, research indicates that the rotifers and acanthocephalans (Exercise 9) are closely related (Wallace and Snell 1991; cf. Garey et al. 1996).

Classification

Three classes are commonly recognized.

1. **Class Seisonidea**. Separate sexes of similar size and morphology. Gonads paired. Single genus (*Seison*), epizootic on marine crustaceans.

2. **Class Bdelloidea**. Only females. Paired ovaries. Mainly swimming and crawling forms. Many capable of becoming desiccated and then rehydrated. Examples: *Habrotrocha* and *Philodina*.

3. **Class Monogononta**. Separate sexes. Single gonads. Males rare and often structurally reduced. Reproduction mainly parthenogenetic, sexual reproduction results in a diapausing embryo. Three orders of benthic, swimming, and sessile forms. Examples: *Asplanchna, Brachionus, Filinia, Keratella, Lecane,* and *Synchaeta*.

■ Observational Procedure:

1. Examine a culture of rotifers using a dissection microscope and describe the locomotory activities you observe. Describe the motion of the "wheel organ" at this magnification. Are the specimens you are studying monogononts (Fig. 8.1A) or bdelloids (Fig. 8.1B)? If both are available, compare their locomotion. Does each move in the same way? Can rotifers be mistaken for protozoans (Exercise 1) or gastrotrichs (Exercise 7) at this magnification? How may they be distinguished?

2. Making a slide preparation of rotifers will depend on the kind available for study: for bdelloid rotifers, suck up a small amount of the debris from the bottom of the culture vessel; for free-swimming monogonont rotifers (two options), blindly suck up a small amount of culture fluid and hope for the best or attempt to pick out a single individual from the culture while observing it using a dissection microscope. Place a ring of methyl cellu-

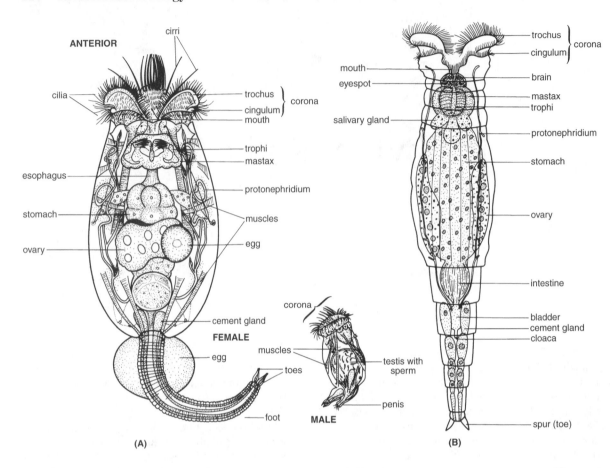

Figure 8.1 Views of two common rotifers. (*A*) *Brachionus plicatilis*, amictic female and male (Monogononta). (B) *Philodina roseola* (Bdelloidea). Each about 300 μm in length.

lose on a slide and add 1–2 drops of the rotifer culture into the center of the ring. Gently mix the methyl cellulose and culture fluid with a toothpick and then carefully apply a cover slip to the preparation, avoiding the introduction of air bubbles.

3. Using low power, examine your slide for rotifers. After locating an animal, identify its anterior and posterior ends (Figs. 8.1 and 8.2). Note that the body may be divided into three general regions: head (with corona), body, and foot with toes (Fig. 8.1). A neck region also may be present or the foot absent. Although highly variable, the corona usually consists of two ciliated rings (trochus and cingulum) at the anterior end.

4. Switch to higher magnification and attempt to distinguish the details of the coronal ciliation, including the trochus and cingulum (Fig. 8.1). How do the cilia appear to move? Are cirri present? Be careful not to let the slide dry out during your examinations. This may

be avoided by periodically adding a drop of culture fluid to the edge of the cover slip.

5. Turn your attention to the internal anatomy of your rotifer specimens. Under high power, locate the stomach, intestine, ovary (ovaries in bdelloids), musculature, protonephridial flame cells, foot with pedal glands, and toes (Fig. 8.1). As you examine the organs, look for the presence of nuclei.

6. If live protozoans or gastrotrichs are available, make a slide of that material and observe these organisms, comparing their general morphology to that of rotifers? Can rotifers easily be confused with ciliated protozoans or gastrotrichs under high power magnification? What sort of descriptive terms might be used to describe the movements of each? What distinctive anatomical features does each phylum possess?

7. Make another slide without methyl cellulose and place a small drop of food suspension (small algae or

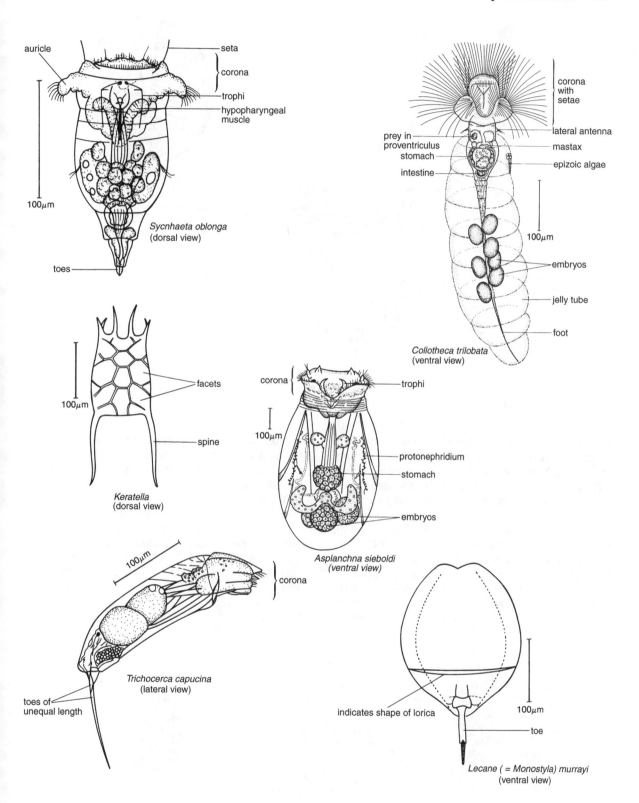

auricle

seta

corona

trophi

hypopharyngeal muscle

100μm

Sycnhaeta oblonga
(dorsal view)

toes

corona with setae

prey in proventriculus

stomach

intestine

lateral antenna

mastax

epizoic algae

100μm

embryos

jelly tube

foot

Collotheca trilobata
(ventral view)

100μm

facets

spine

Keratella
(dorsal view)

corona

trophi

100μm

protonephridium

stomach

embryos

Asplanchna sieboldi
(ventral view)

100μm

corona

Trichocerca capucina
(lateral view)

toes of unequal length

indicates shape of lorica

100μm

toe

Lecane (= Monostyla) murrayi
(ventral view)

Figure 8.2 Several common freshwater rotifers. Bars = 100 μm. (After Koste 1978.)

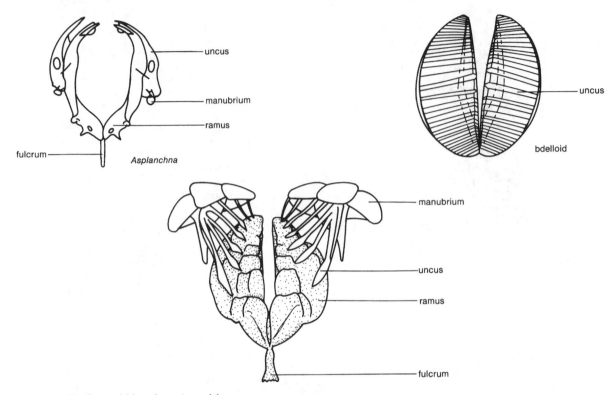

Figure 8.3 Rotifer trophi (not drawn to scale).

yeast) or inert material (carmine powder or latex microspheres, 1–5 μm) at the edge of the cover slip. Draw the suspension across the slide using a piece of toweling and observe the rotifers under low power. Track any particles that become caught in the coronal currents and follow them as they enter the mouth.

Describe the activity of the mastax during feeding. What function do the trophi serve in feeding? Can you see the particles accumulate in the gut? If you watch the animals long enough, you will see them pass the food you provided out of the anus. How long did it take from the start of feeding to first elimination of that food? This is the gut passage time. Were all the cells of the live food completely digested while in the gut? What do these observations tell you about feeding and digestion in rotifers? What are the ecological consequences of these features?

8. If additional species are available for study, examine them and compare their general morphology. What anatomical differences do you observe, especially between monogonont and bdelloid species (Fig. 8.1A and B, and Fig. 8.2)? Zooplankton samples from the field may contain a variety of species, including some with resting eggs (Fig. 8.4). Depending on the species available for study, you may observe the remains of food in their guts.

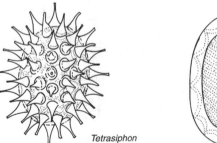
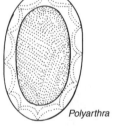

Figure 8.4 Resting eggs of rotifers (not drawn to scale). (After Koste 1978.)

9. If freshwater plants are available from a eutrophic pond or lake, examine them for sessile forms. To do this place a leaf of the plant in a small dish of water and observe it using a dissection microscope at about 30–50×. Large sessile forms (≈500–1,000 μm) such as *Collotheca* (Fig. 8.2) or *Floscularia* may be present. The latter resembles an elongate *Branchionus* (Fig. 8.1), but its corona looks like Mickey Mouse's ears!

10. If time and materials permit, make another slide preparation of one or more of the available species, but add a small drop of bleach to the fluid before putting the cover slip in place. Quickly locate a rotifer on the slide, switch to high power, and observe as the bleach dissolves the soft tissues, leaving only the trophi. Examine the trophi and locate the various parts illustrated in Fig. 8.3.

11. Observe a prepared slide of preserved rotifers and compare them to the live preparations just examined. Locate as many of the features mentioned above as you can. ■

Supplemental Readings

Clément, P., and E. Wurdak. 1991. Rotifera. *In*: F. W. Harrison and E. E. Ruppert (eds.). Microscopic Anatomy of Invertebrates, Vol. 4. Aschelminthes. Wiley-Liss, New York, pp. 219–296.

Edmondson, W. T. 1959. Rotifera. *In*: W. T. Edmondson (ed.). Fresh-water Biology, 2nd ed. John Wiley & Sons, New York, pp. 420–494.

Edmondson, W. T. 1965. Reproductive rate of planktonic rotifers as related to food and temperature in nature. Ecol. Monogr. 35: 61–111.

Epp, R. W., and W. M. Lewis. 1984. Cost and speed of locomotion for rotifers. Oecologia 61:289–292.

Garey, J. R., T. J. Near, M. R. Nonnemacher, and S. A. Nadler. 1996. Molecular evidence for Acanthocephala as a sub-taxon of Rotifera. J. Mol. Evol. (in press).

Gilbert, J. J. 1963. Mictic female production in the rotifer *Brachionus calyciflorus*. J. Exp. Zool. 153: 113–123.

Gilbert, J. J. 1985. Escape response of the rotifer *Polyarthra*: a high-speed cinematographic analysis. Oecologia (Berl.) 66: 322–331.

Gilbert, J. J 1985. Competition between rotifers and *Daphnia*. Ecology 66: 1943–1950.

King, C. 1967. Food, age, and the dynamics of a laboratory population of rotifers. Ecology 48:111–128.

Lubzens, E. 1987. Raising rotifers for use in aquaculture. Hydrobiologia 147: 245–255.

Maly, E. J. 1975. Interactions among the predatory rotifer *Asplanchna* and two prey, *Paramecium* and *Euglena*. Ecology 56: 346–358.

Ricci, C. 1984. Culturing of some bdelloid rotifers. Hydrobiologia 112: 45–51.

Ricci, C., G. Melone, and C. Sotgia. 1993. Old and new data on Seisonidae (Rotifera). Hydrobiologia 225/256: 495–511.

Rieger, R. M., and S. Tyler. Sister-group relationship of Gnathostomulida and Rotifer-Acanthocephala. Invertebr. Biol. 114: 186–188.

Stemberger, R. S., and J. J. Gilbert. 1985. Body size, food concentration and population growth in planktonic rotifers. Ecology 66: 1151–1159.

Starkweather, P. 1980. Aspects of the feeding behavior and trophic ecology of suspension feeding rotifers. Hydrobiologia 73: 63–72.

Starkweather, P. L. 1987. Rotifera. *In*: T. J. Pandian and F. J. Vernberg (eds.). Animal Energetics, Vol. 1. Protozoa Through Insects. Academic Press, Orlando, FL, pp. 159–183.

Wallace, R. L. 1980. Ecology of sessile rotifers. Hydrobiologia 73: 181–193.

Wallace, R. L. 1987. Coloniality in the phylum Rotifera. Hydrobiologia 147: 141–155.

Wallace, R. L., and T. W. Snell 1991. Rotifera. Chap. 8. *In*: J. Thorp and A. Covich (eds.), Ecology and Classifications of North American Freshwater Invertebrates. Academic Press, New York, pp. 187–248.

EXERCISE 9

■

Phylum Acanthocephala

The endoparasitic worms comprising phylum Acanthocephala [a-kan-tho-SEF-a-la; G., *acantho*, spine or thorn + G., *cephala*, head] are distinguished by a protrusible **proboscis** covered with recurved spines (Fig. 9.1). The proboscis anchors adults to the mucosal lining of their vertebrate host (Fig. 9.2). Thorny-headed worms range from about 1 mm to more than 1 m in length depending on the species; most are about 25 mm long. Like cestodes, acanthocephalans, absorb nutrients through the porous body wall or **tegument**, a complex syncytial layer containing fluid-filled canals called **lacunae**. Because of their parasitic nature, acanthocephalans are of economic importance as they affect domestic stock and game animals. Occasionally, there are reports of human infections. In spite of a morphology and life cycle radically different from rotifers (Exercise 8), these two phyla are clearly related for both possess a syncytial epidermis condensed into a sort of cuticle with numerous intracytoplasmic microtubules (See also Garey et al. 1996).

Classification

Four classes are recognized in this small phylum of about 700 species.

1. **Class Archiacanthocephala.** Parasites of birds and mammals. Insects and myriapods serve as intermediate hosts. Examples: *Mediorhynchus* (parasite of birds), *Moniliformis* (parasite of rats), and *Macracanthorhynchus* (parasite of pigs).

2. **Class Palaeacanthocephala.** Parasites of all vertebrate classes. Example: *Plagiorhynchus* (parasite of passerine birds).
3. **Class Eoacanthocephala.** Parasites of fish, amphibians, and reptiles. Example: *Neoechinorhynchus* (parasite of fish).
4. **Class Polyacanthocephala.** Parasites of fish and possibly crocodilians.

■ Observational Procedure: *Macracanthorhynchus*

General Morphology and the Proboscis. Observe a slide (whole mount) or preserved specimen of *Macracanthorhynchus hirudinaceus*, a parasite of pigs. Note the following: (1) recurved hooks on the proboscis, (2) neck, (3) trunk, and (4) genital region. Is there any evidence of segmentation in the thorny-headed worms? Examine a cross or longitudinal section of *M. hirudinaceus* with the proboscis embedded in the intestine of a pig (Fig. 9.2). Are any spines visible in this section? Do you see any evidence of host tissue damage? How much tissue damage would you expect to find, if many of these parasites were attached to the intestinal wall?

Internal Anatomy and Reproduction. Place a specimen in a dissection pan and make an anterior-posterior incision from the region of the genital pore to the neck. Then turn the cut edges laterally and pin them to the base of the pan using fine insect pins. Using Fig.

115

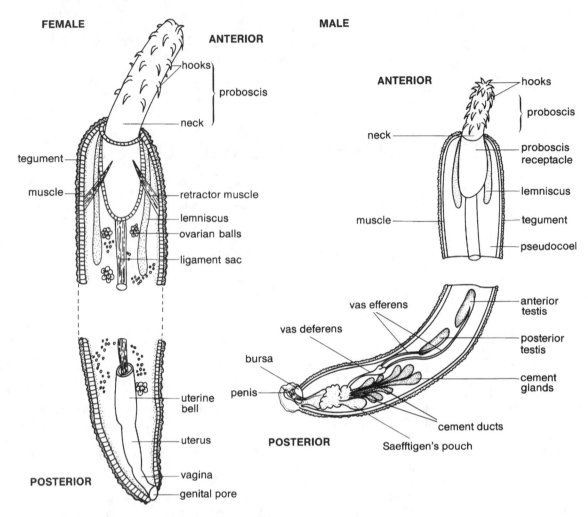

Figure 9.1. Schematic illustration of some external and internal features of male and female acanthocephalans.

9.1 as an aid, locate the following structures: male—(1) lemnisci, (2) reproductive organs, and (3) pseudo-coelom; female—(1) proboscis receptacle, (2) lemnisci, (3) ligamentous sac (which posteriorly communicates with the uterine bell), (4) uterus and vagina, and (5) genital pore.

Observe the muscular complex at the anterior part of the trunk. Cut away a narrow strip of the body wall and examine it under a dissection microscope. It should appear porous. The spaces seen are the lacunar parts of the canal system in the body wall. The canal system also will be evident in prepared slides of cross and longitudinal sections of *M. hirudinaceus*.

In female specimens examine the ligament sac surface. If patches or clumps of cells are present, these probably will be egg clusters (**ovarian balls**). Examine the pseudocoel for eggs and/or put pressure on the uterine bell to force out some eggs for collection. Place some of them on a slide with a drop of water, add a cover slip, and examine. Note the stage of development of several eggs. If available, observe the development of acanthocephalan eggs on a prepared slide (Fig. 9.3). ■

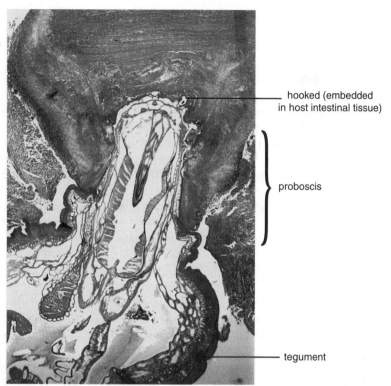

Figure 9.2. Photomicrograph of a longitudinal section of *Macracanthorhynchus hirudinaceus* with its proboscis embedded in the intestine of a pig.

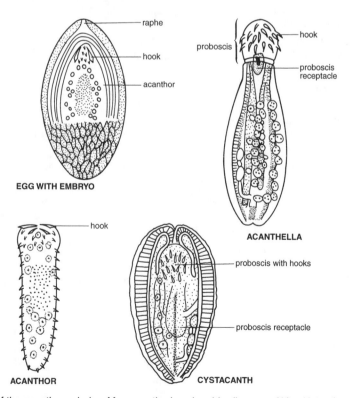

Figure 9.3. Larval stages of the acanthocephalan *Macracanthorhynchus hirudinaceus*. (After Kates from Olsen.)

Supplemental Readings

Bethel, W. M., and J. C. Holmes. 1977. Increased vulnerability of amphipods to predation owing to altered behavior by larval acanthocephalans. Can. J. Zool. 55: 110–115.

Boyd, E. M., 1951. A survey of parasitism of the starling *Sturnus vulgaris* L. in North America. J. Parasitol. 37: 56–84.

Brennan, B. M., and T. C. Cheng. 1975. Resistance of *Moniliformis dubius* to the defense reactions of the American cockroach, *Periplaneta americana*. J. Invertebr. Pathol. 26: 65–73.

Bullock, W. L. 1963. Intestinal histology of some salmonid fishes with particular reference to the histopathology of acanthocephalan infections. J. Morphol. 112: 23–44.

Byram, J. E., and F. M. Fisher. 1973. The absorptive surface of *Moniliformis dubius* (Acanthocephala) I. Fine structure. Tissue Cell 5: 553–579.

Crompton, D. W. T. 1970. An Ecological Approach to Acanthocephalan Physiology. Cambridge University Press, Cambridge, UK.

Crompton, D. W. T., and B. B. Nickol. (eds.) 1985. Biology of the Acanthocephala. Cambridge University Press, Cambridge, UK.

DeGiusti, D. L. 1949. The life cycle of *Leptorhynchoides thecatus* (Linton), an acanthocephalan of fish. J. Parasitol. 35: 437–460.

Dingley, D., and P. C. Beaver. 1985. *Macracanthorhynchus ingens* from a child in Texas. Am. J. Trop. Med. Hyg. 34: 918–920.

Edmonds, S. J. 1965. Some experiments on the nutrition of *Moniliformis dubius* Meyer (Acanthocephala). Parasitology 55: 337–344.

Fisher, F. M., Jr. 1960. On Acanthocephala of turtles, with the description of *Neoechinorhynchus emyditoides* n.sp. J. Parasitol. 46: 257–266.

Garey, J. R., T. J. Near, M. R. Nonnemacher, and S. A. Nadler. 1996. Molecular evidence for Acanthocephala as a sub-taxon of Rotifera. J. Mol. Evol. (in press).

Harrison, F. W., and E. E. Ruppert (eds.). 1991. Microscopic Anatomy of Invertebrates, Aschelminthes, Vol. 4. Wiley-Liss, New York, 424 pp.

Holmes, J. C., and W. M. Bethal. 1972. Modification of intermediate host behavior by parasites. *In*: E. U. Canning and C. A. Wright (eds.). Behavioural Aspects of Parasite Transmission. Academic Press, New York, pp. 123–149.

Kates, K. C. 1943. Development of the swine thorn-headed worm, *Macracanthorhynchus hirudinaceus*, in its intermediate host. Am. J. Vet Res. 4: 173–181.

Lorenzen, S. 1985. Phylogenetic aspects of pseudocoelomate evolution. *In*: The Origins and Relationships of Lower Invertebrates. S. Conway Morris, J. D. George, R. Gibson, and H. M. Platt (eds.). Clarendon Press, Oxford, pp. 210–223.

Miller, D. M., and T. T. Dunagan. 1978. Organization of the lacunar system in the Acanthocephala, *Oligacanthorhychus tortuosa*. J. Parasitol. 64: 436–439.

Moore, J. 1983. Responses of an avian predator and its isopod prey to an acanthocephalan parasite. Ecology 64(5): 1000–1015.

Moore, J. 1984. Parasites that change the behavior of their host. Sci. Am. 250: 108–115.

Whitfield, P. J. 1970. The egg-sorting function of the uterine bell of *Polymorphus minutes* (Acanthocephala). Parasitology 61: 111–126.

Yamaguti, S. 1963. Systema Helminthum. Acanthocephala, Vol. 5. Interscience Publishers, John Wiley & Sons, New York.

■

Phylum Nematoda (Nemata, Nema)

Phylum Nematoda [nem-a-TOD-a; G., *nema*, thread] is a large group of some 15,000 species of threadlike pseudocoelomates known commonly as the roundworms. Free-living species have been found in every habitat examined, e.g., arctic ponds, hot springs, all soils (including mountain and desert), deep sea marine muds, rotting apples, unpasteurized vinegar. Diversity of animal and plant parasitic nematodes is tremendous. Every vertebrate species may serve as a host to one or more nematodes. On a global scale, humans are infected commonly by about 15 species of nematodes, but an additional 15 species may occasionally parasitize humans, a condition called **zoonosis**.

Despite of their breadth in nutritional habits and geographic distribution, all nematodes have the same basic body structure. Typically they are cylindrical worms tapering at both ends (Fig. 10.1) and ranging in size from less than 1 mm to several meters long. Nematodes are covered by a resistant **cuticle**, unique to this group; the cuticle is composed of three layers (**cortical, matrix,** and **fibrous**). Nematodes have a straight digestive tract consisting of a mouth, buccal cavity, muscular esophagus, intestine, rectum, and anus. The buccal cavity, esophagus, and rectum are lined with cuticle. The nematode mouth is surrounded by up to six **lips**, but these may be fused into three lips or may be absent totally. Some nematodes, especially those that parasitize plants, are armed with a long **stylet** which is used in feeding (Fig. 10.1).

Nematodes lay eggs which may be released from the female or may hatch *in utero*. Hatchlings are usually called larvae. However, except for an immature reproductive system, they possess most adult structures; thus they are really juveniles. As the young grow they undergo four molts, at which time the entire cuticle is shed. However, unlike arthropods which also molt, individuals grow in size during intermolt periods. Larvae of some parasitic forms are given specific names: **rhabditiform, filariform**, and **microfilaria**. Parasitic nematodes may have a life cycle which is direct (**monoxenous**), having only one host, or indirect (**heteroxenous**), having two or more hosts.

The main objective of the following study is to direct student attention to the general structure and diversity of free-living and parasitic nematodes, and to review the life cycle of some species important as human parasites. Owing to time constraints, it is probable that only a few of the following studies may be done in your laboratory.

Classification

Two classes and 20 orders of nematodes are recognized.

1. **Class Adenophorea** (Aphasmida) 12 orders. Predominantly free-living nematodes lacking phasmids; amphids located behind the anterior end. Males lacking lateral extensions of the tail. Examples: *Trichinella spiralis* (trichina worm), *Trichuris trichiura* (whipworm), *Metoncholaimus* (free-living in mud on North American coasts).
2. **Class Secernentea** (Phasmida) 8 orders. Predominantly parasitic or free-living terrestrial

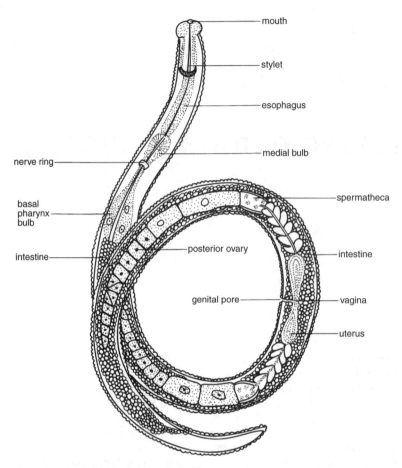

Figure 10.1. Typical female nematode, *Tylenchorhynchus cylindricus.*

nematodes having phasmids; amphids are located forward in the anterior end. Males often have lateral extensions of the tail. Examples: *Ascaris lumbricoides* (intestinal roundworm), *Dracunculus medinensis* (Guinea worm), *Dirofilaria immitis* (dog heartworm), *Enterobius vermicularis* (pinworm), *Necator americanus* (American hookworm), *Rhabditis maupasi* (earthworm parasite), and *Wuchereria bancrofti* (filarial worm), *Turbatrix aceti* (vinegar eel).

■ Observational Procedure:

Free-Living Nematodes

Using a dissection microscope and a Pasteur pipette remove some of the nematodes from a culture containing the vinegar eel, *T. aceti* (Fig. 10.2), or another appropriate species (e.g., *Caenorhabditis, Chiloplacus, Rhabditis*). Place your sample on a concave microscope slide and examine it for specimens. Describe how these animals move. What other animals move in this fashion?

Remove a few specimens, place them on a microscope slide, add a drop of water and a cover slip, and examine under low and high magnifications. Identify the anterior and posterior ends. How does this species feed? Can you see the muscular esophagus or any other internal organs, especially of the gut? Note that the thick cuticle may make identification of internal organs difficult.

Occasionally, you may see copulating pairs of worms. What region of his body does the male use to grasp the female? What region of her body does the male grasp? If the culture is actively growing, all developmental stages should be present. What types of evidence can you provide to suggest that your culture is growing? Are any of the worms molting? Each successive molt stage is larger than the previous one, but does growth occur between molts? How can one go about determining this without observing worms during their entire intermolt period?

Examine a prepared slide of *Turbatrix* or the species you examined alive and locate the esophagus, the

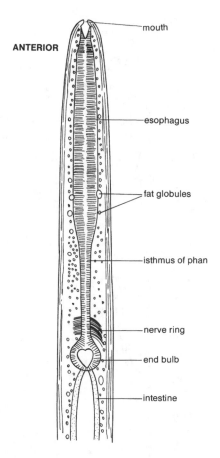

ANTERIOR

- mouth
- esophagus
- fat globules
- isthmus of phan
- nerve ring
- end bulb
- intestine

Figure 10.2. Anterior end of the vinegar eel, *Turbatrix aceti*, a free-living nematode.

remainder of the gut, and reproductive system. Attempt to trace the digestive tract from mouth to anus. Determine the length and diameter of these specimens. This information will help you when you study the other specimens.

Parasitic Nematodes

Nematodes are some of the most important parasites affecting the lives and general welfare of humans, including domestic and game animals and crops. Only few of the better known animal parasites have been selected for this exercise. For each species studied, a brief description of the life history will be given followed by directions for the observations to be made. *If live specimens are to be studied, be very careful not to infect yourself with infective stages (eggs, larvae, adults).*

Pinworm (Fig. 10.3). Pinworm (*Enterobius vermicularis*) or seat worm is a cosmopolitan oxyurid parasite of humans having a direct life cycle. Fertilized females migrate from the colon (usually at night) and lay thou-

sands of eggs in the perianal region. The embryonated eggs contain rhabditiform larvae which may enter the digestive tract, either by inhalation of dust contaminated by the eggs, or through direct ingestion. Once embryos reach the intestine, they hatch and after several successive molts the adults mate and become attached to the colon wall. It is estimated that 500 million people worldwide have **enterobiasis**.

Examine a prepared slide of pinworms and with the aid of Fig. 10.3 identify the anatomical structures of the specimen. Determine the size of this parasite and compare it to the free-living specimen just studied. Examine a slide of pinworm eggs. Note their oval shape and flattened side. Your instructor may discuss the clinical methods used for obtaining and identifying pinworm eggs.

Hookworms (Fig. 10.4). One of the most devastating, cosmopolitan diseases of humans is **ancylosotomiasis,** more commonly known as hookworm disease. Several species of hookworms may be studied with equal educational value; *Ancylostoma duodenale* (Old World hookworm), *Ancylostoma caninum* (domestic dog and cat hookworm), and *Necator americanus* (New World or American hookworm) are examples. New World hookworm is believed to have been brought to the Americas by the slave trade. Most hookworms have a direct life cycle. Adults live attached to the mucosal lining of their host's intestine and feed on blood and tissue fluids. Embryonated eggs are passed in the feces and the first two developmental stages take place in the soil. Third-stage larvae burrow into skin and are transported via the bloodstream to the lungs where the larvae burrow into the alveoli. They are then carried up the trachea and are swallowed. Final molt and sexual maturity are achieved in the small intestine.

Examine a whole-mount slide of a hookworm and note the structures in the mouth region, including a large buccal cavity with **cutting plates** or teeth. Examine a longitudinal section of a hookworm *in situ* and note how the worm is attached to the mucosa. In a whole-mount slide of a male, note the **copulatory bursa** and **spicules**. Determine the size of these intestinal worms. Compare the size and general morphology of *A. duodenale* rhabditiform (free-living) and filariform (infective) larvae. How do hookworms compare in size and general morphology to pinworms?

Filarial Worms (Fig. 10.5). Adults of the human filarial worm, *Wuchereria bancrofti*, live in the lymph nodes of humans where **ovoviviparous** females release thousands of motile embryos known as microfilaria. These make their way to the peripheral blood of humans where female mosquitoes pick them up while taking a blood meal. The larvae migrate to thoracic and leg

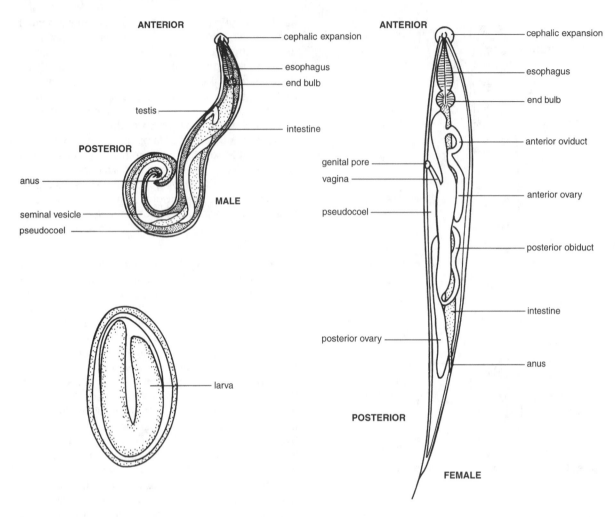

Figure 10.3. Male, female, and eggs of *Enterobius vermicularis* (human pinworm).

muscles of the mosquito where they continue to grow. Upon reaching a certain stage of growth (ca. 2–5 weeks), they migrate to the mosquito's mouthparts and enter the human body when the mosquito takes another blood meal. In humans, the microfilaria find their way to the lymphatic system and attain maturity in about a year. Lymphoid tissues increase in size in response to microfilaria parasitization. If, after repeated attacks, sufficient worms are present in the human host, an enlargement of certain body appendages (e.g., lower limbs, scrotum) occurs producing the disease inappropriately called **elephantiasis**. Other diseases produced by filarial worms include river blindness (*Onchocerca volvulus*) and dog heartworm (*Dirofilaria immitis*) which is common in North America. Specimens of all three species discussed here will be studied.

Observe a preserved specimen of *Dirofilaria immitis*, noting its long, thin form. How can such a long worm develop in a dog's heart? What problems would this cause Fido? Examine slides showing blood smears of *D. immitis* and/or *W. bancrofti* microfilaria. Compare the relative sizes of the microfilaria and red blood cells. Could these worms be mistaken for other blood parasites such as *Plasmodium* or *Trypansoma* (Exercise 1)? Why or why not? Examine a cross section of a nodule produced by the worm, *O. volvulus*. What evidence do you see that the worm is present in the nodule? How can a worm which parasitizes adipose tissue in the human mid-region cause blindness?

Trichina Worm (Fig. 10.6). Trichinosis is acquired by eating insufficiently cooked meat contaminated with the encysted stage of *Trichinella spiralis* (Fig. 10.6).

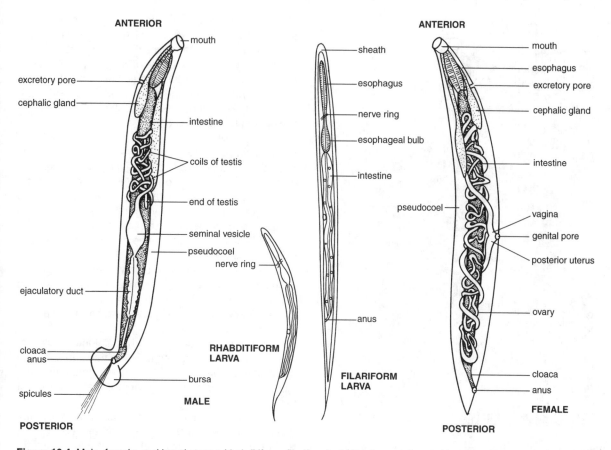

Figure 10.4. Male, female, and larval stages (rhabditiform, filariform) of *Necator americanus* (American hookworm).

Digestive juices in the gut liberate the larvae from the meat and then by removing the cyst wall. Worms achieve adulthood in about 3 days. After mating the male worms die and females penetrate the intestinal mucosa where they release juvenile worms. One female may liberate as many as 2,000 larvae into the host's blood stream before she dies. The larvae are transported thoughout the body and become lodged in skeletal muscle and other muscles with a rich blood supply such as the diaphragm and tongue. Soon after the muscles are invaded, the worms enlarge and encyst, causing harmful tissue damage. In the past, the usual source of human infection was contaminated pig meat. However, bear meat is now a source of infection, but any carnivore, especially rats, may be infected.

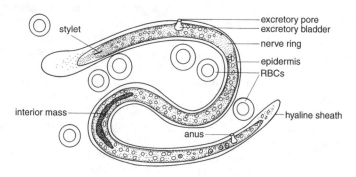

Figure 10.5. Microfilarial stage of *Wuchereria bancrofti* (human filarial worm).

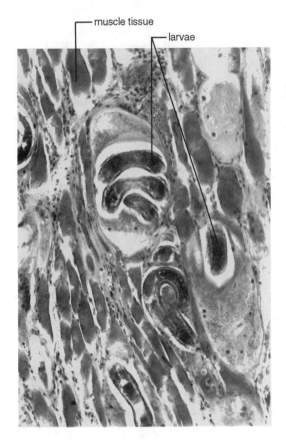

muscle tissue

larvae

Figure 10.6. Photomicrograph of the juvenile stage of *Trichinella spiralis* (trichina worm) encysted in muscle tissue.

Examine a section of muscle with encysted larvae. Note the size of the cysts and their general frequency in the tissue. Estimate the number of trichina cysts per cm^3 of muscle tissue. How can you go about making the appropriate calculations to do this estimation? Is there any evidence of damage to the muscle tissue by the cysts? How might a heavy infection affect host muscle function? Examine a slide of adult *Trichinella* and compare their general anatomy with the other nematodes you have observed. How are they similar; how do they differ?

Whipworm (Fig. 10.7). The life cycle of the human whipworm is direct. Eggs, ingested via fecal contamination, hatch in the small intestine, and in about 2 weeks the juvenile worms migrate to the colon where they attach permanently to the mucosa and feed on blood cells. Whipworms reach sexual maturity in about 3 months. It has been estimated that more than 350 million people worldwide are infected!

Examine a whole mount slide of a whipworm. Describe the shape of the worm and tell why are these worms given their common name? The esophagus is a long, thin-walled, multiglandular tube called the **sti-**

chosome. There are no lips. The buccal cavity is armed with a minute **stylet** which may be seen under high magnification. Determine the size of these worms. Examine a whole-mount slide of whipworm eggs. Note their oval shape with the characteristic bipolar plugs (Fig. 10.7). Compare the eggs of whipworms to pinworms? Take some notes describing how they are similar and how they differ?

Earthworm Parasite (Fig. 10.8). Adults of the species *Rhabditis maupasi* are free-living, soil nematodes, but earthworms play host to juveniles. When the earthworm dies the larvae mature by feeding on bacteria which grow in the decaying tissues. Larvae are believed to enter the earthworm through the nephridiopore or genital openings. If time permits, complete the following exercise.
1. Dissect an anesthetized earthworm. Make a longitudinal cut in the earthworm's muscular body wall and pin back its edges to the dissection pan.
2. Carefully remove the digestive tract from the posterior two-thirds of the body.
3. Remove the viscera anteriorly, cut it into small sections of about 1 cm in length. Place these in separate petri dishes containing 3% sterile agar. Label the plates with the date and the approximate location of the worm section.
4. Cut similar sections from the muscular body wall and prepare similar agar cultures with these.
5. Set the cultures aside at room temperature. In about 3 days examine the cultures using a dissection microscope and estimate the number of nematodes per section. Which part of the worm housed the greatest number of nematodes? Transfer some specimens to a microscope slide; add a drop of water, a cover slip, and observe under higher magnification.

Intestinal Roundworm (Figs. 10.9–10.12). From a morphological point of view, *Ascaris lumbricoides* is not a typical nematode. Nevertheless, it is important to examine more closely because **ascariasis** remains a public health problem in the United States. Estimates suggest a childhood prevalence of about 15% in certain regions (Darby and Westphal 1972)! The size of *Ascaris* makes it suitable for dissection and its reproductive system is often used to demonstrate mitosis and gametogenesis. Specimens used in this study are the pig strain (species) *A. suum*. This species is similar morphologically to the one found in humans, differing only in the minute details of the lips and in aspects of its physiology.

Ascaris has a direct life cycle. When embryonated eggs are swallowed, they hatch in the small intestine where the young worms penetrate the wall and pass into the bloodstream. Upon entering the lung the worms break out into the alveoli where they develop to

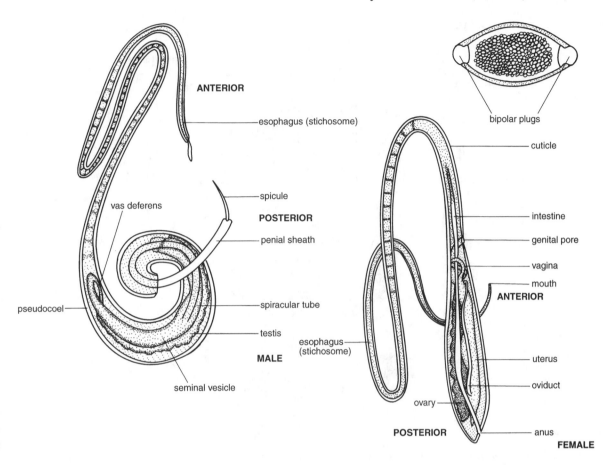

Figure 10.7. The human whip worm *Trichuris trichiura* (male, female, and bipolar eggs).

the fourth stage. Then they are carried up the trachea to the pharnyx where they are swallowed. The worms mature in the small intestine where they mate and begin releasing eggs in the host's feces. Two related species *Toxocara canis* and *Toxocara cati* (parasitizing dogs and cats, respectively) can cause a serious disease of humans known as **visceral larval migrans**.

General Morphology

1. Note the four thin, longitudinal strips of lighter-pigmented cuticle along the length of the worm: two strips are lateral, one dorsal, and one ventral. The lateral strips are broader and seen more easily than the other two. These are external mainfestations of internal structures, the **hypodermal cords**.

2. Examine the cuticular covering under a dissection microscope and note that it appears to be finely segmented. This is not true segmentation, but superficial striations of the cuticle.

3. Locate the terminal mouth and the subterminal and ventral anus.

4. Observe the anterior end in a head-on position under a dissection microscope to see the single dorsal lip and the two subventral lips surrounding the mouth. Note the somewhat dentate condition of the inner surface of the lips.

5. At the periphery of the dorsal lip locate the two cephalic papillae which house special sensory organs.

6. *Ascaris* is sexually dimorphic. In males the posterior end is curved ventrally. Are any of the specimens in the class male? Occasionally a pair of **penial spicules** (copulatory structures) can be seen protruding from the male's cloaca. Females are usually larger than the males.

7. Examine the cuticle surface of a female under the dissection microscope and find her genital pore which is located ventrally about one-third of the body length from her mouth.

Female Reproductive System (Figs. 10.9 and 10.10)

1. Place a female specimen ventral surface down in a dissection pan and orient the specimen and pan so that you may easily observe the worm through a dis-

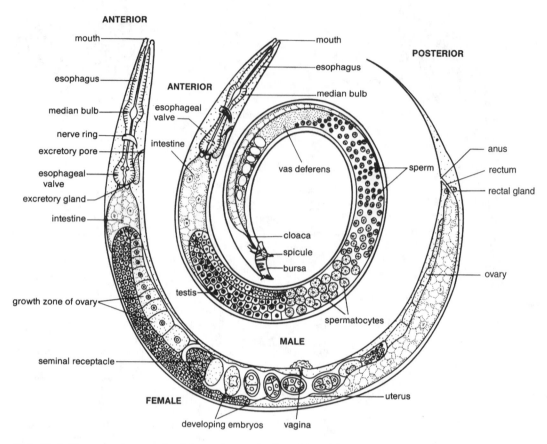

Figure 10.8. *Rhabditis maupasi*, a parasite of the earthworm (male and female).

section microscope. Pin the anterior and posterior ends of the worm to the pan, cover the specimen with water, and dissect the worm as follows.

2. Make a shallow incision in the outer body wall the full length of the worm. Take care not to disturb the internal viscera. After an incision of some 6–8 cm has been made, carefully pin the body wall to the pan. Continue the dissection and pinning until the entire length of the worm has been exposed (Fig. 10.9).

3. Observe the outer body wall just cut. It surrounds a central digestive tract and other organs located in the spacious pseudocoel.

4. Note the highly convoluted pair of tubules that wind about the intestine in the posterior part of the body cavity. The tube varies in dimension and constitutes the female reproductive system (Fig. 10.9).

5. Locate the most anterior part of this structure which appears to penetrate the body wall; this is the **vagina**. The vagina bifurcates forming the two **uteri**.

6. Follow one uterus as it grades into the **oviducts**. Although there is an appreciable decrease in tube diameter, there is no obvious external change in the morphology from oviducts to the ovary which ends blindly.

7. Obtain a prepared slide of a cross section of a female worm showing the reproductive tract. Locate the uterus filled with eggs and the smaller oviducts and ovarian loops in the section (Fig. 10.10). The upper part of the oviduct contains eggs without shells, but the lower portion should have shelled eggs indicating that fertilization has taken place. Some of the eggs may be in early stages of cleavage.

8. Examine slides fertilized and unfertilized of *Ascaris* eggs. Note the egg's thick, bumpy (mammillated) coat (Fig. 10.11). How can one tell the difference between fertilized and unfertilized eggs? Humans infected with *Ascaris* often have only unfertilized eggs present in their stools; what simple fact would account for this situation? Why might it be important for a physician to know whether a person has fertilized or unfertilized in his or her stools? Compare the eggs of pinworm, whipworm, and *Ascaris*. Develop a simple chart or a dichotomous key that will permit you to tell them apart.

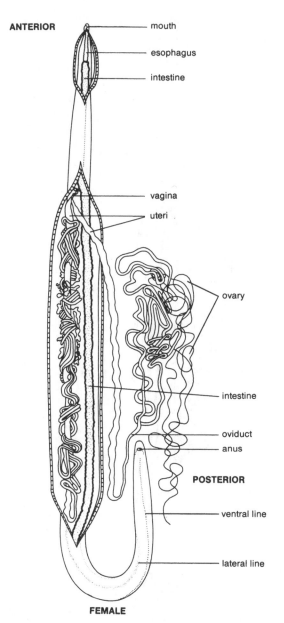

ANTERIOR
mouth
esophagus
intestine
vagina
uteri
ovary
intestine
oviduct
anus
POSTERIOR
ventral line
lateral line
FEMALE

Figure 10.9. *Ascaris lumbricoides*: internal longitudinal view of an adult female.

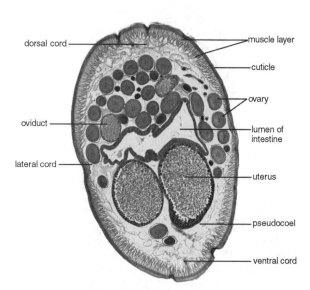

dorsal cord
muscle layer
cuticle
oviduct
ovary
lateral cord
lumen of intestine
uterus
pseudocoel
ventral cord

Figure 10.10. *Ascaris lumbricoides*: cross section of an adult female worm showing parts of the reproductive and digestive systems. (The body form is not circular in this preparation due to distortion during fixation and sectioning.)

3. Locate the single highly contorted tube in the male. The threadlike distal part of the tube constitutes the **testis** and is located forward in the pseudocoel. Like the female oviduct, the gradation from testis to **vas deferens** is not visible externally.

4. As the vas deferens approaches the region of the cloaca, there is a distinct enlargement of the tube. This is the beginning of the **seminal vesicle** which stores sperm.

5. Beyond the seminal vesicle is the ejaculatory duct. It connects the reproductive tract with the cloaca.

6. Obtain a cross section of a male worm showing the reproductive tract (Fig. 10.13). Locate the testis, vas deferens, and seminal vesicle. The seminal vesicle is much larger in diameter than either the testis or vas deferens. It may be possible to find mature sperm cells in slide preparations that show portions of the ejaculatory duct and/or cloaca.

Digestive System and Other Internal Structures of Ascaris (Fig. 10.14)

1. In your dissected specimen, examine the digestive system: esophagus, intestine, rectum, cloaca (male), and anus. It may be necessary to move parts of the reproductive system to expose the digestive tract.

2. Observe a prepared slide showing a cross section of the body of *Ascaris* made through the region of the esophagus (Fig. 10.14). Note the three powerful mus-

Male Reproductive System (Figs. 10.12 and 10.13)

1. Complete a dorsolongitudinal incision of a male *Ascaris* in the same way that was done for the female. Use Fig. 10.12 to assist in the location and identification of the following structures.

2. The male's gential and anal openings open posteriorly into a region called the cloaca. Why is the term cloaca used to describe this cavity? The female lacks a cloaca.

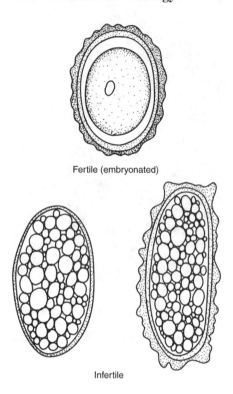

Fertile (embryonated)

Infertile

Figure 10.11. *Ascaris lumbricoides*: fertile and infertile eggs. Eggs are about 60 μm long and 40 μm wide.

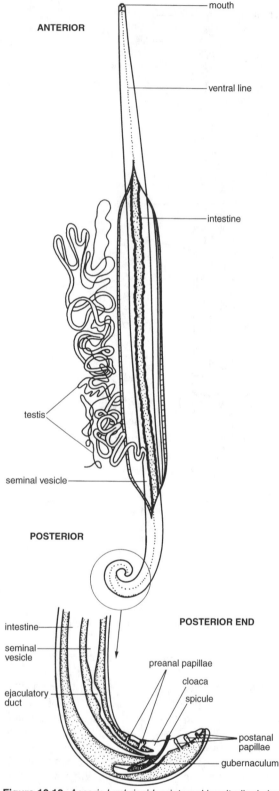

Figure 10.12. *Ascaris lumbricoides*: internal longitudinal view of an adult male.

cle units which produce the sucking action when food is obtained from the host. Note the glandular bodies in the muscle tissue of the esophagus.

3. Strip a section of cuticle from the body and mount it on a slide with a drop of water; add a cover slip. Observe the superficial striation of the cortical (outside) layer and the crosshatched fibers of the fibrous layers. Estimate the angles the fibers make with the longitudinal axis of the worm. How do these fibers aid the muscles in locomotion?

4. Observe a prepared slide showing a cross section of the body wall. Use either sex. Observe the close association between the muscle cell bases and the **syncytial hypodermis**. The muscles at the inner body wall are in four quadrants, divided by the four **cords** (see below). Note that in the cross section being examined, the muscle cells have been cut transversely, which makes it somewhat difficult to visualize their proper arrangement. There are about 150 cells per quadrant and each cell is a very elongated body with a narrow, spindle form.

5. Locate the dorsal, ventral, and lateral cords. The dorsal and ventral cords contain the nerve trunks. Just

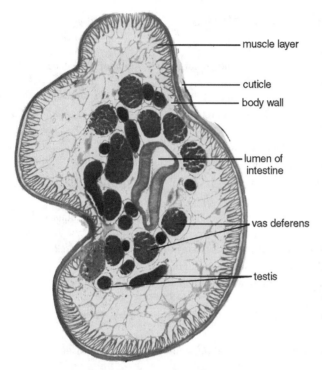

inside the lateral cords you should be able to see the lateral excretory canals.

6. Internal to the muscle cells is the spacious pseudocoel.

7. Examine the intestine. The outermost perimeter of the intestine is called the basement membrane. Just inward from the basement membrane is a layer of columnar epithelial cells. These epithelial cells are elongated and bear distinct nuclei at their bases. On the side opposite the nuclei, a fuzzy coat may be seen on these cells. This coat is actually made up of many microvilli which extend into the lumen of the intestine.

If you are to work with live *Ascaris* eggs, your instructor will give detailed instructions to avoid accidental contamination.

Plant Parasites. If time and materials permit, your instructor will provide information for the study of some plant parasites. As you follow those directions, be sure to compare these parasites to the animals ones just studied. How are they similar and how do they differ? The preparation of a simple chart may help you organize your findings about the nematodes. ∎

Figure 10.13. *Ascaris lumbricoides*: cross section of an adult male worm showing parts of the reproductive and digestive systems. (The body form is not circular in this preparation due to distortion during fixation and sectioning.)

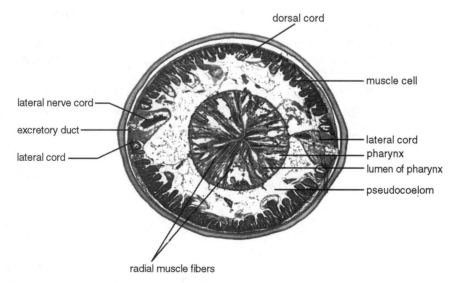

Figure 10.14. *Ascaris lumbricoides*: cross section of a worm at the level of the esophagus.

Supplemental Readings

Beaver, P. C. 1949. Methods of pinworm diagnosis. Am. J. Trop. Med. 29: 577–587.

Brandl, R., W. Mann, and M. Sprinzl. 1992. Mitochondrial tRNA and the phylogenetic position of Nematoda. Biochem. Syst. Eco. 20: 325–330.

Chitwood, M. B. 1970. Nematodes of medical significance found in market fish. Am. J. Trop. Hyg. 19: 599–602.

Crichton, V. F. J., and M. Beverley-Burton. 1975. Migration, growth, and morphogenesis of *Dracunculus insignis* (Nematoda: Dracunculidea). Can. J. Zool. 53: 105–113.

Croll, N. A. (ed). 1976. The Organization of Nematodes. Academic Press, New York.

Croll, N. A., and B. E. Matthews. 1977. Biology of Nematodes. John Wiley & Sons, New York.

Darby, C. P., and M. Westphal. 1972. The morbidity of human ascariasis. J. S. C. Med. Assoc. 68: 104–108.

Dropkin, V. H. 1980. Introduction to Plant Nematology. Wiley-Interscience, New York.

Ehlers, U. 1994. Absence of a pseudocoel or pseudocoelom in *Anophostoma vivipara* (Nematodes). Microfauna Marina 9: 345–350.

Frey, G. F., and J. G. Moore. 1969. *Enterobius vermicularis*: 10,000-year-old human infection. Science 166: 1620.

Gould, S. D. (ed). 1970. Trichinosis in man and animals. Charles C Thomas, Springfield, IL.

Harrison, F. W., and E. E. Ruppert (eds.) 1991. Microscopic Anatomy of Invertebrates, Aschelminthes, Vol. 4. Wiley-Liss, New York, 424 pp.

Lee, D. L., and H. J. Atkinson. 1977. Physiology of Nematodes, 2nd ed. Columbia University Press, New York.

Maggenti, A. 1981. General Nematology. Springer-Verlag, New York.

Malakhov, V. V. 1994. Nematodes: Structure, Development, Classification, and Physiology. Smithsonian Institution, Washington, DC.

Most, H. 1978. Current concepts in parasitology: Trichinosis: preventable yet still with us. N. Engl. J. Med. 298: 1178–1180.

Nicholas, W. L. 1984. The Biology of Free-Living Nematodes, 2nd ed. Oxford University Press, London.

Poinar, G. O. 1994. Nematoda and Nematomorpha. *In*: J. H. Thorp and A. P. Covich (eds.). Ecology and Classification of North American Freshwater Invertebrates. Academic Press, New York, pp. 249–283.

Schad, G. A., and L. E. Rozeboom. 1976. Integrated control of helminths in human populations. Annu. Rev. Ecol. Syst. 7: 393–420.

Sprent, J. F. A. 1952. Anatomical distinction between human and pig strains of *Ascaris*. Nature 170: 627–628.

Sprent, A. F. A. 1969. Nematode *larva migrans*. N. Z. Vet. J. 17: 39–48.

Tripathy, K., F. Gonzalez, H. Lotero, and O. Bolanos. 1971. Effects of *Ascaris* infections on human nutrition. Am. J. Trop. Med. Hyg. 22: 212–218.

Winnepenninckx, B., T. Backeljau, L. Y. Mackey, J. M. Brooks, R. De Wachter, S. Kumar, and J. R. Garey, 1995. 18S rRNA data indicate that Aschelminthes are polyphyletic in origin and consist of at least three distinct clades. Mol. Biol. Evol. 12: 1132–1137.

Wood, W. B. (ed.). 1988. The Nematode, *Caenorhabditis elegans*. Cold Spring Harbor, New York.

Wright, K. A. 1991. Nematode. *In*: F. W. Harrison and E. E. Ruppert (eds.). Microscopic Anatomy of Invertebrates, Vol. 4: Aschelminthes. Wiley-Liss, New York, pp. 11–195.

■

Phylum Nematomorpha

Nematomorpha [NE-mat-o-MOR-fa; G., *nema*, thread + G., *morpha*, form] comprises about 230 species of threadlike, pseudocoelomate worms that lack a functional gut. They range in length from 10 to 100 cm, but are only 1 to a few millimeters in diameter. As juveniles they parasitize arthropods, while the adults are free living. The common name, **horsehair worms**, comes from myths that said these worm arose from horse hairs that had fallen into water. The body wall of nematomorphs consists of a (1) thick **cuticle** often containing plates called **areoles**; (2) single-layered, cellular **hypodermis**; and (3) muscle layer whose fibers are oriented longitudinally as in nematodes. Emergence of adults from the host that housed the juvenile stage takes place in or near water and mating begins soon thereafter.

Classification

Two classes are recognized based on external morphology and aspects of the cuticle: **Nectonematoida** (marine nematomorphs) and **Gordioida** (freshwater, most nematomorphs).

■ Observational Procedure:

Examine a whole worm while keeping it immersed in a shallow dish of water. Determine the length and diameter of the specimen. Locate the anterior and posterior ends. The mouth will be located either terminally or subterminally. The head is called the **calotte** and usually has a darker colored area below it. The posterior end may be unlobed, bilobed, or trilobed, with a subterminal, ventral anus (cloacal aperture). Are any other structures visible on the external surface of the animal?

Observe a slide of a cross section of a horsehair worm (Fig. 11.1). First become familiar with the position of the section on the slide. Determine the dorsal and ventral sides. How can you do that? Once you are familiar with the section, locate the structures indicated in Fig. 11.1.

How do these parasites compare to nematodes? Develop a table that compares the two phyla. How are they similar and how are they different in size, shape, life cycle, and internal anatomy? ■

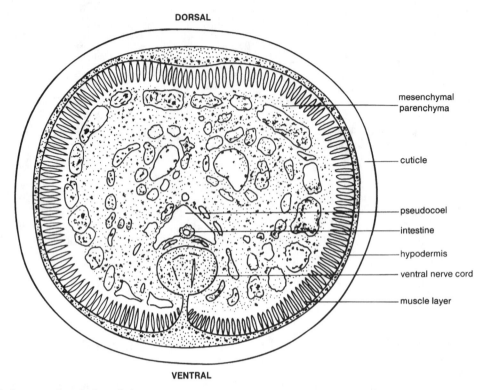

DORSAL

mesenchymal parenchyma

cuticle

pseudocoel

intestine

hypodermis

ventral nerve cord

muscle layer

VENTRAL

Figure 11.1. Cross section of a horsehair worm.

Supplemental Readings

Arvy, L. 1963. Donnèes sur le parasitisme protèlien de *Nectonema* (Nèmatomorphe), hez les Crustacès. Ann. Parasitol. (Paris) 38: 887–892.

Carvalho, J. C. M. 1942. Studies on some Gordiacea of North and South America. J. Parasitol. 28: 213–222.

Lanzavecchia, G., M. de Equileor, R. Valvassori, and G. Scarì. 1995. Body cavities of Nematomorpha. In: G. Lanzavecchia, R. Valvassori, and M. D. Candia Carnevali (eds.). Body cavities: Function and Phylogeny. Selected Symposia and Monographs Unione Zoologica Italiana, 8, Mucchi, Modena, pp. 45–60.

May, H. G. 1919. Contributions to the life histories of *Gordius robustus* Leidy and *Paragordius varius* (Leidy). Ill. Biol. Monogr. 5: 1–118.

Poinar, G. O. 1994. Nematoda and Nematomorpha. *In*: J. H. Thorp and A. P. Covich (eds.). Ecology and Classification of North American Freshwater Invertebrates. Academic Press, New York, pp. 249–283.

Zapotosky, J. E. 1974. Fine structure of the larval stage of *Paragordius varius* (Leidy, 1851) (Gordiodea: Paragordidae). I. the preseptum. Proc. Helminth. Soc. Wash. 41: 209–221.

PHYLUM MOLLUSCA

Members of phylum Mollusca are schizocoelomate, nonsegmented protostomes. However, to some workers, repetition of certain external and internal structures in the class Monoplacophora constitutes evidence of segmentation or that this class arose from the line that led to more definite metamerism (i.e., Annelid-Arthropod line). Two distinct features set the molluscs apart from the other phyla. (1) The ventral surface has been variously modified into a muscular foot or into a group of muscular arms (tentacles). (2) Tissues of the dorsal and lateral surfaces have been modified into a structure called the mantle, which in most classes secretes an external shell. In some forms the shell is internal, much reduced, or absent. In the most primitive forms there is a tendency for the pericardial cavity, the metanephridia, and the gonads to be somewhat continuous. In more advanced groups these systems, especially the reproductive, become more or less separated and independent anatomical units. However, they all represent parts of the embryonic coelom which originates in a schizocoelic manner. The schizocoelom does not function as a hydraulic skeleton in molluscs.

Ghiselin, M. T. 1988. The origin of molluscs in light of molecular evidence. In: Oxford Surveys in Biology, Vol. 5. P. H. Harvey and L. Partridge (eds.). Oxford University Press, Oxford.

Runnegar, B., and J. Pojeta. 1974. Molluscan phylogeny: the paleontological viewpoint. Science 186: 311–317.

Phylum Mollusca

Well over 50,000 described species of unsegmented, schizocoelomate protostomes comprise phylum Mollusca [mo-LUS-ka; L., *molluscum*, soft]. Mollusks range in size from less than 0.5 mm to nearly 15 m long and possess two distinct features that set them apart from other invertebrates. (1) Their organs are enclosed in a visceral mass and dorsal **mantle** (*pallium*, L., a mantle or cloaked) which usually secretes an external calcareous shell. (2) They possess a ventral muscular foot. In each class the foot is variously modified for locomotion and food procurement. Most mollusks have an open circulatory system and in all but one class (Bivalvia) a long, chitinous, rasplike feeding device called the **radula** is present in the buccal cavity. The nervous system is moderately to highly cephalized, primitively with paired dorsal (visceral) and ventral (pedal) nerve cords. Well-developed excretory, respiratory, and reproductive systems are present. Mollusks date to the **Cambrian Period** (Fig. 12.1) and include about 45,000 fossil species. There is a rich diversity of species within most of the classes.

Classification

Seven classes of mollusks are recognized based on morphology of the foot, and presence or absence and type of shell.

1. **Class Monoplacophora.** A small relic class of fewer than twenty species of small (less than 3 cm), non-torted, marine mollusks with a single, dorsal, caplike shell. Many fossil species are represented in this class. Example (extant): *Neopilina*.

2. **Class Polyplacophora.** Mollusks possessing a creeping foot and a shell divided into eight plates. Examples: *Katharina* and *Cryptochiton*.

3. **Class Aplacophora.** Small (less than 1–5 cm), wormlike, marine mollusks, with calcareous spicules in the mantle. These shell-less animals crawl on, or burrow in sediments, or are epizoic on cnidarians in deep, marine waters. They are commonly called **solenogasters** [so-LEN-o-GAS-ters; G., *solen*, channel + G., *gaster*, stomach] because some species have a ventral, longitudinal groove that runs nearly the entire length of the body. Recently a new class, **Caudofoveata** [KAW-do-fo-ve-AT-a; L., *cauda*, tail + L., *fovea*, pit] has been erected comprising animals formerly assigned to Class Aplacophora.

4. **Class Gastropoda.** Snails and snail-like mollusks possessing a well-defined head with tentacles and relatively simple eyes. Most possess a single coiled shell. The foot is adapted for crawling, but some species can swim. Three subclasses are recognized.

   a. **Subclass Prosobranchia.** Shelled marine gastropods with the mantle cavity anterior. Examples: *Busycon* (whelks), *Conus* (cones), *Haliotis* (abalone), *Murex* (drills), and *Strombus* (conchs).

   b. **Subclass Opisthobranchia.** Mostly marine gastropods with much reduced shell or shell lacking.

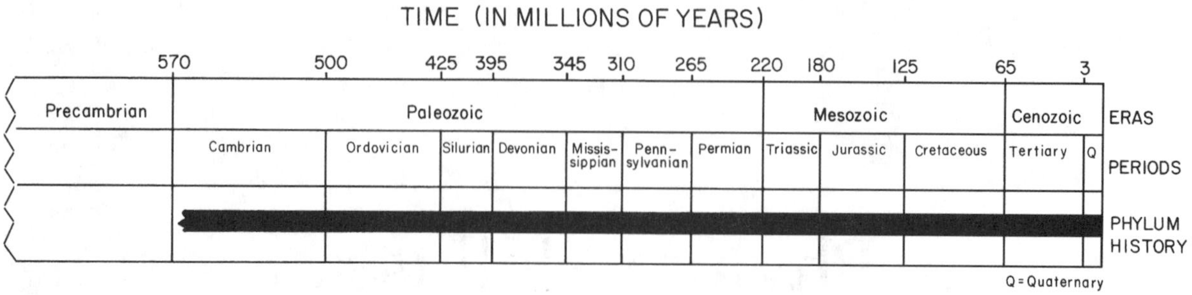

TIME (IN MILLIONS OF YEARS)

570		500		425	395		345	310		265	220	180		125	65	3

Precambrian	Paleozoic								Mesozoic			Cenozoic		ERAS
	Cambrian	Ordovician	Silurian	Devonian	Missis- sippian	Penn- sylvanian	Permian	Triassic	Jurassic	Cretaceous	Tertiary	Q	PERIODS
													PHYLUM HISTORY

Q = Quaternary

Figure 12.1. Geologic history of the phylum Mollusca.

The mantle cavity lacking or is on the right side, or may be partly posterior. Examples: *Aplysia* (sea hare) and *Aeolidia* (nudibranch).

c. **Subclass Pulmonata.** Mostly terrestrial, shelled gastropods with a mantle cavity modified as a lung. Examples: *Lymnaea* (freshwater snail) and *Limax* (land slug).

5. **Class Bivalvia (= Pelecypoda).** Mollusks having two shells (valves) that are hinged dorsally. The head is absent. Large curtain-like gills found in the mantle cavity are used in filter feeding. Six subclasses of freshwater and marine species are recognized, but are not reviewed here. Examples: *Cardium* (cockle), *Mytilus* (common mussel), *Unio* and *Anodonta* (freshwater clam and mussel, respectively), *Ostrea* (oyster), *Pecten* (scallop), *Mya* (soft-shelled clam), and *Mercenaria* (Venus clam or quahog).

6. **Class Scaphopoda.** A small group composed of approximately 350 species of burrowing marine mollusks with a single, dorsoventrally elongate, slightly curved, tubular shell (2–6 cm long) which is open at both ends. The shell resembles an elephant's tusk, so they are known commonly as tusk or tooth shells. A

modified foot and set of delicate tentacles extend from the ventral orifice of the shell. Example: *Dentalium*.

7. **Class Cephalopoda.** Elongated, free-swimming marine mollusks that are highly cephalized, possessing a well-developed nervous system and a foot modified into several highly developed tentacles. Shell external, internal, or absent. Three subclasses are recognized.

a. **Subclass Nautiloidea.** Cephalopods with a multichambered, external shell and many arms. Most extinct species. Example (extant): *Nautilus*.

b. **Subclass Ammonoidea.** Cephalopods with coiled, multichambered, external shells. Subclass is entirely extinct.

c. **Subclass Coleoidea.** Cephalopods with internal shells or shell lacking. Four orders with extant species and one wholly extinct order are recognized. Examples: *Loligo* (squid), *Octopus* (octopus), and *Sepia* (cuttlefish).

Of the seven molluscan classes, three (Aplacophora, Monoplacophora, and Scaphopoda) will not be covered here due to the lack of available commercial suppliers of specimens.

A. Class Polyplacophora

Commonly known as **chitons**, members of the class Polyplacophora [POL-e-pla-KOF-o-ra; G., *poly*, many + G., *placo*, plate + G., *phora*, carry] are a small group (about 800 species) of dorsoventrally flattened marine mollusks, usually ranging in size from 1 to 10 cm long. The ventral foot is a broad, flat, creeping sole and the shell is divided into eight partially overlapping plates or valves that articulate with one another, thus permitting chitons a certain degree of flexibility along the anterio-

posterior axis. Laterally, the mantle extends beyond the plates to form a girdle surrounding the animal.

Chitons have reduced nervous and sensory systems with an indistinct head. They lack eyes, but unique photosensitive structures called **aesthetes** are present dorsally. Chitons are abundant in the intertidal zones along rocky coasts where they attach to rocks and other hard objects. Some species are found at great depths (greater than 5,000 m).

■ Observational Procedure:

External Morphology

1. Obtain a specimen of the chiton *Katharina* and observe its general shape and symmetry (Fig. 12.2). Identify the anterior and posterior ends.

2. Place the specimen dorsal side up. Note that the shell consists of eight similar plates that overlap each other anteriorly. Portions of the plates are embedded in the fleshy mantle. The lateral, thickened part of the mantle is known as the **girdle**.

3. Observe the dorsal surface using a dissection microscope. In young individuals, aesthetes may be seen in the plates. In older specimens of *Katharina*, these and other markings on the plates, are frequently eroded.

4. Place your specimen ventral side up. The most prominent feature of the ventral surface is the broad muscular foot. The head region, anterior to the foot (Fig. 12.2), bears the mouth, but lacks any special sense organs (e.g., tentacles and eyes).

5. Note the distinct groove between the foot and the girdle. This is the mantle cavity or **pallial groove**. It contains a single row of gills on either side of the foot

(Fig. 12.2). Trace the gills of one side from the anterior to the posterior end. Are they the same size throughout? Approximately how many gills are there per side?

6. Remove a gill and place it in a depression slide filled with a drop of water. Hold the base of the gill down with a dissection probe and gently run the point of another probe along the gill surface. Note that the gill is constructed of a series of small, rounded filaments positioned like the pages of a book. What is the significance of the serial filaments? Does this arrangement indicate true segmentation?

7. Find the anus at the posterior end of the pallial groove. It is located just behind the foot. With a dissection microscope, locate the paired **nephridiopores** and **gonopores** that empty into the pallial groove on either side of the foot. The nephridiopores are just anterior to the anus. Locate these pores by scanning the region between the gills and the foot, folding the gills aside with a probe as you proceed anteriorly with the search. In chitons other than *Katharina*, the nephridiopores are found on either side of the anus. The gonopore is located near and slightly anterior to each nephridiopore.

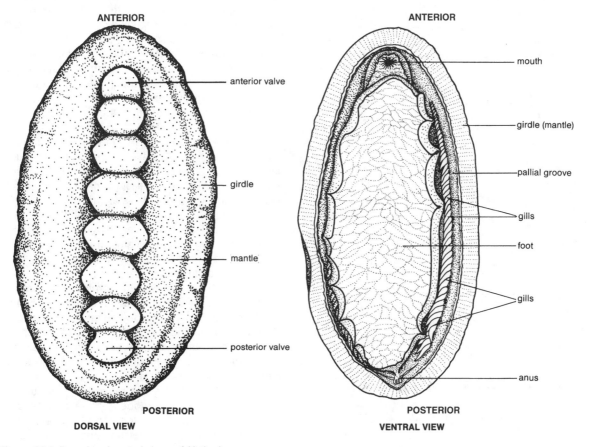

Figure 12.2. Dorsal and ventral views of *Katharina*.

Internal Anatomy

1. To study the internal anatomy of a *Katharina*, the plates must be removed. Begin by **carefully** cutting away the mantle. Do not cut into the mantle between the plates. Cut around the plates laterally until they are nearly exposed. With a sharp scalpel, carefully remove the anterior plate by freeing it from its muscular and connective tissue attachments. When a plate is loose, carefully pry it up using a blunt probe. Working posteriorly, remove the remaining seven plates, placing each aside in order.

2. Examine the plates in their proper order (Fig. 12.3). The medial pigmented layer of the plate is known as the **tegumentum**, and the lateral, nonpigmented layer is called the **articulamentum**. The latter is inserted into the muscular mantle of the chiton. Note that the middle six plates are nearly identical in size and shape, but that the first (head) and last (tail) plates differ.

3. Examine the dorsal surface of the body and observe the thin mantle (Fig. 12.4). A pair of longitudinal muscles are located at dorsally at the midline. What is their function? Locate the oblique and transverse muscles. What are their functions?

4. Pin the specimen to the dissection pan through the anterior and posterior ends and carefully remove the dorsal mantle without damaging the viscera beneath. Submerge the specimen in water and observe it under a dissection microscope.

5. Note the large bundles of muscles near the buccal bulb (Fig. 12.5) They attach to and operate the **odontophore** and **radula**. The salivary glands are located lateral to the muscular bands. Their ducts enter the buccal cavity posteroventrally.

6. Make a short incision through the dorsal surface of the buccal cavity, extending the cut laterally so as not to cut the odontophore muscles. Expose the radula and radular sac (Fig. 12.5); remove the radula intact and observe under the compound microscope. Are the teeth of uniform structure the entire length of the radula? Examine the morphology of the entire radula. Is it uniform from one end to the other? What does this variation indicate about the radula? Find the cartilaginous odontophore over which the radula glides. Determine the action of this complex mechanism. In the next part of this exercise you will compare the form of *Katharina's* radula to that of the gastropod *Helix*.

7. Without disturbing or making further incisions, locate the pharyngeal (sugar) gland, stomach, digestive gland, and intestine (Fig. 12.5). Note the length of the intestine. What does the length of the intestine indicate concerning the diet of this animal?

8. Located the single, mid-dorsal gonad (Fig. 12.5). It is a large, lobed structure that begins in the second third of the body and terminates at the posterior end. Start at the anterior end and trace the gonad until you find the genital ducts that lead to the gonopores.

9. Next locate the **pericardial membrane** and cavity. It is located dorsally in the hindquarter of the body (Fig. 12.5). The cavity surrounds the heart. Locate the triangular median ventricle in the posterior region of this cavity. The aorta leads from the ventricle anteriorly above the gonad. Paired auricles are attached to the lateral sides of the ventricle.

10. Carefully remove the digestive gland and gonad. Locate the thin, many-lobed kidneys extending nearly the entire length of the body. They drain into nephridiopores located just posterior to the gonopores.

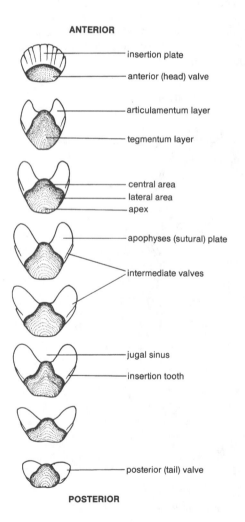

ANTERIOR

insertion plate

anterior (head) valve

articulamentum layer

tegmentum layer

central area
lateral area
apex

apophyses (sutural) plate

intermediate valves

jugal sinus

insertion tooth

posterior (tail) valve

POSTERIOR

Figure 12.3. Dorsal view of isolated plates of *Katharina*.

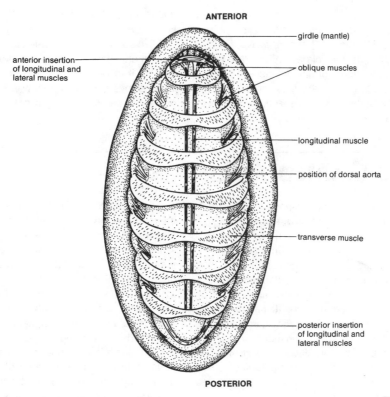

Figure 12.4. Dorsal view of *Katharina* with plates removed.

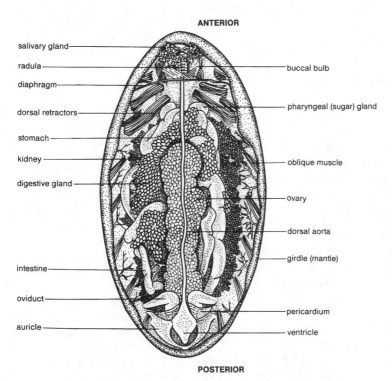

Figure 12.5. Dorsal internal view of the viscera of *Katharina*.

Other Species

Observe a specimen of *Cryptochiton* and compare its external morphology to that of *Katharina*. Note the large size and apparent lack of plates. Where are the plates located? As your instructor directs, inspect other chitons. Look for differences in the size, shape, and ornamentation of the plates and girdle (e.g., granules, scales, spines). Are there any calcareous spicules on the surface of these chitons? What might their function(s) be?

If live specimens are available, observe their general movements, and feeding and homing behavior. How easy is it to remove a chiton from the substratum? If time permits, your instructor will provide instructions for observing the flow of respiratory water through the chiton's mantle cavity. ■

B. Class Gastropoda

Snails are members of class Gastropoda [GAS-tro-POAD-a or GAS-trop-o-dah; G., *gastro*, stomach + G., *podos*, foot], a large group comprising over 40,000 extant and at least 15,000 extinct species of mollusks that vary greatly in size. Three important characteristics are seen in most gastropods. They have (1) a well-developed head, (2) a muscular creeping or swimming foot, and (3) a single asymmetrical shell of calcium carbonate and organic matter. The shell may be absent. Snail shells are usually coiled to the right (clockwise), but left-handed (counterclockwise) shells are known (e.g., *Busycon perversum*, the lightning whelk). The shells of certain sessile gastropods (vermetid worm shells) are irregularly coiled (contorted) and generally elongated.

Students will first study shells and whole specimens of selected species to gain an appreciation of general snail morphology and of the diversity found in the class. The terrestrial pulmonate snail, *Helix*, will be dissected as a class representative.

■ Observational Procedure:

Living Gastropods

If live gastropods are available, make the following observations. Observe locomotion in a living snail. How fast does the animal move? How does the foot appear to move the animal? Describe the movement of the foot when viewed from below. Determine the roles light and gravity play in movement?

Observe feeding in a living snail as it scrapes food from a glass surface. A hand lens may be helpful to make those observations. Note how the radula is applied to the glass. Is all the food removed by the animal? What marks are left by the radula after the animal passes by?

If living pulmonate specimens are available, observe functioning of the pneumostome. Does it remain open or does it periodically open and close? What is the significance of this cycle?

Diversity of Gastropods

There is a rich diversity of shelled forms in this class. Your instructor will provide you with a set of gastropod shells for you to study.

1. Place several different shells on the bench with their **apices** pointing away from you. What common features do these shells share (Fig. 12.6)? Are the **spires** of similar shape? Compare the relative sizes of the body whorls. Are there differences? Note the differences in shell **apertures**, size, and shape (Fig. 12.7).
2. On which side of the shell is the aperture found? Is this consistent in all the specimens present? Shells with the aperture on the right side are right-handed or **dextral** (L., *dexter*, right). Those with the aperture on the left side are known as left-handed or **sinistral** (L., *sinister*, left). View one shell of each type from the apex. Which way do dextral shells coil? Which way do sinistral shells coil? Are there some shells that do not conform to these descriptions?
3. Observe that the outer shell surface may be ornamented with **lamellae, striations, tubercles, spines**, and **ribs**, or may be very smooth (Figs. 12.6 and 12.7). What might be the adaptive significance of these surface features? What types of colors and color patterns are found on the outer shell surface? Are these colors found through the shell? How might these color patterns be produced? (*Hint:* Consider how patterns are produced in a woven rug or mat. Is the pattern printed on at the end or is it incorporated into the rug as it is woven?) What is the significance of color and color patterns to the living snail?
4. Examine a shell that has been cut or broken in half longitudinally. The central column is called the **columella**. Is the shell divided into compartments by the columella? In life, how is the snail positioned in the shell?
5. To gain a better understanding of this point, do the following exercise. Cut a thin strip of construction paper about 10–15 cm × 0.5 cm long. Snake the paper through the whorls of the cut shell and adjust it so that the paper achieves the best possible fit. The paper strip simulates how a long bilateral organism would fit into a helical shell. Does the paper fit particularly well?
6. Repeat this demonstration, but cut the strip in the shape of a spiral. Be sure to keep the width of the spiral strip to about 0.5 cm. This simulates a spiral-

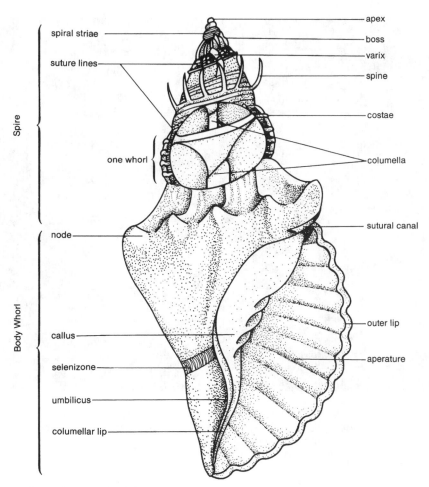

Figure 12.6. Morphology of a generalized gastropod shell.

shaped organism. Which of these presumptive organisms fits the gastropod shell better? Why? Based on these observations, predict the body shape of hermit crabs which inhabit old gastropod shells. Which shells (dextral or sinistral) would they fit into best? Observe the shape of the abdomen of a hermit crab that has been removed from its shell? Which way does the crab's abdomen twist? Which shell type dextral or sinistral will the crab fit best?

7. Not all gastropods have the typical form just studied. Many species are mistakenly identified by novice collectors as members of other molluscan classes or even different phyla. Examine a cowrie shell. Is the spire present? What region of the shell represents the last body whorl? Note the superficial bilateral symmetry of this shell. Describe the growth pattern of this shell.

8. A very different growth pattern is exhibited by *Crepidula* (slipper shell) and *Haliotis* (abalone). Obtain a specimen of each of these species and place them open-side down. Note the growth lines in the shell sur-

face. View these shells from above and determine whether they are dextral or sinistral. In *Crepidula*, note the shelflike **septum** that occupies about one-half of the aperture. What is its function? There is no septum in *Haliotis*, but note the large muscle scar in the central region of the interior. Find the perforations that run near the outer margin of the shell of *Haliotis*. What function do they serve? Also, observe the pearly luster of the inner nacreous or mother-of-pearl layer of a *Haliotis* shell.

9. Compare shell form of *Diodora* (keyhole limpet) to that of other gastropods. Identify the shell aperture and perforation. What function does the perforation serve? Some limpets (e.g., *Acmaea*, true limpet) lack the perforations. In these, excurrent water flow exits the shell from the right side.

10. A very different type of shell occurs in vermetid worm shells. If specimens are available, determine the apex, aperture, and direction of coiling. Can these shells be classified as dextral or sinistral?

Figure 12.7. Diversity of shell form in class Gastropoda (including from left to right, top row: *Mitra* (papal miter), *Conus* (cone), *Vermicularia, Melongena, Busycon*; middle row: *Haliotis* (abalone), *Murex*, limpet; bottom row: *Polinices* (moon snail), *Architectonica*, sundial, *Clanculus* (strawberry top shell), cowrie, keyhole limpets).

11. Obtain a specimen of a nudibranch and place it in a dissection pan. Determine anterior and posterior ends. Note that, except for the location of the genital aperture, the animal is bilaterally symmetrical. Some nudibranchs have numerous finger-like projections on their dorsal surfaces. These are called **cerata**. The cerata apparently function in gaseous exchange by increasing the surface area. Nudibranchs that lack cerata often have other secondary outgrowths that function as gills. Living nudibranchs are often brilliantly colored. This condition is generally considered to be an example of warning coloration to potential predators. As a defense against predators many species produce obnoxious chemicals or use nematocysts obtained from their cnidarian prey. This is sort of an invertebrate recycling program.

12. Examine the sea hare, *Aplysia*. Unless instructed otherwise, only the external features will be examined. Determine the anterior and posterior ends of *Aplysia*. Note the large fleshy foot and winglike **parapodia** used in swimming. At the head, locate the mouth, anterior tentacles, eyespots, and **rhinophores** (olfactory tentacles). On the right side of the head between the right tentacle and rhinophore, locate the small penis. Extending posteriorly and dorsally from the penis is the ciliated **sperm groove** that ends at the gonopore. At the posterior end locate the **anal siphon** and anus. Dorsally, find the opening in the mantle (shell aperture) under which a small, thin shell is located. Does the shell offer much protection? *Aplysia* secretes a purple dye when handled and this is said to help protect it from predators.

13. Observe other gastropods (e.g., *Busycon* and *Limax*) that are available for study. Relate the morphology and anatomy of these specimens to the ones you have just studied.

Anatomy of *Helix*

1. Obtain a freshly killed or expanded, preserved specimen of *Helix* and examine the shell, locating the following major features: apex, spire, suture line, and aperture.

2. Identify the head and foot of the animal (Fig. 12.8). The head bears two pairs of tentacles, the anterior ones being shorter than the posterior. Both pairs can be retracted. The posterior tentacles bear an eye at their distal tip. A small opening to the **common genital**

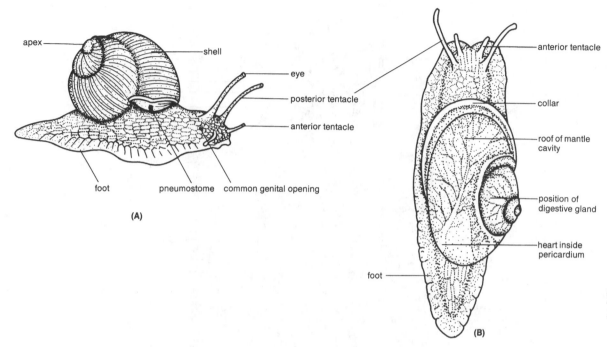

apex

shell

eye

posterior tentacle

anterior tentacle

foot

pneumostome   common genital opening

(A)

anterior tentacle

collar

roof of mantle cavity

position of digestive gland

heart inside pericardium

foot

(B)

Figure 12.8. *Helix*, a pulmonate gastropod. (A) Lateral (with shell). (B) Dorsal (without shell).

duct may be found below and slightly behind the base of the right posterior tentacle. Locate the mouth with its two lateral and one ventral lips. On the right side of the body just below the shell, locate the **pneumostome**, the opening into the mantle cavity that functions as a lung (Fig. 12.8).

3. The shell of *Helix* is thin compared to that of many gastropods. Does *Helix* possess an **operculum**? Carefully break the shell away from the animal. It is not necessary to remove all the shell material completely; the central portion of the shell (**columella**) may be left intact. How does the thickness of the shell of *Helix* compare to the other shells you examined? Speculate on why this difference occurs.

4. Note that the mantle is thicker along the anterior region just behind the head. This is the **collar**; it secretes shell material at the lip of the shell aperture. Carefully make a shallow cut into the lung through the pneumostome. Continue to cut the mantle cavity until the entire lung surface is exposed. Reflect the roof of the mantle cavity to one side to expose the visceral hump. Note that the floor of the mantle cavity is so thin that the internal organs may be seen underneath.

5. Carefully cut into the **pericardial cavity** and expose the heart, identifying the thin-walled auricle and the thicker-walled ventricle. The kidney is located in the roof of the mantle cavity posterior to the heart.

6. Pin your specimen to the dissection tray and immerse it in water. Cut through the floor of the mantle

cavity, being careful not to damage the internal organs. Continue this cut anteriorly through the collar, extending it middorsally to the front of the head. With the aid of Fig. 12.9, locate the following structures.

7. Identify the muscular buccal mass at the anterior end and posterior to it, the nerve ring. A slender esophagus passes from the buccal mass through the ring into the expanded crop region of the gut. Right and left salivary glands will be found on the lateral surfaces of the crop. A duct from each gland extends anteriorly through the nerve ring to the buccal mass. The crop leads into the large stomach, located in the second turn of the visceral hump. Left and right digestive glands can be observed. The left gland is larger and lies alongside the stomach; the smaller right gland occupies the apex of the visceral hump and last whorls of the shell. The intestine leads from the stomach and forms an S-shaped curve in the stomach region.

8. *Helix* has a complex reproductive system. If you are to study it, proceed through steps 8–10 with care. *Helix* is monoecious (hermaphroditic). Carefully tease apart the reproductive organs from the gut (Fig. 12.9) and, if possible, locate the protandrous gonad (**ovotestis**) on the inner coiled face of the right digestive gland. Trace the thin, convoluted duct (**hermaphroditic duct**) that leads from the gonad to the base of an **albumen gland** near the stomach region. Where the duct enters the albumen gland is a small fertilization chamber (or **talon**); here ova are fertilized prior to receiving albumen.

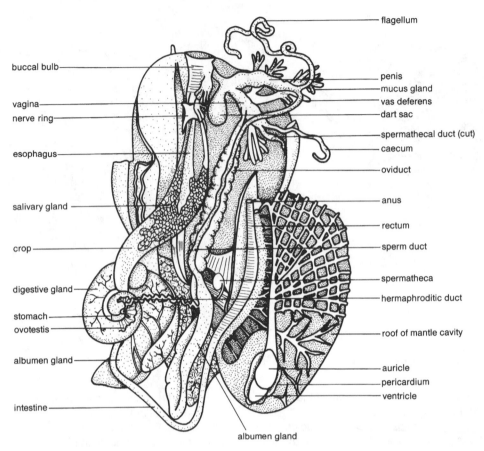

buccal bulb

vagina
nerve ring

esophagus

salivary gland

crop

digestive gland

stomach
ovotestis

albumen gland

intestine

flagellum

penis
mucus gland
vas deferens
dart sac
spermathecal duct (cut)
caecum

oviduct

anus

rectum

sperm duct

spermatheca

hermaphroditic duct

roof of mantle cavity

auricle
pericardium
ventricle

albumen gland

Figure 12.9. Internal anatomy of *Helix*.

9. Locate the broad **spermoviduct** that has separate channels for both male and female reproductive products. It leads anteriorly from the albumen gland to a region where the duct divides into the sperm duct (**vas deferens**) and **oviduct**. The sperm duct leads to the **penis sac** and protrusible penis. Near the junction of sperm duct and penis sac is a long, thin, blind tube called the **flagellum**.

10. The oviduct leads to the vagina and a pair of branched mucus glands enter at about the same point. Nearby, a large bulbous **dart sac** enters too. The muscular dart sac secretes a calcareous needle-like dart that is plunged into the side of the partner before copulation. What purpose might this process serve? Remove the dart sac by cutting its connection to the vagina and observe its muscular structure. A long spermathecal duct joins the vagina at its posterior end. This duct ends in a spherical spermatheca that stores sperm received from another snail during copulation.

11. After you have completed your dissection, remove the buccal mass, make a longitudinal incision down its length, and with the aid of a dissection microscope, locate the radula. Remove it from the buccal mass and examine the teeth under a microscope. How do the teeth of this radula compare to that of the chiton *Katherina*?

Gastropod Larvae

Examine a prepared slide of a gastropod **veliger larva** such as *Crepidula*. Although the internal anatomy of most specimens will be difficult to interpret, attempt to locate the following structures: the double-lobed, ciliated **velum**, visceral mass, foot, operculum, and shell (Fig. 12.10).

The genus *Busycon* (whelks) produces young that complete their larval stages within the egg case. If available, examine a string of egg capsules of a whelk. Some whelks may produce strings of more than 175 parchment-like capsules with 100–200 eggs per cap-

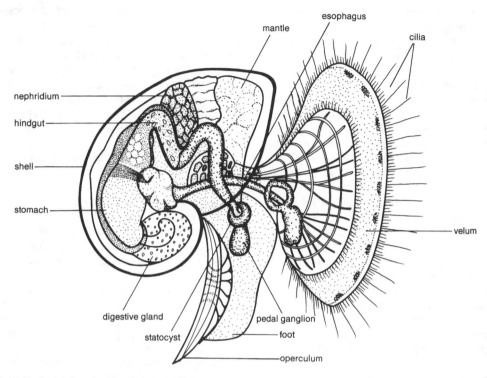

Figure 12.10. Typical veliger larvae of a gastropod.

sule. How many capsules are present on the string you are examining? Note how the capsules are attached to the central cord. Are all of the capsules the same size? The first several capsules of the string are often smaller, more widely spaced, and contain no eggs. What function might the capsules in this region serve?

Examine closely several capsules and determine how young snails emerge from the capsule. If permitted, remove one or more capsules from a string and dissect them. How many young snails are present in each capsule? How big are they? In which direction do the shells of these young snails coil? ■

C. Class Bivalvia

Bivalvia [bi-VAL-ve-a; L., *bi*, two + L., *valva*, folding door] is a large class comprising more than 8,000 extant and about 20,000 extinct species of shelled mollusks. They differ from other mollusks in having a laterally compressed shell of two valves hinged along their dorsal margin. Many bivalves have a hatchet-shaped foot used in digging. This feature accounts for the alternate class name of **Pelecypoda** [pel-e-SIP-o-dah; G., *pelecy*, hatchet + G., *poda*, foot]. Bivalves lack a recognizable head and possess neither radula nor odontophore. Individual species are known commonly by many names including clams, mussels, oysters, cockles, and scallops. Many are economically important food sources and are used in jewelry and, in certain cultures, for money. The wood-boring bivalves, called **shipworms** and **piddocks**,

do considerable damage to pilings, docks, ships, and other wooden structures in marine environments.

Students will study shells of a few selected species to gain an appreciation of the diversity of forms that comprise this class. The marine clam *Mercenaria* (= *Venus*) *mercenaria* will be dissected as a class representative.

■ Observational Procedure:

Live Specimens
If living bivalves are available, observe and describe the various ways they burrow. What behaviors are exhibited when a light or dark shadow strikes the edge of the mantle or siphons? What behaviors are exhibited when the mantle or siphons are touched?

If time and materials are available, your instructor will provide information for observing swimming movements in scallops and flow of water in and out of siphons in filter-feeding clams.

Shell Morphologies

1. As your instructor directs, obtain the shells of several different species of bivalves (e.g., *Anodonta, Arca, Crassostrea, Donax, Mercenaria, Mya, Mytilus,* and *Pecten*). Whenever possible, study specimens with both valves. Note that these shells differ widely from one another in their size and shape, but that they have several features in common. List several of these common features and also several ways in which the shells differ. Do any have more than one type of symmetry? How would you describe shell shape in oysters?

2. Study more closely the shells of *Mercenaria* and the freshwater clam *Anodonta*. Note the thin, dark organic **periostracum** covering the outer surface of *Anodonta*. What functions might the periostracum serve? Observe the concentric lines on the external surface that recede from an elevated point near the hinged margin. These are laid down as growth proceeds. The elevation is the **umbo**, the oldest part of the shell. Umbos of opposite valves are directed toward each other and the approximating points are called **beaks**. In these clams the umbo is directed towards the anterior.

3. Orient both specimens so that the dorsal surface (hinge) is up and the umbo points away from you. Identify the right and left valves (Fig. 12.11). In other species this arrangement is not as clear.

4. Observe the external surface of all specimens you have assembled. Are there any external ridges, spines, or folds on these species? Where are these features located? What purpose might they serve? Examine the surface of the valves closely. Is there evidence of predatory attack (e.g., drill holes or chips along the ventral margin of the shell)? Was this damage repaired before the bivalve died? Do any of your specimens show damage caused by the boring sponge *Cliona*?

5. Turn your attention to the inside of the shells and note any color patterns present. In some species such as *Anodonta* and *Mytilus* the inner surface is made up of **nacre** (mother-of-pearl). It is secreted by the entire outer surface of the mantle. Examine a prepared microscope slide of a section of mantle margin and locate the inner, middle, and outer mantle folds, pallial retractor muscle, and mantle epithelium (Fig. 12.12).

The shell is composed of a number of layers: inner (sometimes nacreous), middle, and outer. The whole shell is covered by periostracum. The middle and outer shell layers and the periostracum are formed at the perimeter of the mantle.

6. Using a hammer and chisel, break the shell, taking care to protect your eyes from shell fragments by wearing safety glasses. View the broken edge of the shell fragment in cross section using a hand lens or dissection microscope. Describe the cross section of the shell. Compare the similarities and differences in shell structure of various species. Are they all the same?

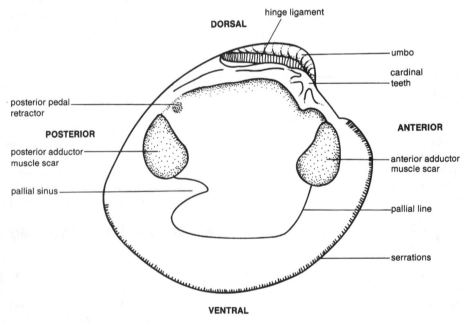

Figure 12.11. Inner surface of the left valve of *Mercenaria*.

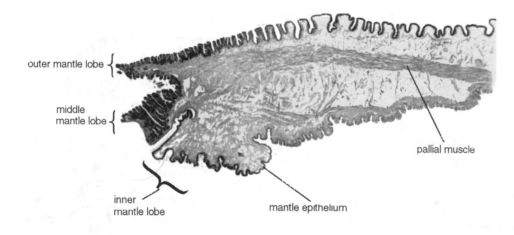

outer mantle lobe {

middle
mantle lobe {

pallial muscle

inner
mantle lobe

mantle epithelium

Figure 12.12. Cross section of the mantle margin of a clam, shell removed.

7. Observe the inner dorsal margins of several species and note the presence of any prominent plates or ridges (Fig. 12.11). These make up the **dentition** or teeth of the valves. Those that run longitudinally along the hinge line are **lateral teeth**. Those grouped near the umbo are **cardinal teeth**. Carefully move the valves, noting how the teeth on opposing valves articulate. Is the fit between valves tight in all species? Note the serrated, inner ventral margin of the valves in *Mercenaria*. How do these serrations fit when the valves close? Suggest ideas that could explain the functional significance of these serrations. Do other species have similar serrations?

8. Observe the anterior and posterior muscle scars and the pallial line on the inner surface of the valves (Fig. 12.11). The pallial line represents the point of attachment of the pallial retractor muscle to the shell.

Internal Anatomy

Preserved or live specimens may be used for this study. *Be sure to heed the warnings given below.*

1. Obtain a preserved specimen of *Mercenaria* in which the valves are held apart by a wooden peg. Using a heavy scalpel, carefully sever the hinge ligament that connects the valves along their dorsal margins. *Use extreme **CAUTION** when doing this procedure; cut away from your body and your fingers! Breaking the edge of the shell with a hammer will provide sufficient space to insert a scalpel. Be careful to protect your eyes from shell fragments when breaking the shell.*

2. With the hinge weakened, carefully insert the scalpel inside the shell at the ventral, posterior margin; sepa-

rate the delicate mantle from the right valve, progressing anteriorly. Be careful not to tear the mantle. Detach the anterior and posterior adductor muscles where they insert at the surface of the right valve. Lift the right valve off the clam, carefully separating any remaining mantle tissues or muscles from the shell. After completing the separation at the hinge, the right valve should come free while the left contains the intact animal.

3. Identify the exposed muscle bundles. The large anterior muscle is the **anterior adductor**, important in keeping the valves closed. Above, and in close proximity, is the smaller **anterior pedal retractor** muscle which withdraws the foot. The large posterior muscle is the **posterior adductor** used also in closing the valves. Above it is the **posterior pedal retractor**.

4. Carefully cut away the right mantle lobe. This will expose the mantle cavity and the organs it contains. Locate the following structures: foot, **ctenidia** (gills), visceral mass, palps, exhalant (excurrent) and inhalant (incurrent) siphons (Fig. 12.13). Note the arrangement of the siphonal openings. The inhalant siphon bears the aperture through which water enters the mantle cavity. The exhalant siphon is the exit channel for water that has passed through the ctenidia, for urine expelled by the kidney, for feces from the adjacent anus, and for gametes. The visceral mass houses the digestive, excretory, and reproductive organs.

5. Examine the arrangement of the ctenidia and their connection to the foot and to the mantle. Observe how they lie in an angular position in the mantle cavity on each side of the body. Note that each ctenidium (right and left sides of the mantle cavity) is composed of two **demibranchs** (inner and outer), the former typically

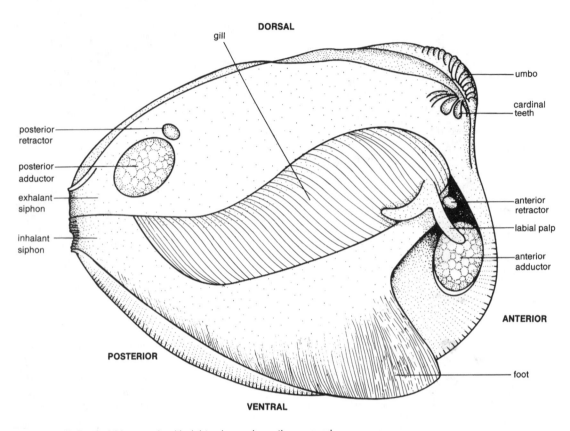

DORSAL

gill

umbo

cardinal teeth

posterior retractor

posterior adductor

exhalant siphon

inhalant siphon

anterior retractor

labial palp

anterior adductor

ANTERIOR

POSTERIOR

foot

VENTRAL

Figure 12.13. Anatomy of *Mercenaria* with right valve and mantle removed.

larger than the latter. Each demibranch is composed of two **lamellae**, ascending and descending.

6. Examine a prepared microscope slide of a section of a demibranch (Fig. 12.14) and locate the numerous **ostia** in the lamellae through which water enters. Water is transported dorsally within the ctenidia through numerous **water tubes**. A complex arrangement of cilia on the vertically aligned filaments that make up the surface of the lamellae creates the flow of water. All tubes empty into the dorsal **suprabranchial cavity**, which in turn empties into the exhalant siphon. Blood vessels are profusely distributed throughout the ctenidia. Note the minute chitinous rods present to the inside of the ciliated surface. What is the significance of these rods? Water entering the mantle cavity also carries food particles, that are retained on the surface of the ctenidia. Ciliary action moves material to food grooves located on the ventral margin of each demibranch. Within the groove, food materials are moved towards the labial palps and then into the mouth.

7. With a pair of scissors cut the mantle free from the outer demibranch. This will expose the suprabranchial cavity; some of the water tubes may be visible. Carefully remove the remaining gill parts on the right.

Posteriorly there is a union of both chambers, forming a **branchial-cloacal** chamber.

8. Hold the animal with the anterior-most point of the foot directed toward you. Locate the anterior adductor muscle. Just below the muscle are the labial palps and slitlike mouth with outer and inner lips (Fig. 12.15). With scissors make a lateral incision and follow the mouth cavity posteriorly for a short distance. The cavity exposed is the esophagus which opens up into an enlarged stomach. Surrounding, and opening into the stomach, are the dark **digestive diverticular (gland)** (Fig. 12.15). Also opening into it from a cecum may be a long, flexible, translucent **crystalline style** that rotates against the chitinous gastric shield of the dorsal stomach wall, releasing enzymes used in extracellular digestion. The intestine is a long, coiled tube within the visceral mass. What does the length of the intestine tell us about the food of this organism? The later part of the intestine ascends dorsally, where it enters the anterior end of the pericardial cavity. From here the rectum passes posteriorly through the pericardial cavity, passes through the heart, and finally ends in an anus on the face of the posterior adductor muscle (Fig. 12.15).

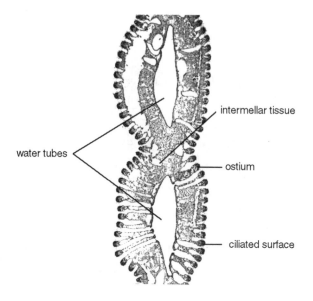

Figure 12.14. Photomicrograph of a section of a bivalve gill.

9. To observe the pericardial cavity and its contents, it may be necessary to remove some of the overlying mantle tissues, if they have not been removed earlier

in the dissection. Locate the region just dorsal to the suprabranchial cavity and carefully remove the mantle tissue; the pericardial cavity should then be exposed (Fig. 12.15). Locate the rectum surrounded by the single, median ventricle of the heart.

10. Locate the two auricles laterally attached to the ventricle and the anterior and posterior aortae. In fresh preparations the heart still may be beating. If your dissection was done properly, you may be able to see the **nephrostome** and the genital aperture, both located just ventral to the pericardial cavity.

11. Follow the rectum to the anus located on the posterior face of the posterior adductor muscle and near the exhalant siphon. Because the circulatory and nervous systems, and reproductive and renal ducts are somewhat difficult to differentiate in a general dissection, they will not be studied here.

Other Specimens

The Mussel *Mytilus*. Carefully open the shell of a mussel of the genus *Mytilus*. Note the reduced foot and cluster of threadlike filaments arising from the heel of the foot anteriorly. These are **byssal threads** secreted by a byssal gland. The mussel uses the threads for

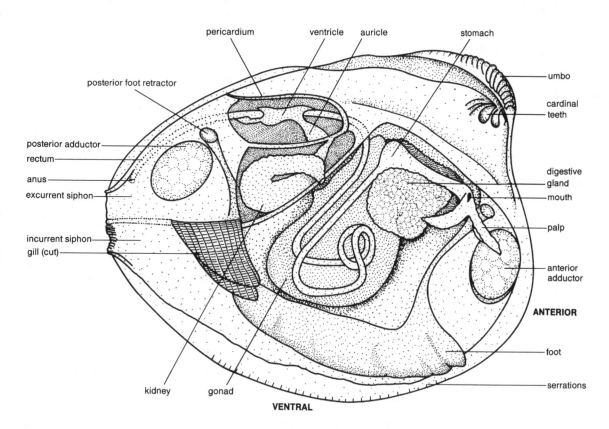

Figure 12.15. Partially dissected *Mercenaria*, depicting internal anatomy.

attachment to hard substrata. Note the difference in size of the adductor muscles.

Scallops. Along the edge of the mantle are numerous **ocelli**. Carefully open the shell of a scallop. Note the single, large, posterior adductor muscle. The adductor is divided into two, approximately equal sections of smooth and striated muscle tissue. The latter is responsible for the rapid clapping of the valves when the scallop swims. The muscle is the part most often consumed. Note the orientation of the muscle fibers within the cylindrical adductor. Occasionally one is served a dish made with scallops but the muscle fibers of the flesh runs from side-to-side and not top-to-bottom. How do you know that this food item is not an adductor muscle from a scallop? (Think skates.)

Wood-Boring Bivalves. Examine pieces of wood that have been colonized by wood-boring bivalves. If the bivalves are present, note the size, shape, and modifications to their shells from the basic bivalve plan. Why are these bivalves called shipworms? How is the shell modified to allow these mollusks to burrow? How is the burrow lined?

Larval Stage. Examine specimens or prepared slides of bivalve larval stages. The first larval stage is a ciliated, somewhat barrel-shaped **trochophore**. Note the position of ciliary bands. The **veliger** stage is more complex. Although the internal anatomy of most specimens will be difficult to interpret, attempt to locate the following structures of the veliger stage: ciliated **velum**, visceral mass, adductor muscles, and valves. Compare the anatomy of a free-living veliger with that

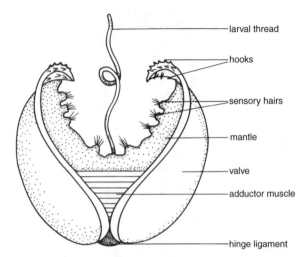

Figure 12.16. Glochidium, parasitic veliger larval stage of freshwater clams.

of the parasitic **glochidium** larva of the freshwater clam, *Anodonta*. Attempt to locate the visceral mass, adductor muscle, hinge ligament, valves (with hooks possessing small spines), sensory cilia, and attachment thread (Fig. 12.16).

Fossil Specimens
Bivalves have a long geological history, dating back to the **Cambrian Period** (Fig. 12.1). If fossil specimens are available, examine their general shell morphology and compare them to extant species. ■

D. Class Cephalopoda

Cephalopoda [SEF-a-lo-POAD-ah or sef-a-LOP-o-da; G., *cephalo*, head + G., *poda*, foot] comprises about 600 extant and 8,000 extinct species of marine mollusks. Most cephalopods are less than 1 m long, but some very large forms are known. In general, cephalopods are adapted for swimming, although many species are bottom-dwellers, crawling about on their arms. The shell is generally reduced or even absent (e.g., octopuses) in living cephalopods, but well-developed shells were possessed by members of the extinct subclasses Ammonoidea and Nautiloidea (represented today by members of the genus *Nautilus*).

Cephalopods have four major features. (1) All possess a well-developed brain and complex nervous system. (2) The anterior portion of the foot is modified into a circle of prehensile, muscular arms and tentacles that surround the mouth. Except in *Nautilus*, these appendages bear

numerous, muscular suckers that aid in prey capture. (3) Cephalopods have a muscular buccal apparatus containing beaklike jaws and a radula. (4) Locomotion in this group is unique. Water taken into the mantle cavity can be expelled in a forceful contraction of the mantle walls. This exhalation furnishes a type of jet propulsion.

The common Atlantic-coast squid, *Loligo pealeii*, will be studied in detail as the class representative. Given the diversity within the class, *Loligo* should not be considered to be a typical cephalopod.

■ Observational Procedure:

External Anatomy
1. Orient the squid in the dissection pan so that the arms and tentacles are directed away from you and the broad flat surface of the fins is directed upwards. In

terms of biological function, the posterior end of the animal is now nearest you, the anterior end farthest away, and the dorsal surface is upward. (Fig. 12.17). Now identify anterior, posterior, dorsal, and ventral surface in terms of the other molluscan classes. Why do these two ways of viewing the animal differ?

2. Note the finely striated appearance of the surface of the fins. These striations are muscle bundles that control fin movement. The spotted condition of the body surface is due to the presence of pigmented cells called **chromatophores**.

3. Gently scrape the surface of the body removing the transparent outer cuticle. Each chromatophore is a small membranous sac to which tiny muscle fibers are attached. Contraction of muscle fibers enlarges the chromatophore and makes the pigment visible.

4. Arrange the squid with the ventral side up and the head nearest you. The body form is elongate and arrow-shaped (Fig. 12.18). The head region appears as though it is inserted in a central tube. This is the tough, muscular mantle.

5. Pick up the squid and view the animal from the oral aspect. While applying lateral pressure to the tube, look into the mantle cavity. The blunt lateral projections are extensions of the **mantle locking ridges**. Cartilaginous locking grooves are located on the inside surface and will be observed later. The pointed extension on the dorsal surface of the mantle marks the location of the internal shell (**gladius**) or **pen** (Fig. 12.17). The cone-shaped structure extending beyond the mantle on the ventral surface is the **funnel**.

6. Locate the eyes on either side of the head in the region of the funnel. They will be considered in detail later.

7. Two kinds of appendages are located at the anterior end around the mouth; shorter ones are arms and

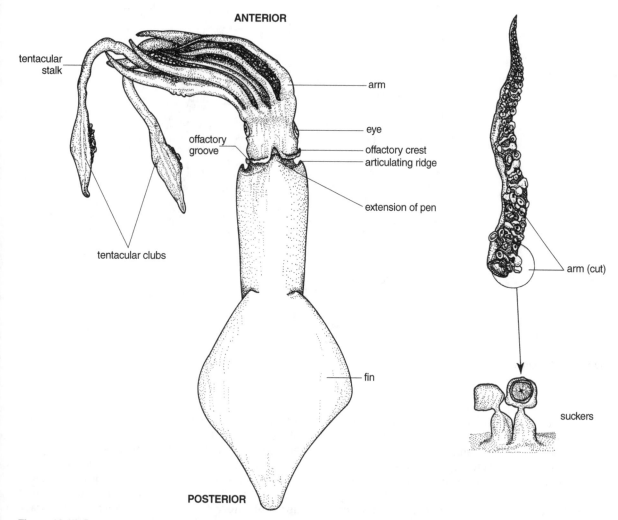

Figure 12.17. Dorsal view of the squid *Loligo pealeii,* with enlargement of an arm and suckers.

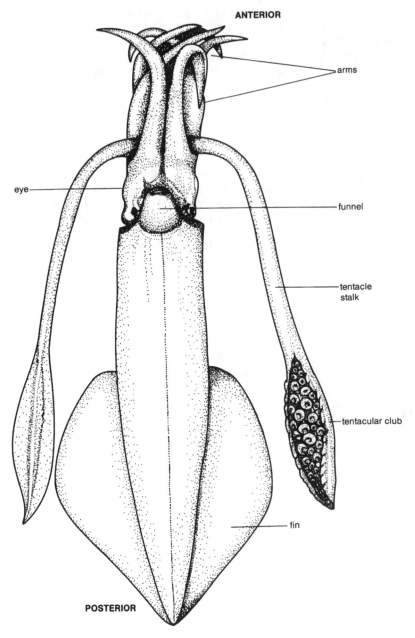

ANTERIOR

arms

eye

funnel

tentacle stalk

tentacular club

fin

POSTERIOR

Figure 12.18. Ventral view of the squid *Loligo pealeii*.

the longer ones with club-shaped ends are tentacles. How many arms and how many tentacles are present?

8. Examine the arms and tentacles. How do they differ as to size and shape? Note any differences in size and position of sucker cups on the arms and tentacles (Fig. 12.17).

9. Remove several suckers and determine how they function. What do they resemble?

10. Locate the **aquiferous pore** at the anterior edge of the eye (Fig. 12.19). This pore is probably used for equalizing pressure in the eye chamber.

11. Observe the articulation of the mantle ridges with the funnel locking cartilages on each side of the funnel. Near the eye, the integument is folded and thickened as the **olfactory crest**, under which is found the **olfactory groove** (Figs. 12.17 and 12.19).

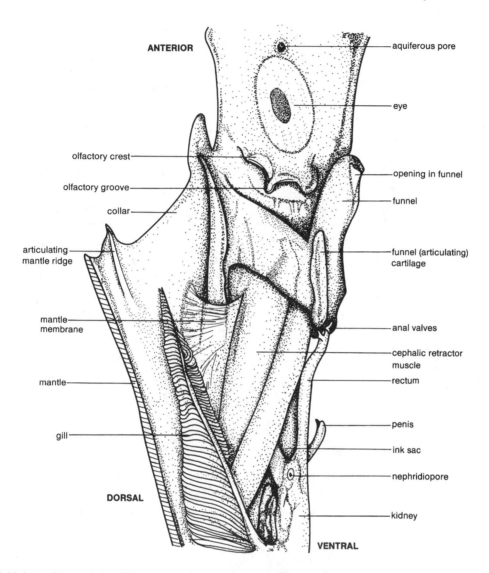

ANTERIOR

aquiferous pore

eye

olfactory crest

olfactory groove

collar

opening in funnel

funnel

articulating
mantle ridge

funnel (articulating)
cartilage

mantle
membrane

anal valves

cephalic retractor
muscle

mantle

rectum

gill

penis

ink sac

nephridiopore

DORSAL

kidney

VENTRAL

Figure 12.19. Lateral internal view of the trunk-head region of a male squid (*Loligo pealeii*).

Internal Anatomy

1. Orient the squid with its ventral side up (Fig. 12.18). With a sharp, heavy-duty scalpel, make an incision from the mantle aperture to the posterior end. First make a *shallow cut;* follow this by a second cut that completely separates the mantle. Make sure this cut is not so deep it disturbs the internal organs. Gently fold the two sides of the mantle laterally and, with large dissection pins, firmly attach the outer edges of the mantle to the base of the dissection pan (Fig. 12.20). If the mantle cannot be secured with dissection pins, make a shallow longitudinal cut the length of the mantle on either side.

2. With the aid of Fig 12.20, determine the sex of your specimen. Sever the funnel lengthwise, exposing its cavity. The large paired glands inside the funnel are the **organs of Verrill**, whose function is not known. The liplike structure at the forward edge of the funnel is the funnel valve that regulates water flow out of the mantle cavity. Note the fold extending from the funnel to the other side of the head. This is one of the **lateral valves** that prevents water from exiting the mantle cavity along the same path from which it came. Water exits only through the funnel. Describe the functional morphology of the structures involved with circulation of water through the mantle cavity.

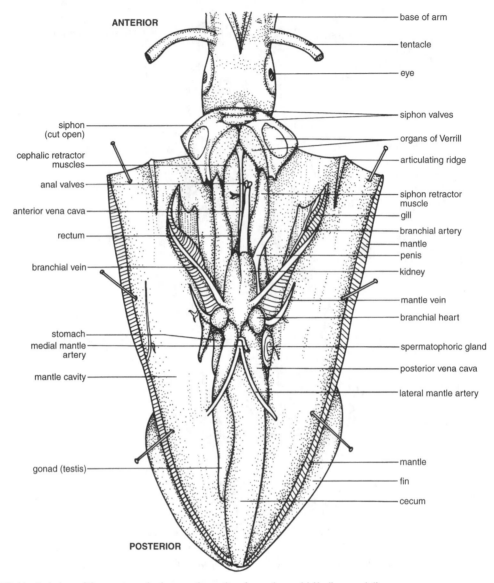

Figure 12.20. Ventral view of the anatomy in the mantle cavity of a male squid (*Loligo pealeii*).

3. Locate the **rectum** and the anal opening at the end. The anus bears two flaplike valves. How do they function; are they active or passive? Why does the anus possess valves of this sort?

4. Below the rectum is the ink sac (Fig. 12.20). This small sac is filled with a black pigment called **melanin**. Observe the membranous tissue binding the ink sac to the rectum. Using sharp, fine-pointed scissors and a scalpel, carefully separate and remove the ink sac from the rectum.

5. Open the sac in a small dry dish. Remove some of the thick, dark ink with a pipette or syringe and place a drop into some water. Note how deeply the ink stains the water.

6. Two pairs of large muscles are found in the area of the ink sac. Locate the **funnel retractor muscles** that extend dorsally from under the funnel and dip strongly under the central organs. Beneath these muscles and still larger in size are the paired **cephalic retractor muscles** (Fig. 12.20).

7. The paired, large, pinnate structures in the mantle cavity are the gills (Figs. 12.19 to 12.21). In specimens injected with latex, the gills will be colored, usually red with some blue. Dorsally the gills are attached to the mantle by the gill mantle membrane.

8. Follow the gills to their proximal junction with the other viscera. Along the outer edge is a large blood vessel (**efferent branchial vein**). Oxygenated blood is

ANTERIOR

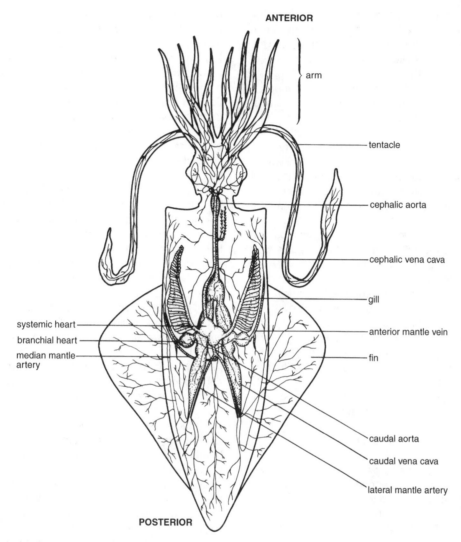

arm

tentacle

cephalic aorta

cephalic vena cava

gill

systemic heart

branchial heart

median mantle artery

anterior mantle vein

fin

caudal aorta

caudal vena cava

lateral mantle artery

POSTERIOR

Figure 12.21. Circulatory in *Loligo*. (Modified after Williams.)

returned through this vessel from the gill to the **systemic heart**.

9. Between the bases of the gills is a pair of elongate, sponge-textured organs. This is the region of the kidney. It appears as if the branchial veins are directly connected to the kidney. Actually they pass under this organ and connect to the systemic heart (Fig. 12.21). This connection will be better observed later. At the base of each gill is a bulblike **branchial heart**. The branchial hearts collect venous return and pump blood into the vessels of the gills.

10. Extending back from the branchial hearts are two elongated conelike structures with broadened bases. They lie next to the branchial hearts. These vessels vary in size according to the amount of blood contained. They are the posterior **vena cavae** (posterior

mantle veins). It will be revealed by later dissection that the posterior vena cava gathers venous blood from the far part of the body opposite the head and empties it into the branchial heart.

11. Coming from the opposite area of the body is another venous vessel that forks in the region of, and courses through, the kidney. Because of the close association of the forks, the vessels are hard to see. They are described as the nephridial portions of the anterior vena cava. Carrying venous blood, it too empties into the branchial heart. With blood from both the anterior and posterior vena cava, the branchial heart functions as an accelerating pump that forces blood into the gills for aeration.

12. In addition to the larger vessels that empty into the branchial heart, there are three smaller vessels that bring blood directly from the mantle (**anterior** and **medial mantle veins**) and from the gonad (**gonadal vein**). The latter vein will not be seen at this time. The anterior and medial mantle veins may be seen lateral to the branchial hearts.

13. Arrange the squid so that the head is nearest you. Take the head in your hand and spread the arms and tentacles. Observe a ring of small finger-like projections, **buccal lappets**, surrounding the mouth. Note their arrangement and the suckers they bear. The buccal lappets are united with the buccal membrane which they support. Inward to and below the buccal lappets surrounding the mouth, is another membrane, the **peristomial membrane**. The margin of this membrane forms a rugose collar (Fig. 12.22).

14. Grasping the free edge of the funnel with a pair of forceps, lift it and sever the two small funnel protractor muscles extending between the head and funnel. Also cut the lateral funnel valves.

15. Separate the pair of arms immediately below the funnel by making a medial lengthwise incision at the base of the arms. Again note the buccal and peristomial membranes. As the arms are separated, notice their arrangement around the buccal mass. The general muscular organization of the specialized foot is complex and will not be studied.

16. Apply slight pressure to this region of the body to flatten the arms and tentacles thereby revealing the **buccal mass** (Fig. 12.22).

17. Examine the buccal mass. How muscular is this structure?

18. Cut the peristomial and peripheral buccal membranes, exposing a compact body with the end displaying dark chitinous jaws (mandibles) that are similar to an inverted parrot's beak. The upper jaw is overlain by the lower jaw (Fig. 12.22). What function do these mandibles serve?

19. Using a sharp scalpel make a midsagittal cut through the buccal mass. Locate the jaws and, in the floor of the mouth, a structure called the **tongue**. A duct from the median posterior salivary gland opens at the tip of the tongue.

20. With forceps, grasp the distal end of the tongue and bend it backwards until the radular area is exposed. The radula is surrounded by the upper jaw and is easily identified by having numerous teeth. Note the flaplike plates at the side of the radula. The lower (ventral) edges of the paired flaps lead to the esophagus. With the dissection microscope, study the dentition of the radula. How does the radula of *Loligo* compare to those of the other mollusks you studied? The space enclosed between the jaws is the buccal cavity; it communicates with the esophagus.

21. To locate the esophagus, it will be necessary to make a deep, medial incision on the surface of the head at the base of the arms. The incision must be exact as the esophagus is a relatively small tube. As you dissect, a hardened structure, the **cephalic cartilage**, will be encountered at about the level of the eyes. Also exposed are whitish, pulpy masses, on either side of the tubular esophagus. These two tissue masses are the **cerebral** and **pedal ganglia**. The pedal ganglion is on the same side of the esophagus as the funnel.

22. Follow the esophagus from the buccal mass attachment to the place where it penetrates the cream-colored **digestive gland** (Fig. 12.23).

23. Remove the digestive gland and expose the esophagus, which lies in a groove between the digestive gland and the body wall. The digestive gland also surrounds the elongated median posterior salivary gland. Anterior salivary glands are embedded in the buccal mass and may be found with careful dissection. The anterior glands produce a lubricating secretion and the posterior salivary gland a poison and hyaluronidase. What are the functions of the poison gland and the enzyme **hyaluronidase**? The small vessel, located alongside the esophagus and connected with the systemic heart, is the **cephalic aorta** (Fig. 12.21).

Study of the digestive system will be temporarily discontinued and attention given to the circulatory and the reproductive systems. It is difficult to dissect the digestive canal at this point and still leave the circulatory and reproductive systems intact.

24. Two types of hearts (branchial and systemic) have been mentioned (Fig. 12.21). The branchial hearts are venous and the systemic heart is arterial. The systemic heart receives aerated blood from both gills. Blood moves from the systemic heart to the body via the anterior (cephalic) aorta, posterior aorta, and genital artery. Valves in each vessel prevent blood from flowing back into the heart. Blood returns to the central circulatory system through a complex system of veins. The anterior (cephalic) vena cava and posterior vena cava are enlarged portions of this system. Blood in the vena cava passes to the gills via the branchial hearts. Blood vessels leaving the branchial hearts possess valves that prevent return of the blood.

25. Locate the small vessel that extends between and slightly posterior to the branchial hearts. This is the posterior aorta, which communicates with the systemic heart. It is three-pronged, with two lateral vessels and one medial vessel. The central one is the median mantle (pallial) artery. The other two forks are the lateral mantle (pallial) arteries. That portion of each posterior vena cava in close association with the kidney before entering the branchial heart is known as the nephridial portion of the posterior vena caval vein. The right and left lateral mantle arteries run along the inner margins of each posterior vena cava (Fig. 12.21).

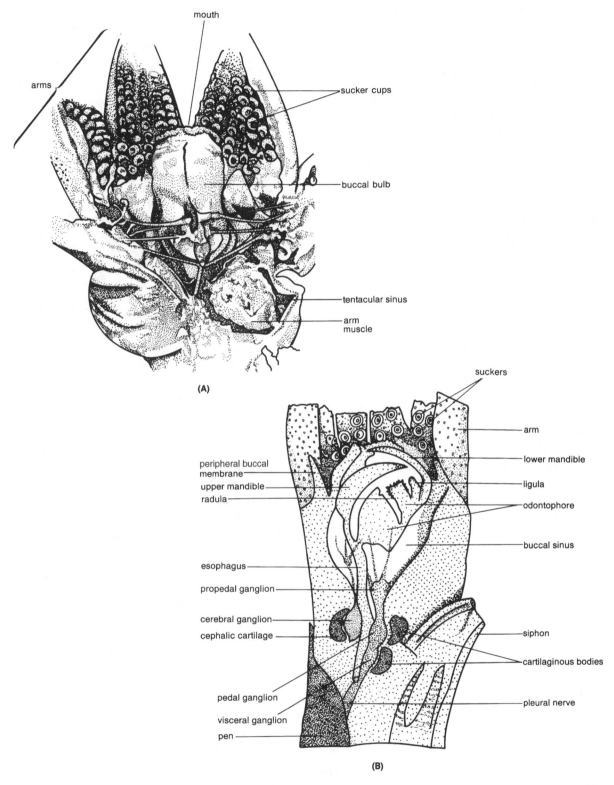

Figure 12.22. *Loligo.* (A) Base of arms sagittally separated to expose the buccal mass. (B) Midsagittal diagrammatic section of the head region.

26. Using a broad-blade scalpel, gently scrape away the nephridial tissue between the bases of the gills. A somewhat convoluted mass of tissues beneath the kidney is the digestive duct appendages. When additional pancreatic tissue is removed the quadrate-shaped systemic heart can be seen (Fig. 12.21). The forwardmost corner of the systemic heart is the part communicating with the anterior aorta, supplying the head and foot. Large lateral vessels running to the inside of the gills are the efferent branchial veins. Smaller blood vessels may be found attached to the systemic heart. They are very small and only careful dissection will insure their being located. These vessels carry arterial blood to the branchial hearts.

27. Locate the genital artery arising from the systemic heart. The dorsal corner of the heart leads into the conspicuous short posterior aorta that subsequently branches into the three mantle arteries mentioned previously.

28. Turn your attention to the remainder of the digestive system and trace the esophagus posterior to the large, muscular stomach. Attached to the stomach is a thin-walled, saclike organ, the **spiral cecum** or **gastric pouch** (Figs. 12.23 and 12.24). Both stomach and cecum may vary in size depending on the amount of food eaten before the animal was preserved. Inside the cecum, located on the proximal wall, is a large, spiral system of folds. These **ciliary leaflets** sort minute food particles sent from the stomach.

Reproductive System: Male

1. Using the handle of your scalpel, gently lift and reflect the gastric pouch laterally.

2. The large, somewhat firm but flattened, light colored organ in the dorsal part of the cavity and under the gastric pouch is the testis (Fig. 12.25). It is enclosed by the perivisceral (peritoneal) membrane. Sperm are passed freely into the cavity surrounding

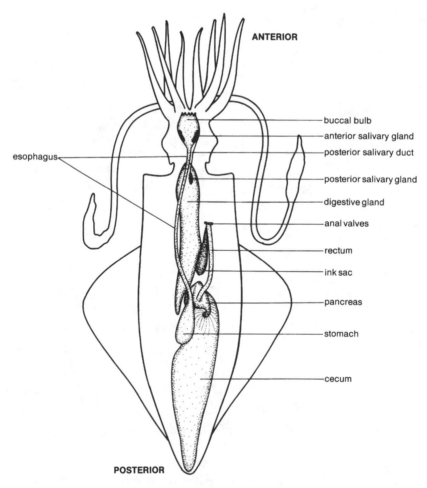

Figure 12.23. Schematic view of the digestive organs of *Loligo*.

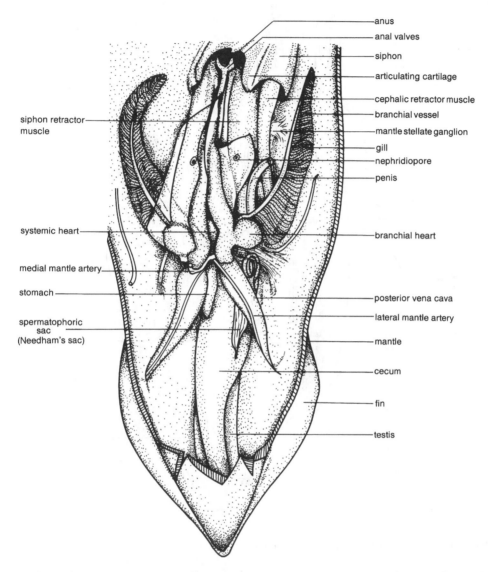

anus
anal valves
siphon
articulating cartilage
cephalic retractor muscle
branchial vessel
mantle stellate ganglion
gill
nephridiopore
penis

siphon retractor muscle

systemic heart
medial mantle artery
stomach

branchial heart

posterior vena cava
lateral mantle artery

mantle

cecum

fin

testis

spermatophoric sac (Needham's sac)

Figure 12.24. Central visceral mass with periviseral membrane turned back to expose nephridial and circulatory organs of a male *Loligo*.

the testis. The sperm exit from the testis through a slit-like opening on its surface.

3. Locate the spindle-shaped organ, the **spermatophoric sac** (Fig. 12.25). Alongside the sac is the coiled vas deferens. The free end of the vas deferens is enlarged as the **sperm bulb** (ciliated funnel). Sperm are swept by ciliary currents into the bulb. After entering the bulb they pass into the convoluted vas deferens.

4. The vas deferens connects to an enlarged convoluted tubule, the **spermatophoric glands**. These glands provide secretions that enclose the sperm in spermatophores (Fig. 12.25). Spermatophores exit the

spermatophoric glands by way of the spermatophoric duct to the spermatophoric sac or **Needham's sac** where they are stored. Exclusive of the testis, Needham's sac, when filled with spermatophores, is the largest structure of the male reproductive system. The spermatophores exit to the mantle cavity through the muscular penis that extends anteriorly from Needham's sac.

5. Remove some of the spermatophores from the sac or from the penis and examine them under the low power of the compound microscope. Study the shape and size of the spermatophores.

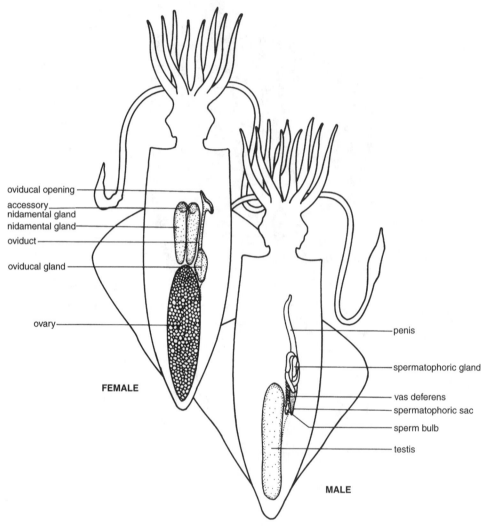

oviducal opening
accessory nidamental gland
nidamental gland
oviduct
oviducal gland
ovary

FEMALE

penis
spermatophoric gland
vas deferens
spermatophoric sac
sperm bulb
testis

MALE

Figure 12.25. Female and male reproductive systems of *Loligo*.

Reproductive System: Female

1. Examine a female squid. If your specimen is a mature female, note the prominent paired **nidamental glands** (Fig. 12.25). They secrete the coverings for the egg capsules. The capsule covering is elastic and hardens somewhat upon contact with water.

2. The nephridia, rectum, ink sac, and part of the heart region are covered by these glands. The anterior accessory nidamental glands are enlarged when the female is mature. However, they are usually small, rounded bodies located just below (as you view the specimen) the nidamental glands on the anterior side. The accessory glands secrete an elastic membrane about each egg.

3. Remove the nidamental glands by separating them with the blunt end of a scalpel, and lifting them from the body. Be careful not to disturb the ink sac if it has not

been removed. The large oviduct, lying parallel to the rectum should be evident.

4. The ovary lies in the coelomic cavity in the same position as the testis. Note how the perivisceral membrane encloses the ovary.

5. Locate the large oviducal opening of the oviduct by following the course of the oviduct forward.

6. In the region of the branchial heart, the oviduct is dilated and is called the oviducal gland. It secretes the egg covering. The oviduct at this point turns down for a short distance and then reverses direction toward the ovary.

The Eye and Shell

If time permits, examine the squid's eye. Remove one eye from its cavity by severing the nerve connection at the base of the cavity. With forceps remove the outer covering or false cornea of the eye. The true cornea lies underneath the false one. Remove the cornea. The opaque, ball-like structure is the lens. Surrounding the lens is the iris. Under the iris the surface of the eye is dark in color. Remove this membrane. The ciliary muscles controlling the lens may be seen as a concentric ring around the lens. By holding the lens and its band of ciliary muscles to the light, one will discover that what appears as a single band is really made up of several colored bands. Remove the inner lining or retina of the eye and examine it with the aid of a dissection microscope. Your instructor will provide additional information if you are to compare the eye of a squid to that of a vertebrate.

Push the viscera to one side and carefully dissect free the translucent, amber-colored gladius or pen (internal shell) embedded within the mantle.

Other Specimens

Sepia. Cuttlefish (*Sepia* and its allies) are not as streamlined and as rapid swimmers *Loligo*. The European cuttlefish (*Sepia officinalis*) is found in shallow waters where it lies buried just under the surface of the sediment during the day. At night *Sepia* becomes a free-swimming predator. A brief study of the external anatomy of *Sepia* will be made and compared to that of *Loligo*.

1. Obtain undissected specimens of *Sepia* and *Loligo* and determine the anterior-posterior axis of each specimen. How does the body shape of these two cephalopods differ? Compare and contrast other features of the two specimens.
2. Observe how the head of *Sepia* is inserted into the muscular mantle tube.
3. Locate the fins of both specimens. How do they compare? Do they have the same shape? How might these structural differences be manifest as functional differences in swimming?
4. As in *Loligo*, the extension of the mantle margin on the dorsal surface is the extension of an internal shell.
5. Examine a specimen of the internal shell of *Sepia*. This structure is known as the **cuttlebone**. Describe its shape. How does this shell compare to that of *Loligo*? Of what is it composed? Using a dissection microscope or hand lens, examine the inside of a piece of cuttlebone that has been cut in cross section at an oblique angle. Note the thin septa. How could this shell add buoyancy to the animal? (*Hint*. Examine the form of the shell of *Nautilus*.)
6. Locate the eyes and the chitinous jaws (mandibles).
7. Note that *Sepia* has short arms and long club-shaped tentacles. How many arms and how many tentacles does *Sepia* have? Examine closely the arms and tentacles. How do they compare to the squid?

Octopus. Octopuses are very different types of cephalopods from those studied so far. Though capable of swimming by jet propulsion, octopuses more often crawl along the bottom in search of prey or hide, waiting for prey. *Octopus* is capable of learning simple tasks and may be one of the most intelligent of all invertebrates. A brief comparison will be made of the general morphology of *Octopus* to that of the other cephalopods studied.

Obtain undissected specimens of *Octopus, Sepia,* and *Loligo*, and compare their general external anatomy. Identify the anterior and posterior ends of *Octopus*. Locate the funnel of the specimen. Where is water taken into the mantle cavity in this organism? How is it expelled? Is there any noticeable internal shell as in *Loligo*? (NB: *Octopus* has paired internal shell vestiges called **stylets**.) How many arms does *Octopus* have? Compare the suckers to those of *Loligo* and *Sepia*. How are they different? Locate the eyes and funnel on *Octopus*.

If available, examine a female specimen of the pelagic octopod, *Argonauta*. Note the paired dorsal arms with their broadly expanded membranes. The membranes secrete the shell that functions as a buoyancy organ and also as an egg case. Examine the shell of *Argonauta*. Note that it has a general spiral shape.

Nautilus. *Nautilus* is the only extant cephalopod with a well-developed external shell. This species lives in the tropics of the western Pacific ocean, ranging from the surface to depths of about 500 m. *Nautilus* is an active diurnal hunter, grasping prey with its 38 sucker-less tentacles.

1. Obtain shells of *Nautilus* and a few gastropods that have been cut in a medial longitudinal section. Compare the external morphology of the *Nautilus* to that of the gastropods. Is shell coiling the same in both of these shells? What type of coiling is found in *Nautilus*? Note that each successive whorl is in the same plane as previous ones.
2. Examine the outer surface of *Nautilus* and note any external color patterns that may be present. Also note the central depression called the **umbilicus**. This represents the axis of coiling that develops because the body whorls do not fully reach the axis. Is there external evidence of internal septa?
3. Examine the inner cut face of the *Nautilus* shell (Fig. 12.26) and locate the aperture septa, **camerae** (chambers), nacreous layer, and connecting rings of the **siphuncle**.
4. What do the different-size camerae represent? Where was the major portion of the body of the living organism located?
5. Where is the perforation of the siphuncle located on the septal face?

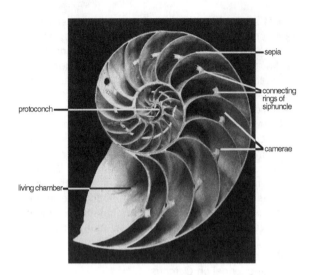

protoconch

living chamber

sepia

connecting rings of siphuncle

camerae

Figure 12.26. Photograph of the internal structure of the shell of *Nautilus*.

6. In the axis of coiling, locate the umbilical perforation. This is the internal manifestation of the umbilicus noted above.

7. Also locate the **protoconch**, a small camera that is the initial chamber of the shell.

8. *Nautilus* regulates its buoyancy by altering the relative volumes of liquid and gas in the camerae through action of the siphuncle. Given this information, describe how *Nautilus* is oriented in the water. In which direction does it swim? How does this bouyancy system relate to the one in the cuttlefish?

9. If available, examine a preserved *Nautilus* and observe the position of hood, eyes, tentacles, funnel, and shell.

10. Examine the surface of the tentacles noting the lack of suckers.

11. Determine the thickness of the hood. What is the function of this structure?

12. Compare the general anatomy of *Nautilus* to that of the three other cephalopods studied. Note that the funnel in *Nautilus* is not a closed tube, the edges only overlap.

Fossil Forms

Examine the fossilized cephalopods on display. In coiled forms, what type of coiling is exhibited? In ammonoids note the suture pattern formed by the septa where they meet the outer margin of the shell. How would you describe these patterns? In a sectioned specimen locate the camerae. Compare shells of ammonoids with that of *Nautilus*. Are there any similarities or differences? Do any of the fossil specimens exhibit evidence of the siphuncle (i.e., septal neck)? The internal anatomy of these forms is believed to be similar to that of *Nautilus*.

Not all fossil cephalopods had coiled shells, some were loosely coiled, arc-shaped, or even had straight shells (Fig. 12.27). Examine a fossil species which had a straight shell (e.g., *Geisenoceras*). Where were the camerae located with respect to the living animal in these forms? How did straight-shelled species maintain a balanced horizontal position in the water? Or did they? Is there any evidence of **cameral** or **siphonate deposits** that could help counterbalance the animal? How would that make the animal like to teeter-totter? ∎

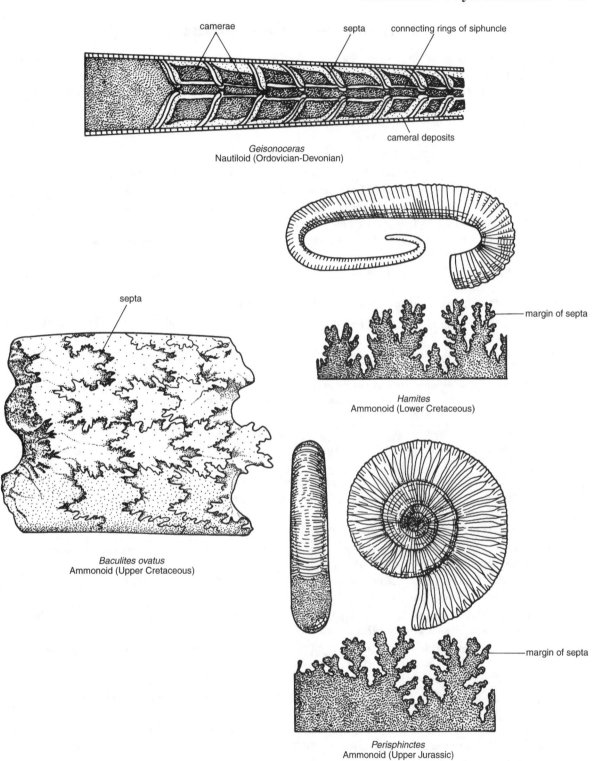

Figure 12.27. Examples of fossil cephalopods.

Supplemental Readings*

Abbott, N.J., R. Williamson, and L. Maddock (eds.). 1995. Cephalopod Neurobiology. Neuroscience Studies in Squid, Octopus, and Cuttlefish. Oxford University Press, New York.

Barnes, J.R., and J.J. Gonor. 1973. The larval settling response of the lined chiton *Tonicella lineata*. Mar. Biol. 20: 259–264. (P)

Bertness, M.D., and C. Cunningham. 1981. Crab shell-crushing predation and gastropod architectural defense. J. Exp. Mar. Biol. Ecol. 50: 213–230. (GA)

Bilyard, G.R. 1974. The feeding habits and ecology of *Dentalium entale stimpsoni*. Veliger 17: 126–138. (S)

Boyle, P.R. 1977. The physiology and behavior of chitons. Ann. Rev. Oceanog. Mar. Biol. 15: 461–509. (P)

Boyle, P.R. (ed.) 1983/1986. Cephalopod Life Cycles. Vols. 1–2. Academic Press, New York. (C)

Britton, J.C., and B. Morton. 1982. A dissection guide, field and laboratory manual for the introduced bivalve *Corbicula fluminea*. Malacol. Rev. Supple 3: 1–82. (B)

Denton, E.J., and J.B. Gilpin-Brown. 1973. Flotation mechanisms in modern and fossil cephalopods. Adv. Mar. Biol. 11: 197–268. (C)

Emerson, W.K. 1976. The American Museum of Natural History Guide to Shells. Alfred A. Knopf, New York. (GN)

Feder, H.M. 1972. Escape responses in marine invertebrates. Sci. Am. 227: 92–100. (GN)

Fretter, V. (ed.). 1968. Studies In The Structure, Physiology and Ecology of Mollusks. Academic Press, New York. (GN)

Gilles, R. 1972. Osmoregulation in three molluscs: *Acanthochitona discrepens* Brown, *Glycymeris glycymeris* L. and *Mytilus edulis* L. Biol. Bull. 142: 25–35. (P, B)

Hadfield, M.G. 1978. Metamorphosis in marine molluscan larvae: an analysis of stimulus and response. *In*: F.-S. Chia and M.E. Rice (eds.). Settlement and Metamorphosis of Marine Invertebrate Larvae, pp. 165–175. Elsevier, New York. (GN)

Harman, W.N. 1972. Benthic substrates: their effect on freshwater Mollusca. Ecology 53: 271–277. (GN)

Hughes, R.N. 1986. A Functional Biology of Marine Gastropods. Johns Hopkins University Press, Baltimore, MD. (GA)

Lane, F.W. 1960. Kingdom of the Octopus: The Life History of the Cephaopoda. Sheridan House, New York. (C)

Lemche, H. 1957. A new living deep-sea mollusk of the Cambro-Devonian, Class Monoplacophora. Nature 179: 413–416. (M)

Linsley, R.M. 1978. Shell form and the evolution of gastropods. Am. Sci. 66: 432–441. (GA)

Lutz, R.A., D. Jablonski, and R.D. Turner. 1984. Larval development and dispersal at deep-sea hydrothermal vents. Science 226: 1451–1454. (GA)

Meinhardt, H. 1995. The Algorithmic Beauty of Sea Shells. Springer-Verlag, New York. (GN)

Morris, P.A. 1975. A Field Guide To Shells, 3rd ed. Houghton Mifflin Co., Boston, MA. (GN)

Palmer, A.R. 1977. Function of shell sculpture in marine gastropods: hydrodynamic destabilization in *Ceratostoma foliatum*. Science 197: 1293–1295. (GA)

Palmer, A.R. 1979. Fish predation and the evolution of gastropod shell sculpture: experimental and geographical evidence. Evolution 33: 697–713. (GA)

Purchon, R.D. 1977. The Biology of Mollusca, 2nd ed. Pergamon Press, New York. (GN)

Rhoads, D.C., R.A. Lutz, E.C. Revelas, and R.M. Cerrato. 1981. Growth of bivalves at deep-sea hydrothermal vents along the Galápagos Rift. Science 214: 911–913. (B)

Runnegar, B., and Pojeta, J. 1974. Molluscan phylogeny: the paleontological viewpoint. Science 186: 311–317. (GN)

Sabelli, B. 1979. Simon and Schuster's Guide to Shells. Simon and Schuster, New York. (GN)

Saunders, W.B. 1984. *Nautilus* growth and longevity: evidence from marked and recaptured animals. Science 224: 990–992. (C)

Scheltema, A.H. 1978. Position of the class Aplacophora in the phylum Mollusca. Malacologia 17: 99–109. (A)

van der Spoel, S., A.C. van Bruggen, and J. Lever. (eds.). 1979. Pathways in Malacology. Dr. W. Junk Publishers, The Hague. (GN)

Stanley, S.M. 1975. Why clams have the shape they have: an experimental analysis of burrowing. Paleobiology 1: 48. (B)

Trueman, E.R. 1968. The burrowing activities of bivalves. Symp. Zool. Soc. Lond. 22: 167–186. (B)

Vermeij, G.J. 1975. Evolution and distribution of left-handed and planispiral coiling in snails. Nature 254: 419–420. (GA)

Vermeij, G.J. 1978. Biogeography and Adaptation. Harvard University Press, Cambridge, MA. (GN)

Vermeij, G.J. 1993. A Natural History of Shells. Princeton University Press, Princeton NJ. (GN)

Ward, D.V. 1972. Locomotory function of squid mantle. J. Zool. Lond. 167: 437–449. (C)

Wilbur, K.M. 1983–1988. The Mollusca. Vols. 1–12. Academic Press, New York. (GN)

*The supplemental literture listed here is coded to indicate the general topic covered by the paper: GN = geneal Mollusca; A = Aplacophora; B = Bivalvia; C = Cephalopoda; GA = Gastropoda; M = Monoplacophora; P = Polyplacophora; S = Scaphopoda.

PHYLUM ANNELIDA

Members of phylum Annelida are schizocoelomate, segmented protostome worms. The most striking feature of annelids is their body construction, consisting of a series of similar compartments called **segments** or **metameres**. This feature is called **segmentation** or **metamerism** [me-TAM-er-ism; G., *meta*, after + G., *mere*, a part]. Annelida is the first phylum to be covered here that exhibits this characteristic; the repetition of body units in cestodes is **strobilization**, not segmentation. Metamerism is probably the greatest advancement displayed by annelids. In two of the three classes (Polychaeta and Oligochaeta), the external rings generally correspond to an internal division of the body. In the third class (Hirudinea or leeches), external annulation does not correspond to internal segmentation. Most major organ systems are affected by metamerism (e.g., body wall musculature, nervous, excretory, and vascular). Metamerism is important in burrowing activities of annelids. The schizocoelom acts as a hydostatic skeleton when circular and longitudinal muscles of the body wall contract alternately. This is because the volume of each segment remains relatively constant despite variations in their linear proportions. Another phylum, **Pogonophora**, may be related to the polychaetes.

Brinkhurst, R.O. 1982. Evolution in the Annelida. Canadian J. Zool. 60: 1043–1059.

Cutler, E.B. 1975. The phylogeny and systematic position of the Pogonophora. Syst. Zool. 24: 512–513.

Eernisse, D.J., J.S. Albert, and F.E. Anderson. 1992. Annelida and Arthropoda are not sister taxa: a phylogenetic analysis of spiralian metazoan morphology. Syst. Biol. 41: 305–330.

Harrison, F.W., and M.E. Rice (eds.). 1993. Onychophora, Chilopoda, and Lesser Protostomata. Vol. 12: Microscopic Anatomy of Invertebrates. Wiley-Liss, New York, pp. 327–369 [Pogonophora]; pp. 371–460 [Vestimintifera].

Trueman, E.R. 1975. The Locomotion of Soft-Bodied Animals. Edward Arnold, London.

■

Phylum Annelida

Phylum Annelida [an-NEL-ee-da; L., *annellus*, ring] is a diverse group of over 12,000 marine, freshwater, and terrestrial worms having a body that is composed of a longitudinal series of similar cylindrical compartments called **segments** or **metameres**. There is a spacious schizocoelom that is compartmentalized by septa in polychaetes and oligochaetes. In class Hirudinea (leeches) there are many external rings called **annuli** and groups of annuli correspond to internal segmentation, but not in a simple fashion. Also, in the leeches the schizocoelom is filled with connective tissue and has been reduced to a complex system of sinuses. Many members of the class Polychaeta have well-developed, lateral locomotory appendages called **parapodia** that exhibit a high degree of neuromuscular organization. Parapodia are lacking in the other classes. In the polychaetes and oligochaetes, bristle-like **setae** usually are present on many segments; setae are absent in most leeches.

Classification

Three classes of annelids are commonly recognized (Polychaeta, Oligochaeta, and Hirudinea), although in some classification systems the latter two are grouped together (e.g., subphylum **Clitellata**), because both possess a clitellum and other similarities. The polychaetes are then placed in the subphylum **Aclitellata**. At one time, Archiannelida was employed as another class.

However, it is now considered to be a catch-all group and is no longer recognized.

1. **Class Polychaeta**. Mainly marine but with a few freshwater annelids distinguished by either paired, lateral appendages called parapodia bearing numerous setae or by numerous anterior tentacles. Most possess a distinct head with eyes, palps, and cirri. Some have antennae. Examples: *Aphrodita* (sea mouse), *Lepidonotus* (the scaleworm), *Nereis* (= *Neanthes*) (sand or clam worm), *Eunice* (Samoan palolo worm), *Chaetopterus* (parchment tube worm), *Arenicola* (lugworm), *Sabella* (fan worm or feather-duster).

2. **Class Oligochaeta**. Terrestrial and aquatic annelids with segmentation both external and internal. Parapodia are absent, but setae are present. A distinct, bandlike clitellum is formed on several anterior segments in sexually mature individuals. Examples: *Tubifex* (aquatic tubificids) and *Lumbricus terrestris* (the night crawler).

3. **Class Hirudinea**. Predominantly freshwater leeches with some terrestrial and marine forms having a conspicuous, superficial external annulation that is not reflected internally as segmentation in any simple fashion. Parapodia are lacking and setae are usually absent, but a distinct clitellum is present. Examples: *Hirudo medicinalis* (medicinal leech) and *Glossiphonia complanata*.

167

Fossil Forms

Because they have few hard parts, annelids do not fossilize well. Most putative fossils consist of trails and burrows. However, several good specimens of the class Polychaeta are represented in the **Burgess Shale Fauna.** Therefore, this class dates at least to the mid-**Cambrian Period** and probably earlier (Fig. 13.1).

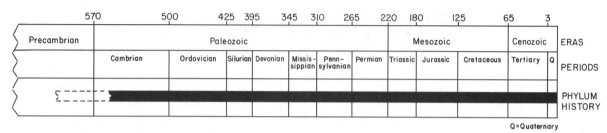

Figure 13.1. Geologic history of the phylum Annelida.

A. Class Polychaeta

The largest class of annelids is Polychaeta [POL-ee-KE-tah; G., *poly*, many + G., *chaeta*, hair or bristle] comprising more than 8,000 species that range in length from a few millimeters to more than 3 m. Two general groups were previously recognized in this diverse class: free-moving forms, the **Errantia** [er-RAN-she-ah; L., *erran*, wander], and sedentary forms, the **Sedentaria** [sed-en-TAR-re-ah; L., *sedent*, sit], although the distinction is not completely clear-cut. Modes of food procurement also show great diversity and include raptorial predators, herbivores, scavengers, and sediment and filter feeders. Few polychaetes are parasitic.

The body comprises a series of similar cylindrical segments, each having a pair of lateral appendages called parapodia, that are important in locomotion and gaseous exchange (Fig. 13.2). The schizocoelom is compartmentalized by intersegmental septa that are perforated to allow the coelomic fluid to pass from one segment to another. Many of the same body functions are performed in each segment. Not all polychaetes are uniformly segmented and the parapodia may be grouped into regions that differ in size, shape, and function. These specialized, functional regions are called **tagmata** [TAG-ma-ta; G., *tagma*, a division]. Tagmatization is often quite pronounced in some forms.

■ Observational Procedure: Polychaeta

Nereis

Members of the genus *Nereis* (= *Neanthes*), commonly known as rag, sand, or clam worms, are found in shallow marine and estuarine waters worldwide.

Live Specimens

1. Obtain a live *Nereis* or similar species. Do you see any color patterns? Gently run your fingers down the side of the worm. What do you feel? These leglike structures are the **parapodia**. Place a *Nereis* on a piece of wet paper that you have previously marked with a grid of lines 1 cm apart; this will help you observe the movement of the animal better. How does the animal move? What role do the parapodia play in locomotion? Is their movement synchronous? Do the parapodia move in a coordinated fashion? If so describe that coordination. Are the parapodia in contact with the surface at all times? How do they grip the surface? How do the parapodia function as a series of lever arms? Determine how fast *Nereis* can move. Stimulate the animal with a gentle prod from the rear. Does the pattern of locomotion change when *Nereis* is crawling fast? Determine how *Nereis* moves around obstacles?

2. Place *Nereis* in a clear glass tube filled with seawater having a diameter only slightly wider than the width of the worm. Put the tube in a dish filled with seawater. How does the worm react to this habitat? Does the worm remain in the tube? What function do the parapodia play under these circumstances?

3. Remove the worm from the tube and place it in a tank of seawater. Can *Nereis* swim? Describe the swimming movements. How do the movements of *Nereis* compare to those of planarian and nematode worms (Exercises 5 and 10)?

External Morphology

The following dissection of *Nereis* may be done with a freshly anesthetized worm or one that has been pre-

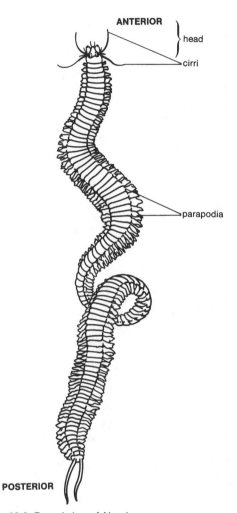

ANTERIOR

head

cirri

parapodia

POSTERIOR

Figure 13.2. Dorsal view of *Nereis.*

served. In either case, the worm should be kept submerged during the dissection. In general razor blades do not work well on a fresh worm; very fine scissors should be used. If an anesthetized worm is used for dissection, be sure that it is killed at the end of the exercise. If a preserved worm is to be used for the following observations, return the live *Nereis* to the holding tank.

1. Obtain a specimen of *Nereis* and examine both surfaces. There is a slight, midventral depression that runs the length of the worm. Observe the distinct segmentation with paired, lateral parapodia, and the appendages on the head region (Fig. 13.3). Is this segmentation uniform? Examine the anterior region and locate the **prostomium**, eyes, **palps**, and antennae, **peristomium** (segments 1 and 2), and peristomial **cirri**. Note the terminal position of the anus.

2. If you are examining a preserved worm, determine whether the pharynx is extended or retracted? If mus-

cles can only contract, how can a worm extend its pharynx? Examine an extended pharynx and locate the jaws and pharyngeal **paragnaths**.

3. Observe the worm under a dissection microscope. Do the segments vary in size or morphology along the length of the worm? The first parapodia-bearing segment (number 3) is called the first **setiger**.

4. Using a scalpel, remove several parapodia from different places along the worm's body. Make sure they are cut close to the body. Place the parapodia in a watch glass, cover with water, and examine them using a dissection microscope. Locate the **notopodium, neuropodium** with extra processess or **ligules**, setae, and dorsal and ventral cirri (Fig. 13.4). Do the various parapodia differ greatly from one another? Is there evidence for tagmatization in *Nereis*? What would you have to see to conclude that tagmatization was present?

5. Compare a prepared whole mount of a parapodium with your specimen (Fig. 13.4). Species may be differentiated on the basis of morphological details of their parapodia. How does your specimen differ from the preparation from the biological supply house?

6. Note that both the notopodium and neuropodium are supported internally by a stiff, chitinous rods called the **acicula**.

7. Remove a few setae from several parapodia, place them in a drop of water on a slide, add a cover slip, and observe under a microscope. Note that the setae are not just uniform rod-shaped structures. How do the setae function in locomotion? Are they birefringent?

Internal Anatomy

1. With a scalpel carefully make a parasagittal, shallow incision from the dorsal side, about one-third of the way from the anterior end.

2. With a small water dropper, siphon some liquid from the coelomic compartments of several segments. Examine this fluid by placing several drops on a glass slide, cover with a cover slip, and view under low power of the compound microscope. Gametes may be present, depending on the season the worm was collected. Abundant corpuscles should be seen. Following your examination of the coelomic fluid, cover the worm with water and proceed with the dissection.

3. Continue the dissection forward until the peristomium (segments 1 and 2) is reached. Pin the body wall to the bottom of the dissection pan as the incision is made.

4. Observe the numerous **septa**. These may have been torn during the parasagittal incision.

5. Count the segments posteriorly from the peristomium to the seventh segment (fifth setiger). Note the enlarged pharyngeal region (segments 3–6, setigers 1–5) and the absence of septa there. Locate the dorsal

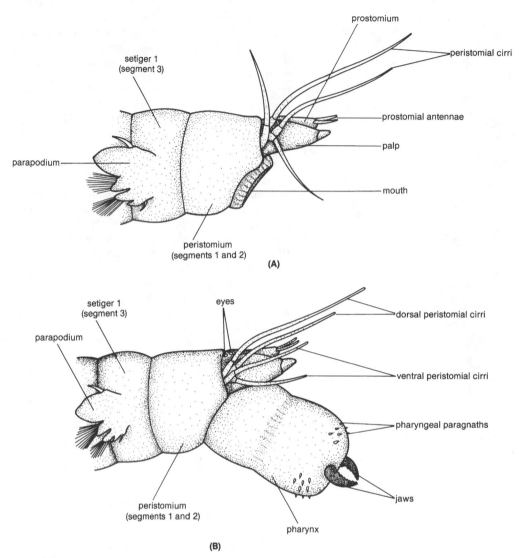

Figure 13.3. Lateral anterior view of the anterior end of *Nereis*. (A) Pharynx retracted. (B) Pharynx extended.

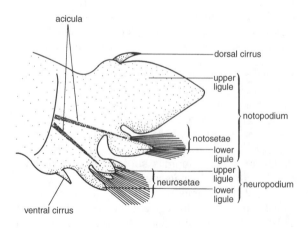

Figure 13.4. Typical *Nereis* parapodia.

blood vessel, that runs longitudinally over the gut (Fig. 13.5). Coming off the vessel are paired lateral vessels that communicate with the parapodia and body wall. The lateral vessels may be difficult to observe in your specimen.

6. Trace the gut toward the posterior end. The esophagus occurs in segments 7–11 (setiger 5–9). In the region of the esophagus, paired **esophageal cecae** are attached laterally to the gut. Posterior to the esophagus the gut is designated as the **stomach-intestine**. In each segment are paired nephridia located ventrolaterally (Fig. 13.5).

7. Sever the gut between the 15th and 20th segments. Carefully remove the gut up to the pharynx and sever again. Make sure that you do not damage the ventral nerve cord.

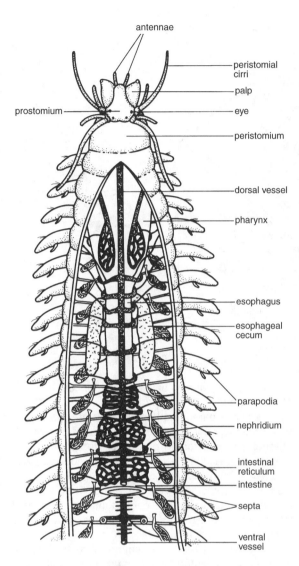

Figure 13.5. Internal dorsal view of *Nereis*.

The image labels (top to bottom, left to right):
antennae, peristomial cirri, palp, eye, peristomium, prostomium, dorsal vessel, pharynx, esophagus, esophageal cecum, parapodia, nephridium, intestinal reticulum, intestine, septa, ventral vessel

ventral longitudinal muscle masses, and ventral nerve cord (Fig. 13.6).

Aphrodita

Observe a specimen of the sea mouse, *Aphrodita*. The dorsal side appears hairy because of fine setae arising from the notopodia. Turn the specimen over and identify the anterior and posterior ends. Locate the terminal anus. Does *Aphrodita* exhibit tagmatization?

Arenicola

This worm, commonly known as the lugworm, lives on both U.S. coasts at about the low-tide mark where it burrows in sandy to muddy sediments.

1. Observe a specimen of *Arenicola*. There is a shallow, narrow, linear depression running along the entire ventral length of *Arenicola*. Locate the terminal anus.

2. Note that the parapodia are modified and reduced in size in comparison with the other worms just studied (Fig. 13.7).

3. *Arenicola* also has distinct **annulation** between segments. Body segments extend from the first annulation behind a parapodium to the first annulation behind the next parapodium.

4. At the anterior end, a fleshy pharynx and buccal mass, covered with small papillae, may be extended from the mouth (Fig. 13.7). Just behind the pharynx is the minute prostomium that lacks appendages. Behind the prostomium, the peristomium may be seen; it is the first annulus.

5. Note that the body appears to be divided into three distinct tagmata: (a) anterior, pre-branchial; (b) branchial; and (c) posterior, post-branchial (Fig. 13.7). The first tagmata past the peristomium may be identified by the presence of reduced notopodia. They increase slightly in size towards the posterior end. The neuropodia are very difficult to see in this region. Segments in the next tagmata have biramous parapodia and possess gills, branched outgrowths of the body wall. In this region the neuropodia appear as slightly raised welts with slits. Appendages are lacking beyond the last gill-bearing segment.

Chaetopterus

Chaetopterus, the parchment tube worm, lives in U-shaped tubes that are opened at both ends. These marine worms may be found in marine mudflats, but their tubes are only exposed during the lowest tides.

1. Obtain a specimen of *Chaetopterus* which has been removed from its tube. Locate the spatulate anterior end. There is a small, dorsal, ciliated groove at this end. The outer lip of the mouth is the **peristomal collar**.

2. Beginning anteriorly, observe that the segments and parapodia are not of a uniform morphology. Rather, there is a differentiation of the animal into three

8. Once the pharyngeal region is reached, continue to remove the gut and note the position of the nerve cord. Just above the nerve cord the ventral blood vessel may be found. Locate the **circumpharyngeal connectives**; these encircle the pharynx and communicate with the dorsal ganglia or brain.

9. Find the nerves which extend anterior from the brain to the prostomium.

10. Observe the nerve cord along its length. How are the ganglia arranged in each segment?

11. Obtain a prepared slide of a cross section of *Nereis* and locate the gut, dorsal and ventral gut suspensors, dorsal and ventral blood vessels, dorsal and

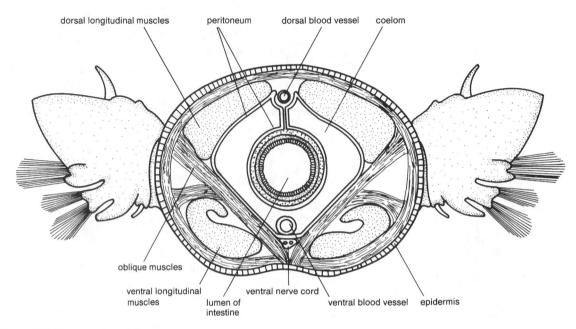

Figure 13.6. Cross section of *Nereis*.

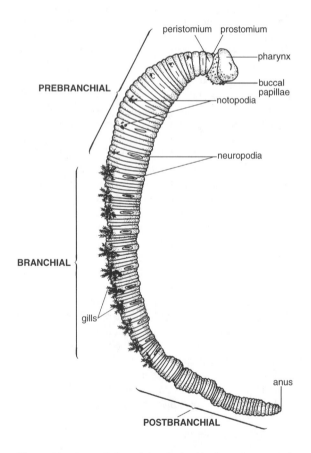

Figure 13.7. Lateral view of *Arenicola* with pharynx extended.

tagma: (a) anterior with uniramous parapodia, (b) middle with greatly modified parapodia, and (c) posterior with biramous parapodia (Fig. 13.8). Water containing food and oxygen is drawn into the tube at the worm's anterior end and forced out the posterior vent. This movement is accomplished by special, modified parapodia called **fans** (segments 14, 15, and 16). Notopodia of segment 12 collect food. They are long and winglike (**aliform**) and contain mucus glands that produce a film of mucus that stretches between the notopodia forming a baglike structure.

3. At segment 13 is a cup-shaped structure where food is concentrated and formed into small balls. These balls are passed anteriorly along the small, dorsal ciliated groove that extends to the mouth, a shallow, funnel-like depression.

4. Note that parapodia posterior to the fans are well developed and similar in shape, although they decrease in size at the posterior end. The anus is terminal.

Other Polychaetes

Diversity in the Polychaeta is very great. One should not have the impression that the few specimens reviewed here give a complete view of the diversity of this class. Other polychaetes may be available for your study, and you should examine additional specimens as your instructor directs. For example, free-moving forms such as *Glycera* (beak-thrower or blood-worm) and *Lepidonotus* (scale worm) and sedentary species such as *Pectinaria* (= *Cistenides*) (trumpet worm),

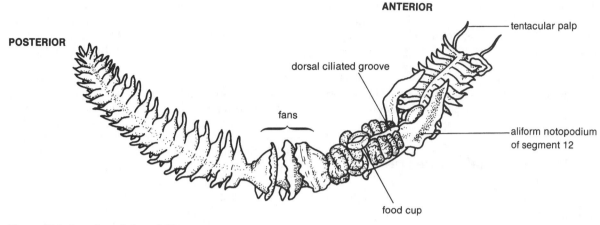

ANTERIOR

POSTERIOR

tentacular palp

dorsal ciliated groove

fans

aliform notopodium
of segment 12

food cup

Figure 13.8. Dorsolateral view of *Chaetopterus*.

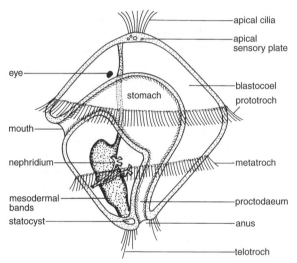

apical cilia

apical
sensory plate

eye

stomach

blastocoel

prototroch

mouth

nephridium

metatroch

mesodermal
bands

statocyst

proctodaeum

anus

telotroch

Figure 13.9. Typical polychaete trophophore larva.

Diopatra (plume worm), *Sabella* (fanworm), *Sabellastarte* (feather-dusters), and *Spirorbis* (calcareous tube worms) may be available for observation.

Larvae

Sexes are separate in most polychaetes, though distinct gonads are not usually present. Seasonal outgrowths of the coelomic lining in the ventral part of many segments form the gametes that are released into the body cavity where they mature, and then to the environment, either when the body wall bursts, or through the excretory system, or through gonoducts which develop when the worm is sexually mature. In many species the embryo develops into a **trochophore larva** (Fig. 13.9), whereas in other species development is direct. After metamorphosis the young worm settles on the substratum and develops into an adult. If available, examine a prepared slide of a trochophore larva. ∎

B. Class Oligochaeta

The Oligochaeta [o-LEE-go-KEY-tah; G., *oligo*, few + G., *chaeta*, long hair or bristle] is the second largest class of annelids, containing more than 3,000 species. Although oligochaetes are commonly called earthworms, the term is technically incorrect because many species are aquatic. Most inhabit freshwaters, but there are many microscopic marine species too. Most feed on dead and decaying materials, but earthworms ingest soil feeding on the microfauna and microflora. The night-crawler represents a group that browses on grass and leaves on the surface of the soil. Oligochaetes range in length from less than 0.5 mm to more than 7 m.

The oligochaete body consists of a series of uniform, cylindrical segments. As in polychaetes, many of the same bodily functions are performed in each segment. However, the extensive tagmatization of segments present in some polychaetes, is not found in the oligochaetes. Also parapodia are not present in the oligochaetes. However, setae often occur on the lateral surface of each segment in many members of the class. Some possess numerous setae, perhaps up to 200 around the equator of each segment. One prominent region of the epidermis in sexually mature individuals is the **clitellum**, a bandlike, glandular structure that may be one to many cells thick. The size of the clitellum varies depending on the species and the time of the year. The clitellum produces mucus during copulation and later secretes a **cocoon** or capsule where embryonic development takes place.

In this exercise *Lumbricus terrestris* will be studied in detail as the class representative. Although it is not a typical oligochaete, it is a convenient organism with which to work.

■ Observational Procedure: Oligochaeta

Live Specimens

If possible begin your studies with a live specimen. Do not be put off by the slimy feel of the organism, and do not waste the opportunity of learning new things about *Lumbricus* by wearing latex gloves.

1. Obtain a live *Lumbricus*. Note any differences in the color pattern on the dorsal and ventral surfaces. Gently hold opposite ends of the worm in each hand noting the strength the worm exhibits upon contraction. Are parapodia present on the sides of the worm? Gently run your fingers down the length of the worm while holding on to one end. What do you feel? Are the setae as numerous as in *Nereis*? Run your fingers down the worm again, first in one direction then in the other. Is there a difference in the feel of the setae? What accounts for the difference and what function might this feature serve in locomotion?

2. Place the worm on a piece of wet paper that you have previously marked with a grid of lines 1 cm apart; this will help you observe the movement of the animal better. How does *Lumbricus* move? Does locomotion in *Lumbricus* resemble in any way that of *Nereis*? Describe how these movements are like peristalic waves.

3. Move the worm to a sheet of wet glass. Does movement still occur? How is this organism better suited to crawl through something (e.g., soil or mulch) than over a flat surface?

4. Drop *Lumbricus* into a tank of freshwater. Can it swim? Compare the locomotory abilities of the earthworm to *Nereis*. How are they similar? How are they different? How does movement in *Lumbricus* compare to other wormlike organisms that you have studied?

External Morphology

The following dissection of *Lumbricus* described may be done with a freshly anesthetized worm or one that has been preserved. Some anatomical details are best seen with a live worm, including other stages of the parasite, *Monocystis*. An excellent view of the functioning of the dorsal vessel and hearts also is possible. In both cases, the animal should be kept sub-merged during the dissection. In general razor blades do not work well on a fresh worm; very fine scissors should be used. If an anesthetized worm is used for dissection, be sure that it is killed at the end of the exercise.

1. Obtain a preserved specimen of *Lumbricus* or one that has been recently anesthetized. Note the lack of any tagmata in this species. Identify the worm's anterior and posterior ends. At the anterior end is the mouth, located on the ventral portion of segment 1. In this region are two parts: a mid-dorsal projection, the prostomium, and the peristomium (segment 1) that surrounds the mouth (Fig. 13.10).

2. At the posterior end locate the **pygidium** [p-JID-e-um; G., *pyg*, rump] bearing the anus.

3. Gently pull the animal between your fingers to feel the setae. Eight setae in four pairs are found on every

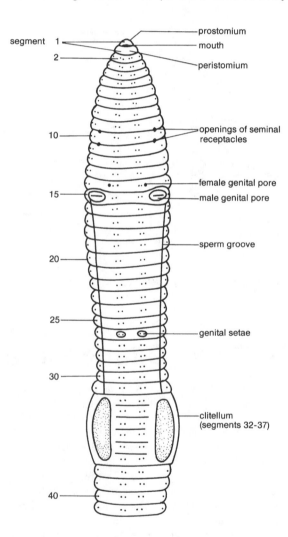

Figure 13.10. Ventral anterior view of *Lumbricus terrestris*.

segment except the first and last. Are the setae randomly scattered around the animal or is there a pattern to their placement?

4. *Slide*: Select a cross section of *Lumbricus* which shows setae. Note the position of the setae and the muscles attached to them (Fig. 13.11). Setae are highly birefringent and may be easily identified in the section by examining it for birefringent structures.

5. Arrange the worm so that the dorsal surface is uppermost. Starting at segment one, count posterior to segment number 31. Beyond this may be found a swelling on the body surface. This is the clitellum, which covers segments 32–37 and forms the cocoon from secretion by large gland cells.

6. *Slide*: Examine a cross section of *Lumbricus* at the level of the clitellum. Note that the clitellum is a thickened region of glandular cells. Is the clitellum of uniform thickness around the worm? Which region is the thinnest? Under high magnification, observe the large, globular secretory cells that make up the bulk of the clitellum. The darkly staining cells closest to the surface secrete mucus and cocoon material; those cells

with less stain secrete albumin for nourishing the developing embryos while in the cocoon.

7. Turn the worm ventral surface uppermost and count the segments from the prostomium until you reach the 15th segment. In this segment find the paired male **genital pores**. The male pores are tiny, but around them there is a distinct swelling resembling a pair of lips.

8. On the 14th segment locate a pair of extremely small openings in the same position as the male pores. These are the external pores of the **oviducts** (Fig. 13.10). The paired openings of the **seminal receptacles** are located between segments 9 and 10, and 10 and 11, but they are very small and difficult to find.

9. On the ventral side locate the two **sperm grooves** and trace them from each male gonopore to the clitellum. Review the process of copulation in the earthworm in your textbook.

10. Two other types of pores that may be seen on the surface of *Lumbricus* are the **nephridiopores** and **coelomic pores**. Nephridiopores are paired and located by the segmental furrows just anterior and slightly above the ventrolateral setae.

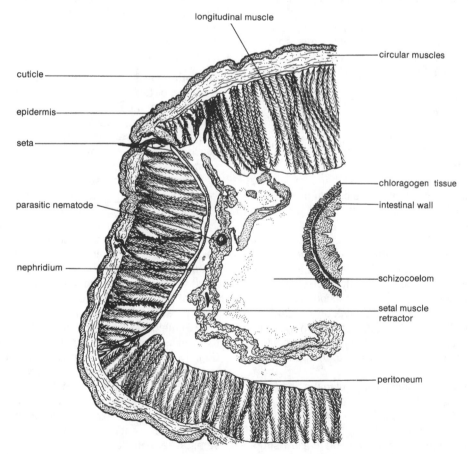

Figure 13.11. Cross section of *Lumbricus terrestris* showing the setae and other internal structures.

11. Coelomic pores are single and release fluid which helps keep the worm's surface moist. They are located middorsally in the intersegmental furrows in the middle and posterior body.

Body Wall

The outermost layer of the body wall is the noncellular **cuticle**. This is secreted by special cells in the epidermis. Microscopic pores present in the cuticle allow epidermal secretions to pass to the outside of the body.

1. Scrape some cuticle from the epidermis. Place the sample on a glass slide, add a drop of water, and apply a cover glass. Observe under low power of the compound microscope. What do you observe? The smooth texture of the cuticle plus mucoid secretions give the characteristic slippery feel to the worm's surface. Immediately below the cuticle is the epidermis, composed of several kinds of columnar epithelial cells.

2. *Slide*: Obtain a median longitudinal section of *Lumbricus* and examine the body wall using high magnification. Note the different cell types found in the epidermis. Inward from the epidermis are two muscle-tissue layers. The outermost of these is the circular muscle layer. Scattered among the fibrils are brownish, pigment granules.

3. Note that the circular muscles in any segment are not continuous with the muscles in other segments.

4. The innermost muscle layer consists of longitudinal muscle cells that are continuous from segment to segment.

5. Just below the longitudinal muscles is the peritoneum. This lines the schizocoelom and is continuous with the septa.

6. Between each pair of segments, locate a septum.

Internal Anatomy

1. Extend the worm full length with the ventral surface down. Fasten the specimen to the bottom of a dissection pan by placing pins through the prostomium and the pygidium.

2. Using a razor blade or very fine scissors, make a *shallow*, midsagittal incision through the anterior third of the body wall. Be very CAREFUL not to cut too deeply thereby disturbing the underlying viscera. *Work slowly and patiently; a rushed dissection can render the specimen useless for detailed anatomical study.*

3. About halfway along the cut, start to turn the body wall over laterally and pin it to the pan. Work posteriorly, continuing to pin the body wall to the pan as you cut.

4. Note the septa that connect the gut to the body wall. The coelom is found between each pair of septa.

5. Add water to the pan so that the worm is covered completely. All observation should be made with the worm submerged to prevent drying and to allow structures to float freely for better observation.

6. Several organ systems have been exposed by your dorsal incision. Study your worm with a dissection microscope. Some structures, such as the seminal funnels and the vas deferens, may be difficult to locate. However, with care you should be successful at locating them. These structures also should be observed in prepared microscopic sections.

Circulatory System (Fig. 13.12)

1. The dorsal blood vessel is located on the upper surface of the gut. In preserved worms this vessel usually appears as a light brown line. Close examination will reveal the small vessels joining the main vessel at right angles in each segment. The dorsal vessel is a collecting tube and pulsates, moving the blood from posterior to anterior.

2. In segments 7 through 11 locate five pairs of hearts that extend dorsoventrally around the esophagus, connecting the dorsal and ventral vessels (Fig. 13.12). These hearts pulsate, contracting in coordinated rhythm with the dorsal vessel. The ventral blood vessel will be observed when the gut is removed later in the dissection.

3. *Slide*: Obtain a cross section of *Lumbricus* at the level of a heart. Note the position and size of the heart(s) and the thin muscular wall. In most sections, connections to the dorsal and ventral vessels will not be evident. Why?

Reproductive System (Fig. 13.12)

1. Posterior to the second pair of hearts in segment 8 locate the three pairs of large, light-colored structures that almost fill the whole area of the worm in this region. These are the seminal vesicles. Their bases are attached ventrally, but they expand dorsally to occupy space in segments 9–12.

2. *Slide*: Obtain a cross section of *Lumbricus* at the level of the seminal vesicles and observe the large size and general construction of these structures. Note that they are not just large sacs. What is their function?

3. Using a scaplel, remove a small portion of one seminal vesicle and macerate it on a microscope slide in a small amount of water. Add a cover slip and observe this preparation using a compound microscope. Scan the seminal vesicles for gametocysts of the gregarine protozoan parasite *Monocystis* (Phylum Protozoa, Exercise 1B). Gametocysts resemble a small sphere filled with many minute spindle shaped structures, the sporocysts (Fig. 13.13). If you macerated your preparation very well, most of the sporocysts will probably be free of their gametocyst. You also should examine the prepared cross section of *Lumbricus* for *Monocystis*.

4. Under the lobes of the two anterior pairs of seminal vesicles (segments 9, 10) are the paired, spherical seminal receptacles (Fig. 13.12) where sperm are stored.

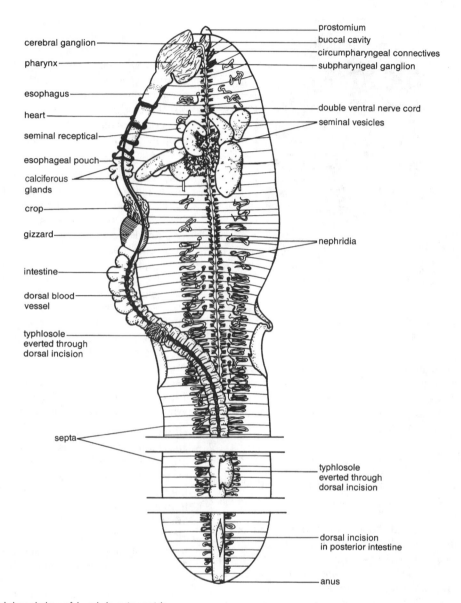

cerebral ganglion
pharynx
esophagus
heart
seminal receptical
esophageal pouch
calciferous glands
crop
gizzard
intestine
dorsal blood vessel
typhlosole everted through dorsal incision
septa

prostomium
buccal cavity
circumpharyngeal connectives
subpharyngeal ganglion
double ventral nerve cord
seminal vesicles
nephridia
typhlosole everted through dorsal incision
dorsal incision in posterior intestine
anus

Figure 13.12. Internal dorsal view of *Lumbricus terrestris*.

5. *Slide*: Obtain a cross section of the **seminal receptacles (spermatheca)**. The vesicles are located ventrolaterally and may contain many spermatozoa. Testes are found near the nerve cord in the anterior of segments 10 and 11, and the ovaries similarly in segment 13.

6. Near the bases of the middle seminal vesicle and located midventrally are the first pair of testis sacs (segment 10). The second pair is located in segment 11. The **sperm funnels** are buried under a membrane, the testis sac, and may be difficult to locate.

7. If this is the case, turn your attention to segment 15 and locate the vasa deferentia which run anteriorly

along the body wall. Follow that tube anteriorly as it leads to a funnel.

8. *Slide*: View a cross section at the level of the sperm funnels. Sperm funnels may be located easily as the sperm stain a dark color. In longitudinal section, sperm funnels appear to be highly convoluted.

9. Ovaries are located in the anterior part of segment 13. Eggs set free in the coelomic cavity are collected by the **ovarian funnels**. Oviducts open to the outside of the body in segment 14. The oviduct will not be easily seen, but the area of the ovarian funnel may be identified by the button-like opening on the septum directly posterior to the ovary.

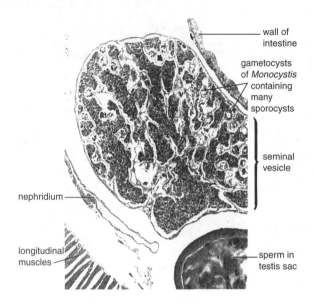

wall of intestine

gametocysts of *Monocystis* containing many sporocysts

seminal vesicle

nephridium

longitudinal muscles

sperm in testis sac

Figure 13.13. Cross section of *Lumbricus terrestris* at the level of the seminal vesicles.

10. *Slide*: Examine cross sections of *Lumbricus* at the levels of the testes and ovaries and locate the gonadal tissues. Note their relatively small size in comparison to other reproductive organs (i.e., seminal vesicles).

Digestive System (Fig. 13.12)

Buccal Cavity and Pharynx. The digestive system is a straight tube with a few specialized regions.

1. Locate the peristomium of your specimen. At this level a buccal cavity opens into the pharynx (segments 4 and 5).

2. Note the size and shape of the pharynx, and locate the numerous muscle fibers that connect the pharynx to the body wall. Take care not to damage the brain that is located dorsally at the anterior junction of the buccal cavity and pharynx (Fig. 13.12).

3. Using your finger, gently press on the pharynx and note the general texture of this organ.

4. *Slide*: Examine a cross section of the pharynx and note the distribution of the powerful pharyngeal muscle. What is the function of this powerful muscle?

5. Observe a medial section through the anterior of the worm and locate the pharynx.

6. In this section note the **pharyngeal dorsal diverticulum**. What function might this structure perform?

7. Note the muscle fibers that connect the pharynx dorsally to the body wall.

Esophagus. The esophagus is a long tube (segments 6–13) that carries food from the pharynx to the crop (segments 14 and 15).

8. Follow the pharynx posteriorly where it opens into the esophagus (Fig. 13.12).

9. On either side of the esophagus are two pairs of bulbous evaginations (segments 11 and 12) called **calciferous glands**. One of their functions is regulation of calcium concentration in bodily fluids, and consequently ionic and pH balance.

10. *Slide*: Locate these glands in a cross section of the esophagus. Note the cavities between elongate cells of the gland.

Crop and Gizzard. From the esophagus food passes into a thin-walled storage organ called the **crop** (segments 14 and 15) and then into a muscular, cuticle-lined **gizzard** (segments 16–19) which is a food-grinding organ (Fig. 13.12).

11. Locate these organs in your specimen. Using your finger, gently press on the crop and gizzard and compare the general texture of these organs. What does this indicate about these two organs?

12. *Slide*: With an anterior medial section of *Lumbricus*, locate the crop and gizzard. How can you tell these organs apart in this section?

13. Also observe these two organs in cross section. Are muscles present in the walls of the crop or gizzard? How are the muscles distributed in the crop?

14. Observe the inner lining of the crop and gizzard. Why would the gizzard have a thick cuticular lining?

Stomach-Intestine. After being processed in the gizzard food passes into the stomach-intestine.

15. Continue the midsagittal incision of the body wall for the full length of the worm, pinning the body wall to the dissection pan as you proceed. Trace the intestine from the gizzard to the anus (Fig. 13.12).

16. Observe the external morphology of the intestine along its length. Note that externally there is only minor regionalization for the gut past the gizzard.

17. Remove 1-cm sections from anterior, mid, and rectal portions of the intestine. Cut these sections open lengthwise, place them in small glass dishes, and observe the lining with the aid of a dissection microscope.

18. In the anterior section, note the presence of the **typhlosole**. Is the typhlosole found in the other sections of the intestine? What does it look like in each section? What is the function of the typhlosole?

19. *Slide*: Examine a cross section of the intestine. The intestine is just below the dorsal blood vessel. Both are surrounded by a special tissue called **chloragogen**. Suspended from the dorsal side of the intestine is the typhlosole (Fig. 13.14). From what region of the intestine was your section made? How can you tell?

20. Note that chloragogen tissue also is found inside the typhlosole. Locate the thin layer of cells between the intestinal epithelium (gastrodermis) and the

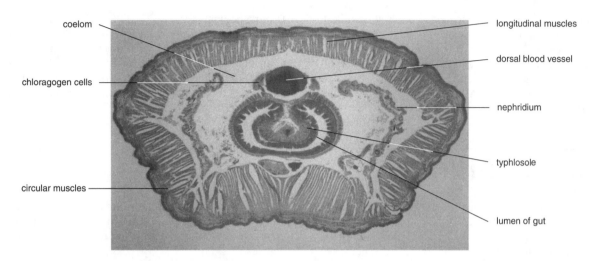

Figure 13.14. Cross section of *Lumbricus terrestris* at the level of the intestine.

chloragogen tissue. This layer is sometimes called the **submucosa**, but it is actually a layer of circular muscle.

Excretory System (Fig. 13.12)

A pair of **metanephridia**, one on each side, is located in the lateral area of each body segment, with the exception of the first two segments and pygidium.

1. In the region of the gizzard and posterior to it, examine the inner body wall surface of each segment and locate the metanephridia.

2. Using a dissection microscope examine a metanephridium in situ. The nephrostome and nephridiopore will be difficult to locate.

3. *Slide*: Compare this structure to a whole mount slide of a metanephridium. Locate the convoluted tubules, bladder, and the vessels of the circulatory system.

4. Locate the nephridiopore in a prepared cross section. What other structures of the metanephridium do you observe? *Lumbricus* is a host to a parasitic nematode, *Rhabditis maupasi*, which is believed to enter the worm via nephridiopore or genital openings (Phylum Nematoda, Exercise 10).

Nervous System (Fig. 13.12).

1. Dorsal to the buccal cavity, at a point where the buccal cavity and the pharynx meet, locate the brain or **suprapharyngeal ganglia**. Projecting anteriorly into the prostomial region are the prostomial nerve fibers. Extending laterally from each cerebral ganglion are the **circumpharyngeal connectives** that lead to the **subpharyngeal ganglia**.

2. Using scissors, carefully cut through the pharynx and esophagus and remove that section to expose the circumpharyngeal connectives and the subpharyngeal ganglia.

3. Observe the size difference between the supra- and subpharyngeal ganglia. Note the ring form made by the circumpharyngeal connectives.

4. *Slide*: Locate the suprapharyngeal and subpharyngeal ganglia in cross sections made at the appropriate levels.

5. Trace in your specimen the ventral nerve cord as it extends posteriorly from the subpharyngeal ganglia. Just above the nerve cord is the ventral blood vessel.

6. In each body segment locate a ganglionic enlargement of the nerve cord.

7. Find the three pairs of lateral nerves which arise from the nerve cord in each segment. Lateral nerves are more or less independent in all their sensory and motor functions, allowing localized action to one segment or several widely separated or even contiguous segments.

8. *Slide*: Examine several cross sections of *Lumbricus* below the level of the subpharyngeal ganglion and locate the ventral nerve cord.

9. Just above the nerve cord locate the ventral blood vessel and within the nerve cord find the **midventral subneural blood vessel** and two **lateral neural blood vessels**.

10. In the dorsal area of the nerve cord note a centrally placed tubelike body with two smaller ones subequal in size to either side. These are **giant fibers**. Giant fibers are large neurons that transmit impulses unimpeded the full length of the worm, thereby permitting a sudden withdrawal of the worm down its burrow.

11. *Slide*: Examine the structure of the supra- and subpharyngeal ganglia to the ventral nerve cord using cross sections. Below the giant fibers are two circular structures that are composed of motor and sensory neurons. In some cross sections you may be able to locate lateral nerves that leave the ventral nerve cord in each segment.

12. *Slide*: In medial sections, the ventral nerve cord may be traced posteriorly for some distance. Occasionally segmental ganglia and lateral nerves may be seen.

Other Species

Lumbricus terrestris is not a typical oligochaete, but it is a fairly easy one with which to work. If time permits, your instructor may provide specimens of other oligochaetes, and observational instructions for their study (e.g., *Aeolosoma, Enchytraeus, Dero, Stylaria, Tubifex*). ∎

C. Class Hirudinea

Members of the class Hirudinea (or Hirudinoidea) [hi-ru-DIN-e-a; L., *hirudin*, leech], commonly known as leeches, are a small group (ca. 600 species) of highly specialized annelids. They are mainly freshwater, but terrestrial and marine forms are known. Most leeches range in length from less than 1 mm to over 5 cm. The very largest may reach more than 0.25 m. Leeches are popularly identified as blood-sucking parasites. The large leeches, *Hirudo medicinalis* (Europe) and *Macrobdella decora* (native to the United States), were used in bloodletting during the eighteenth and nineteenth centuries. Under some rather special circumstances they still are used today. Heavy infestations of leeches are not uncommon and may cause anemia or even substantial mortality to fish in hatcheries and in natural habitats. Most leeches are temporary parasites, abandoning their host after becoming engorged with blood. However, as a class, leeches are not strictly ectoparasitic; many are predaceous or scavengers.

Leeches have fewer segments than do other annelids. What appear to be segments in leeches are called **annuli** and are actually surface subdivisions of individual segments. This is similar to what was observed in the polychaete *Arenicola*. Unlike other annelids, the number of segments in leeches is fixed at 34 (some accounts, which do not count the prostomium, report 33). The body is rounded to flattened and may be divided into five regions. Traditionally the segments within these regions are given Roman numerals: (1) head (segments I to VI), (2) preclitellum (segments VII to IX), (3) clitellum (segments X to XII), (4) trunk (segments XIII to XXIV), and (5) a terminal region (segments XXV to XXXIV) with an anus (segments XXV to XXVII) and posterior sucker (segments XXVIII to XXXIV).

∎ Observational Procedure: Hirudinea

Live Specimens

Leeches may be capable of sucking your blood, but if you handle them with a quick hand they will not have an opportunity to make a meal of you.

1. What types of color patterns are present on the dorsal and ventral slides of the animal? How do they compare to those of *Nereis* and *Lumbricus*? Are they cryptic?

2. Gently run your fingers down the side of the worm. Are setae present? Gently pick up the leech and permit it to attach to you hand or, if you are timid, to a glass slide. How strong are the suckers in making the attachment? Wait until the leech is attached with both suckers and gently prod its anterior end so that it releases its hold from that end. How does the leech react? Describe how it re-establishes its hold. Now do the same thing, but to the posterior sucker. How does the leech react? Are these reactions different? Describe how it re-establishes its hold in each case. Place the leech onto a sheet of glass and observe its locomotory abilities. How does the leech move when out of water?

3. Drop the leech into a tank of freshwater. How does it swim? Does it swim in the same manner as *Nereis*? Is the entire body involved in this movement? Notice in particular the plane of undulations of the body. Compare the locomotory abilities of the leech with the other annelids you have observed. How are they similar? How are they different?

External Morphology

1. Obtain one or two preserved leeches such as *Hirudo medicinalis* and *Haemopis grandis*. Observe the elongated body with the numerous annuli. Note any patterns in body pigmentation.

2. Place the leech on a dissection tray and identify the dorsal and ventral surfaces (Fig. 13.15). The dorsal surface is usually darker. Why might this be advantagous to the animal in its natural habitat?

3. Locate the anterior and posterior **suckers**. The anterior sucker is the smallest and surrounds the mouth. Note the general shape of the posterior sucker.

4. Several minute eyespots are located on the dorsal surface at the anterior end just above the mouth. Observe the eyespots with the dissection microscope or hand lens.

5. On the mid-dorsal surface of the worm, just before the posterior sucker, is the anus. It is often helpful to

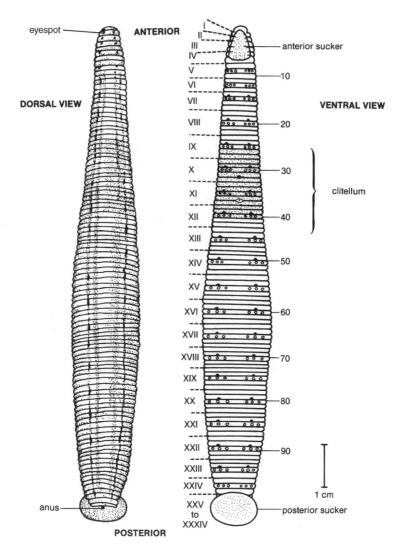

Figure 13.15. Dorsal and ventral views of *Hirudo medicinalis*.

dry the surface of the animal with a paper towel to see the anus. Why is the anus located here and not at the very posterior part of the animal, i.e., within the posterior sucker?

6. Turn the worm over on its dorsal side. About one-fifth to one-fourth of the way from the anterior end in the ventral midline are two small **gonopores**; the male gonopore is anterior to the female gonopore. In some specimens, the penis may be everted from the male pore. It is easier to locate these pores if the surface is dry.

7. Although sensory papillae are difficult to see on preserved worms, attempt to locate them. They occur on the dorsal and ventral sides of the body. The annulus on which sensory papillae are found marks the middle of a segment that extends through several annuli (Fig. 13.15).

8. On the second annulus of each segment, almost in the depression between the annuli, are the paired nephridiopore openings (Fig. 13.15). There are five annuli per segment in the midsection of *H. medicinalis*.

Internal Anatomy

Glossiphonia complanata

(Fig. 13.16). Stained prepared microslide mounts of *G. complanata* will now be observed. Under both low and high powers of the dissection microscope locate the following anatomical features: annuli, eyespots, proboscis and proboscis cavity, salivary gland, crop, gastric ceca, intestine, intestinal ceca, rectum, anus, posterior sucker, ovary, and testes. Note the paired eyespots and the highly muscular proboscis. Although the various anatomical parts shown in the illustration

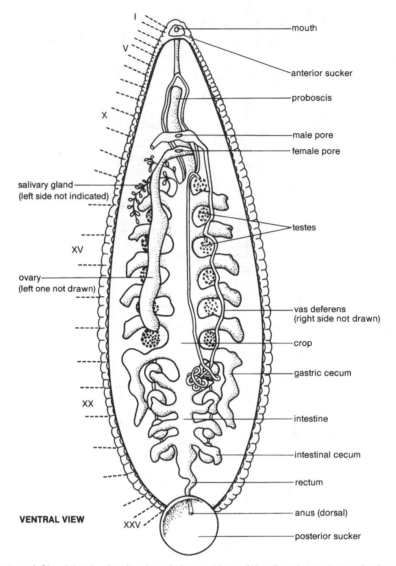

Figure 13.16. Ventral view of *Glossiphonia*, showing the relative positions of the digestive and reproductive systems.

are not exactly the same for all leech species, the basic structure is similar.

Hirudo medicinalis

The dissection of a leech is a difficult task, but care and patience will provide you with a greater appreciation for these complex organisms.

1. Place a specimen of *H. medicinalis* in a dissection pan on its dorsal surface. Firmly pin the leech to the pan by inserting a dissection pin through the tip of the anterior sucker and another pin through the tip of the posterior sucker. Cover the specimen with water.

2. *Carefully* cut away the lower lip to expose the three jaws, arranged as a triangle; two jaws are lateral and one is dorsal (Fig. 13.17). Observe the position of the

jaws and the numerous dilator muscles connected to the pharynx. Carefully remove one jaw with a fine-pointed scalpel. Place the jaw in a drop of water in a depression slide and examine under a compound microscope. Note the numerous minute serrations along the cutting edge of the jaw.

3. Set aside your specimen and obtain a cross section of a leech at the level of the pharynx and observe the very muscular, triangular lumen of the pharynx. What is the functional significance of a triangularly shaped pharynx?

4. Return to your specimen and turn it over on its ventral side. Pin the specimen firmly to the dissection pan. Make a shallow medial incision from about the level of the anterior sucker posteriorly for about 25 annuli. At the posterior and anterior limits of this incision, make

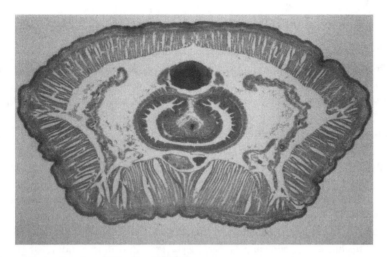

Figure 13.17. Interior view of the oral region of *Hirudo medicinalis*.

two lateral cuts. Using a sharp scalpel, cut the body wall from the fibrous internal mass of tissue. Pin the flap to the dissection pan. Note that complete, internal septa and a spacious body cavity observed in the other annelids are absent. Throughout the coelom is a loose mesenchymal mass called **botryoidal tissue**. This is a highly vascularized mass of tissue whose main com-

ponent is pigmented cells. Spaces, called **coelomic sinuses**, between the loose-connected mass of fibers are the remains of the coelom.

5. Obtain a cross section of a leech at the midbody level and note the position of the botryoidal tissue. Also note the relative positions of the gut and sinuses (Fig. 13.18). ■

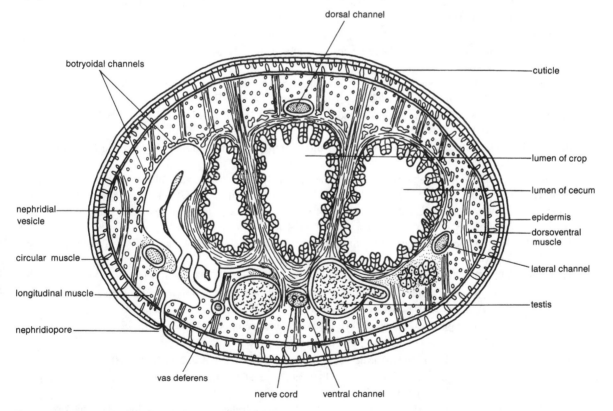

Figure 13.18. Cross section of a leech at the midbody level.

Supplemental Readings*

Aston, R.J. 1973. Tubificids and water quality; a review. Environ. Pollut. 5: 1–10. (O)

Barnes, R.D. 1965. Tube-building and feeding in chaetopterid polychaetes. Biol. Bull. 129: 217–233. (P)

Bonomi, G., and C. Erse'us (eds.). 1984. Aquatic Oligochaeta. Developments in Hydrobiology 24. Dr. W. Junk Publishers, Boston, MA.

Brinkhurst, R.O. 1982. Oligochaeta. In: S.P. Parker (ed.), Synopsis and Classification of Living Organisms, Vol. 2. McGraw-Hill, New York, pp. 50–61. (O)

Brinkhurst, R.O. 1982. Evolution of the Annelida. Can. J. Zool. 60: 1043–1059. (G)

Brinkhurst, R. 1984. Comments on the evolution of the Annelida. Hydrobiologia 109: 189–191. (O)

Brinkhurst, R.O., and S.R. Gelder. 1991. Annelida: Oligochaeta and Branchiobdellida. In: J.H. Throp and A.P. Covich (eds.). Ecology and Classification of North American Freshwater Invertebrates. Academic Press, New York, pp. 401–435. (O)

Chapman, G. 1958. The hydrostatic skeleton in the invertebrates. Biol. Rev. 33: 338–371. (G)

Clark, R.B. 1962. Structure and function of polychaete septa. Proc. Zool. Soc. Lond. 138: 543–578. (P)

Dales, R.P. 1963. Annelids. Hutchinson University Library, London. (9G)

Darwin, C. 1881. The Formation of Vegetable Mould, Through the Action of Worms, with Observations on Their Habits. John Murray, London. (G)

Davies, R.W. 1991. Annelida: Leeches, Polychaetes, and Acanthobdellids. In: J.H. Throp and A.P. Covich (eds.). Ecology and Classification of North American Freshwater Invertebrates. Academic Press, New York, pp. 437–479. (P, H)

Edwards, C.A., and J.R. Lofty. 1977. Biology of Earthworms, 2nd ed. Chapman and Hall, London.

Fauchald, K. 1975. Polychaete phylogeny: a problem in protostome evolution. Syst. Zool. 23: 493–506. (P)

Foster, N. 1972. Freshwater polychaetes (Annelida) of North America. Biota of Freshwater Ecosystems Identification Manual 4: 1–15. (P)

Gerard, B.M. 1967. Factors affecting earthworms in pastures. J. Anim. Ecol. 36: 235–252. (O)

Glaessner, M.F., and M. Wade 1966. The late Precambrian fossils from Ediacara, South Australia. Palaeontology 9: 599–628. (G)

Greene, K.L. 1974. Experiments and observations on the feeding behavior of the freshwater leech Erpodbella octoculata (L.) (Hirudinea: Erpobdellidae). Arch. Hydrobiol. 74: 87–99. (H)

Harrison, F.W., and S.L. Gardiner. 1992. Microscopic Anatomy of Invertebrates, Vol. 7. Annelida. Wiley-Liss, New York. (G)

Hendrix, P.F. (ed.). 1995. Earthworm Ecology and Biogeography in North America. Lewis Publishers, Boca Raton, FL. (O)

Klemm, D.J. (ed.) 1985. A Guide to the Freshwater Annelida (Polychaeta, Naidid and Tubificid Oligochaeta, and Hirudinea) of North America. Kendall/Hunt Publishing Co., Dubuque, IO. (G)

Knight-Jones, E.W., J.E. Bailey, and M.J. Isaac. 1971. Choice of algae by larvae of Spirorbis, particularly of Spirorbis spirorbis. 4th Eur. Mar. Bio. Sym. 1: 89–104. (P)

Mann, K.H. 1962. Leeches (Hirudinea), Their Structure, Physiology, Ecology, and Embryology. Pergamon Press, New York. (H)

MacGinitie, G.E. 1939. The method of feeding of Chaetopterus. Biol. Bull. 77:115–118. (P)

Mettam, C. 1967. Segmental musculature and parapodial movement of Nereis diversicolor and Nepthys homberghi (Annelida: Polychaeta). J. Zool. Lond. 153: 245–275. (P)

Mikulic, D.G., D.E.G. Briggs, and J. Kluessendorf. 1985. A Silurian soft-bodied biota. Science 228: 715–717. (H)

Mill, P.J. (ed.). 1977. Physiology of the Annelids. Academic Press, New York. (G)

Nicholls, J.G., and D. Van Essen. 1974. The nervous system of the leech. Sci. Am. 230: 38–48. (H)

Pettibone, M.H. 1982. Polychaeta. In: S.P. Parker (ed.), Synopsis and Classification of Living Organisms, Vol. 2. McGraw-Hill, New York, pp. 3–42. (P)

Rupp, R.W., and M.C. Meyer, 1954. Mortality among brook trout, Salvelinus fontinalis, resulting from attacks of freshwater leeches. Copeia (1954): 294–295. (H)

Sawyer, R.T. 1984. Arthropodization in the Hirudinea: evidence for a phylogenetic link with insects and other Uniramia. Zool. J. Linn. Soc. Lond. 80: 303–322. (H)

Sims, R.W., and B.M. Gerard. 1985. Earthworms. Synopses of the British Fauna No. 31, Brill, Leiden, Netherlands. (O)

Vetvicka, V., P. Sima, E.L. Cooper, M. Bilej, and P. Roch. 1993. Immunology of Annelids. CRC Press, Boca Roton, FL. (G)

Wallwork, J.A. 1983. Earthworm Biology. Studies in Biology No. 161, Edward Arnold, London. (O)

Wells, G.P. 1949. The behavior of *Arenicola marina* L. in sand and the role of spontaneous activity cycles. J. Mar. Biol. Ass. U.K. 28: 465–477. (P)

Woodin, S.A. 1974. Polychaete adundance patterns in a marine soft-sediment environment: importance of biological interactions. Ecol. Monogr. 44: 171–187. (P)

*The supplemental literature listed here is coded to indicate the general topic covered by the paper: G = general Annelida; P = Polychaeta; O = Oligochaeta; H = Hirudinea.

ARTHROPODOUS PHYLA

Members of phylum Arthropoda are schizocoelomate, metameric (segmented), protostomes. Annelids also exhibit these characteristics, but arthropods possess a unique chitinous **cuticle** or exoskeleton that is often strengthened with calcium deposits. The exoskeleton is not equally thickened in all regions of the body. Separate plates and associated with each body segment or **somite** (G., *soma*, body) and these are connected to one another by flexible, thinner regions called articulating membranes. Periodically the arthropod forms a new exoskeleton underneath the old which is then shed in a **molt**. The new exoskeleton is then stretched to a larger size before hardening. In primitive arthropods each segment possesses a pair of appendages, but this condition does not hold for most species. There is a strong tendency for reduction in number of appendages and segments, and for **tagmatization** of segments. The schizocoelom is much reduced and does not function as hydraulic skeleton as occurs in annelids. The large body cavity in arthropods is actually a hemocoel (a modified blastocoel).

Some workers believe that the Arthropoda may have arisen several different times (i.e., polyphyletic arthropodization) from the annelid-arthropod line. However, many others argue that the phylum is monophyletic. Although recent studies have placed both the Onychophora and Tardigrada within the Arthropods, their exact lineage has not been determined. Until the taxonomy is resolved, we will treat these groups as a separate phylum. The Pentastomida are probably arthropods closely related to class Branchiura (Abele et al. 1989; Riley et al. 1978).

Abele, L.G., W. Kim, and B.E. Felgenhauer. 1989. Molecular evidence for inclusion of the Phylum Pentastomida in the Crustacea. Mol. Biol. 6: 685–691.

Ballard, J.W., G.J. Olsen, D.P. Faith, W.A. Odgers, D.M. Rowell, and P.W. Atkinson. 1992. Evidence from 12S ribosomal RNA sequences that onychophorans are modified arthropods. Science 258: 1345–1348.

Boudreaux, H.B. 1979. Arthropod Phylogeny with Special Reference to Insects. John Wiley & Sons, New York.

Eernisse, D.J., J.S. Albert, and F.E. Anderson. 1992. Annelida and Arthropoda are not sister taxa: a phylogenetic analysis of spiralian metazoan morphology. Syst. Biol. 41: 305–330.

Garey, J.R., M. Krotec, D.R. Nelson, and J. Brooks. 1996. Molecular analysis supports a tardigrade-arthropod association. Invertebr. Biol. 115: 79–88.

Manton, S.M. 1973. Arthropod phylogeny—a modern synthesis. J. Zool. 171: 111–130.

Manton, S.M. 1977. The Arthropoda: Habits, Functional Morphology and Evolution. Clarendon Press, Oxford, UK.

Riley, J., A.A. Banaja, and J.L. James. 1978. The phylogenetic relationships of the Pentastomida: the case for their inclusion within the Crustacea Int. J. Paras. 8: 245–254.

■

Phylum Onychophora

Phylum Onychophora [o-NEE-kof-or-ah; G., *onychos*, claw + G., *phora*, to bear] comprises some 70 species of small (2–5 cm long) wormlike organisms that resemble caterpillars. Onychophorans called velvet worms or walking worms, have features similar to both arthropods and annelids. Because of this, some authors have regarded onychophorans to be a sort of missing link between Arthropoda and Annelida. Arthropod characteristics include an open circulatory system with a dorsal tubular heart bearing ostia, a large hemocoel, a reduced schizocoelom that is limited to small sacs associated with the gonads and nephridia, a respiratory system of tracheae and spiracles, and a pair of modified appendages (e.g., **mandibles**) for feeding. The presence of ciliated **nephridia** (coelomoducts), lack of jointed appendages, and structural configuration of the body wall are annelidan in character. However, research argues that they are arthropods (Ballard et al. 1992). Although the body appears to be unsegmented, the nervous system and nephridia are segmented and the position of the legs indicates segmentation. All onychophorans are dioecious, with males being smaller and having fewer legs than do females. A larval stage is absent; development is direct.

Distribution of onychophorans is discontinuous in both tropical and southern temperate regions. The animals are nocturnal and most are tropical, living in moist areas under leaves, rocks, and logs. A fossil specimen from the **Burgess Shale Fauna** of the **Cambrian Period** indicates the phylum is ancient (Fig. 14.1). Present distribution of the phylum can be explained by continental drift.

Classification

Onychophora is divided into two families. Peripatidae contains members having primarily an equatorial distribution. Example is *Periparatus*. Peripatopsidae, exemplified by *Peripatopsis*, is limited to the southern hemisphere.

■ Observational Procedure:

Peripatus

1. Observe one or more preserved specimens of *Peripatus* under the dissection microscope. The elongate body has two indistinct regions: a head and a trunk of 14 to 43 fused annulated regions. Gently pressure applied to the outside of the body with a blunt probe will demonstrate that the animal is covered with a thin, flexible chitinous **cuticle**. Note the short, unjointed, paired legs on the trunk (Fig. 14.2). How many pairs of legs are on your specimen? The number varies with the species and sex. Examine closely the pair of curved terminal claws and **transverse pads** on each leg. Note the **tubercles** and their arrangement on the legs and body. Using high power of the dissection microscope, attempt to see the **nephridiopore** on the ventral surface near the inner base of each leg. The anus opens on the ven-

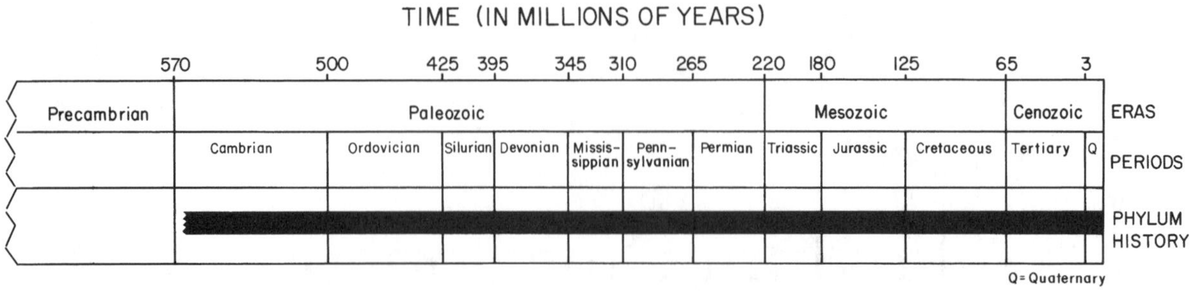

TIME (IN MILLIONS OF YEARS)
570
500
425
395
345
310
265
220
180
125
65
3
Precambrian
Paleozoic
Mesozoic
Cenozoic
ERAS
Cambrian
Ordovician
Silurian
Devonian
Missis-sippian
Penn-sylvanian
Permian
Triassic
Jurassic
Cretaceous
Tertiary
Q
PERIODS
PHYLUM HISTORY
Q= Quaternary

Figure 14.1. Geologic history of phylum Onychophora.

tral side of the posterior end. Attempt to see the gonopore just anterior to the anus (Fig. 14.2).

2. Locate the two annulated **antennae** (preantennae) on the head (Fig. 14.2). Sensory bristles cover the antennae. A small ocellus may be seen near the base of each antenna. Posterior to each antenna and on the ventral surface are two **oral papillae**. Glands in each papilla discharge a secretion that hardens, forming a net of adhesive threads that catch prey such as worms, insects, and other small invertebrates. Between the oral papillae is the mouth, containing paired mandibles. Note the peribuccal lobes or lips surrounding the mouth.

(A)
preantenna
eye
oral papilla
leg
claw
transverse pads
oral papilla
leg
preantenna
tongue
lips
teeth
(B)
nephridiopore
gonopore
anus
anal gland opening
(C)

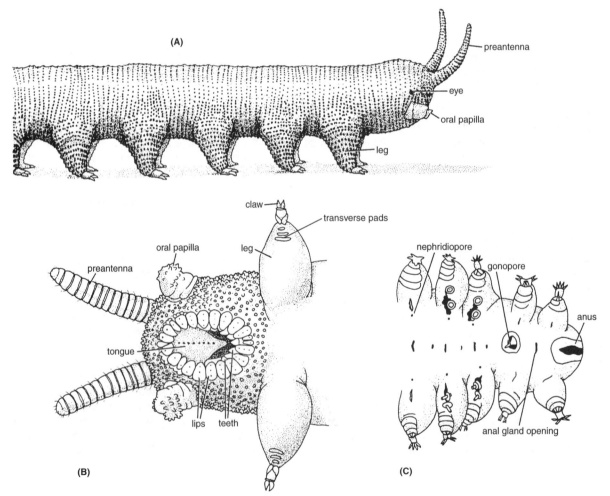

Figure 14.2. *Peripatus.* (A) Side view of body. (B) Anteroventral area. (C) Posteroventral area. (After Sherman and Sherman.)

3. Obtain a slide of a cross section of *Peripatus* and examine it under low power to become familiar with the orientation of the animal on the slide. Switch to high power and locate the cuticle. It will be darkly stained. Note that the papillae are evident and that they are tipped with numerous spines. Find the following layers in order of their position: cuticle, epidermis, dermis, and circular and longitudinal muscles (Fig. 14.3).

4. Depending on the manufacture of slide, your preparation may have more than one section. Scan all sections in an attempt to understand the anatomy of *Peripatus*; it may take the examination of several slides to find all of the structures discussed below (Fig. 14.3).

On the legs, look for evidence of claws and transverse pads. Locate the gut, noting that the intestine is composed of columnar cells. The nervous system consists

of a **brain, two circumpharyngeal connectives**, and two ventrolateral **nerve cords**. Observe the latter in the cross section. The paired ovaries or testes also may be seen in a cross section. Located near the gut, attempt to find the slime glands (dorsolateral to lateral in position) and salivary glands (lateral to ventrolateral in position). You also may be able to make out the dorsal heart and, running tangentially, dorsoventral muscles strands.

Now that you have observed several annelids (Exercise 13) and an onychophoran, compare the overall morphology and internal organization of both kinds of organisms. What features do onychophorans share with annelids? What characteristics are unique to each? You may wish to organize your conclusions into tabular form and save it so that you may add similar observations concerning the arthropods (Exercise 15). ■

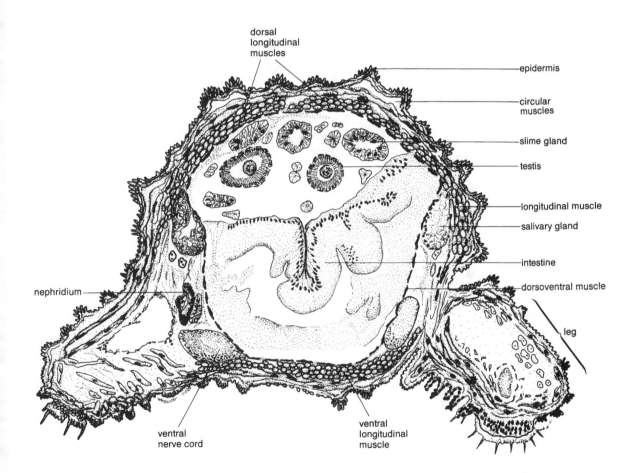

Figure 14.3. Cross section of *Peripatus* through the level of the testes.

Supplemental Readings

Anderson, D.T. 1966. The comparative early embryology of the Oligochaeta, Hirudinea and Onychophora. Proc. Linn. Soc. New South Wales 91: 10–43.

Ballard, J.W.O., G.J. Olsen, D.P. Faith, W.A. Odgers, D.M. Rowell, and P.W. Atkinson 1992. Evidence from 12s ribosomal sequences that onychophorans are modified arthropods. Science 258; 1345–1348.

Boudreaux, H.B. 1979. Anthropod Phylogeny with Special Reference to Insects. John Wiley & Sons, New York.

Jamieson, B.G. 1986. Onychophoran-euclitellate relationships: evidence from spermatozoal ultrastructure. Zool. Scrip. 15: 141–155.

Manton, S.M. 1937. Studies on the Onychopora. II. Feeding, digestion, excretion, and food storage of *Peripatopsis*. Philos. Trans. R. Soc. Lond. B 227: 411–464.

Manton, S.M. 1949. Studies on the Onychophora. VII. The early embryonic stages of *Peripatopsis*, and some general considerations concerning the morphology and phylogeny of Arthropoda. Philos. Trans. R. Soc. Lond. B 223: 484–580.

Manton, S.M. 1950. The locomotion of *Peripatus*. J. Linn. Soc. Zool 41: 529–570.

Manton, S.M. 1973. The evolution of arthropodan locomotory mechanisms. II. Habits, morphology and evolution of the Uniramia (Onychophora, Myriapoda, Hexapoda) and comparisons with Arachnida, together with a functional review of uniramian musculature. J. Linn. Soc. Lond. Zool. 53: 275–375.

Manton, S.M. 1977. The Arthropoda: Habits, Functional Morphology, and Evolution. Clarendon Press, Oxford.

Poinar, G. Jr. 1996. Fossil velvet worms in Baltic and Dominican amber: Onychophoran evolution and biogeography. Science 273: 1370–1371.

Robson, E.A. 1964. The cuticle of *Peripatus*. Q. J. Microsc. Sci. 105: 281–299.

■

Phylum Arthropoda

Arthropoda [ar-THROP-o-dah; G., *arthro*, joint + G., *pod*, foot] is an enormous phylum of segmented (metameric), schizocoelomic protostomes possessing an exoskeleton with jointed appendages. Arthropods range in size from microscopic to several meters in length. Estimates of the number of described species range from 750,000 to 1 million. However, it is predicted that up to 10 million or more species of insects are extant and perhaps one species per day becomes extinct due to clear-cutting the Brazilian rain forest. Arthropods are found from the ocean depths to mountain peaks, and even several thousand meters in the air. They have enormous medical, veterinary, ecological, and economic importance. Some of the most virulent diseases to plague humans are carried by arthropod vectors: the mosquito *Aedes aegypti* transmits the virus causing yellow fever, the flea *Xenopsylla cheopis* carries the bacterium causing black death, and the mosquito *Anopheles* spp. carries the protozoan causing malaria. The immense size of this phylum precludes anything but a brief overview to be presented in the following exercises.

Arthropods possess four important features. (1) A unique **exoskeleton** or **cuticle** made of **chitin** and **scleroprotein** completely covers the body. The exoskeleton is often strengthed by deposits of calcium salts. (2) Primitively, arthropods possess one pair of jointed appendages per body segment (**somite**), but as segments have been lost or fused in the evolution of advanced species this condition is lost. (3) The arthropod body can usually be divided into two or three distinct **tagmata**. (4) Cilia are nearly absent.

Additional features of the phylum include: (1) presence of a reduced coelom, (2) presence of a modified blastocoel (a **hemocoel**) that functions as an open circulatory system, (3) absence of a nephridial excretory system, (4) a high degree of cephalization, and (5) sexual reproduction with development characterized by distinct morphological stages. Arthropods have a geologic history dating from the **Cambrian Period** (Fig. 15.1).

Classification

Arthropoda is comprised of four subphyla: Trilobita, Chelicerata, Crustacea, and Uniramia.

1. **Subphylum Trilobita.** Extinct arthropods possessing three longitudinal lobes (two pleural and one axial) and three tagmata (head, thorax, and abdomen). Trilobite systematics is based on external features alone. Examples: *Elrathia, Diacalymene, Peronopsis,* and *Placops.*

2. **Subphylum Chelicerata.** Arthropods possessing two body regions, or with the entire body fused into a single unit, and no antennae. The first pair of appendages are called chelicerae. This subphylum is divided into three classes: Merostomata, Pycnogonida, and Arachnida.

   a. **Class Merostomata.** Two subclasses of mostly extinct aquatic chelicerates having book gills and a posterior telson: Order Xiphosura (containing the horseshoe or king crab, *Limulus polyphemus*); Order Eurypterida (entirely extinct group of aquatic chelicerates known as giant water scorpions).

191

LIVERPOOL JOHN MOORES UNIVERSITY
LEARNING SERVICES

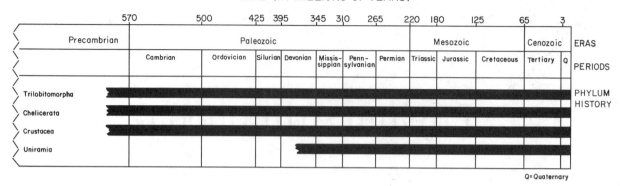

Figure 15.1. Geologic history of the four subphyla in phylum Arthropoda.

b. Class Pycnogonida (Pantopoda). Spider-like marine chelicerates with 4–6 pairs of very long legs. Examples: *Achelia, Pycnogonum,* and *Nymphon.*

c. Class Arachnida. Most chelicerates are arachnids. There are 12, predominantly terrestrial, orders. Examples: Scorpiones (scorpions), Araneae (spiders), Opiliones (harvestmen or daddy longlegs), and Acarina (mites and ticks).

3. Subphylum Crustacea. Mainly aquatic arthropods with biramous appendages and two pairs of antennae. One classification scheme (Schram 1986) presents four classes: Remipedia, Phyllopoda, Maxillopoda, and Malacostraca. Other schemes divide crustaceans into about a dozen classes. Collectively all but one of these classes are sometimes referred to as the lower crustaceans or **entomostracans** [en-to-MOS-tra-kans; G., *entom,* insect + G., *ostracum,* shell]. Examples: Branchiopoda—*Artemia* (brine shrimp), *Branchinecta* (fairy shrimp), and *Bosmina, Daphnia* (water fleas); Ostracoda—*Cypris;* Copepoda—*Calanus;* Cirripedia—*Balanus* and *Lepas* (barnacles). The remaining single class, **Malacostraca,** [mal-a-KOS-tra-ka; G., *malac,* soft], is then referred to as the higher crustaceans. Examples: Isopoda—*Oniscus* (pill bugs) and *Limnoria* (wood lice); Amphipoda—*Gammarus* (freshwater

scuds) and Orchestoidea (beach fleas); Euphausiacea—*Euphausia* (krill); Decapoda—*Penaeus* (shrimp), *Callinectes, Cancer, Paralithodes, Pagurus, Uca* (crabs), *Cambarus* (crayfish), and *Homarus* (lobsters).

4. Subphylum Uniramia. Arthropods with appendages of one branch and a single pair of antennae. The subphylum contains three major classes (Insecta, Chilopoda, and Diplopoda) and the two minor ones, Symphyla and Pauropoda.

a. Class Insecta (Hexapoda). Arthropods having three body regions (head, thorax, and abdomen), a single pair of antennae (rarely absent), and wingless or with one or two pairs of wings borne on the thorax. Two subclasses are recognized: **Apterygota** (primitive wingless insects, including silverfish and springtails) and **Pterygota** (winged or wingless insects divided into two divisions, Exopterygota and Endopterygota, based on the type of metamorphosis). The first division includes dragonflies, grasshoppers, cockroaches, termites, and bugs. The second division includes beetles, butterflies, flies, fleas, and ants.

b. Myriapodous Arthropods. Myriapods comprise four classes, two of which are of some importance: Chilopoda (centipedes) and Diplopoda (millipedes).

A. Subphylum Trilobita

Trilobita [tril-o-BI-tah; L., *tri,* three + G., *lob,* lobe] is a moderately sized group of over 10,000 extinct species of primitive, dorsoventrally flattened arthropods with a chitinous exoskeleton strengthened with calcium and phosphate deposits. These marine animals were most abundant during the **Cambrian** and **Ordovician Periods** of the **Paleozoic Era**, but were extinct by its end (Fig.

15.1). Although a few large species are known to reach nearly 1 m, most adult trilobites were small (less than 10 cm long). The name trilobite refers to their oval body being longitudinally divided into three lobes: two lateral **pleural lobes** and one medial **axial lobe**. The body consisted of three tagmata: **cephalon** (head), **thorax, pygidium** (abdomen).

Although extinct for over 200 million years, trilobites possessed many features common to the phylum Arthropoda. As in other phyla with a rich fossil history, a complex nomenclature has been developed by paleontologists. Here students will study only the basic morphology of these remarkable arthropods.

■ Observational Procedure:

1. Observe specimens of several species and identify the anterior and posterior ends, the three lobes (axis and two pleural), and the three tagmata (cephalon, thorax, and pygidium) (Figs. 15.2 and 15.3).

2. The cephalon of several fused segments forms a large dorsal head shield that is divided into various parts by **facial sutures** and **furrows**. Trace the central **axial furrow** that delimits the axis from the pleural lobes through all three tagmata.

3. Examine specimens to see if their cephalons are divided into a central region, the **cranidium** (Fig. 15.3) and lateral **cheeks** (Fig. 15.2). In the center of the cranidium, often elevated above the cephalon, locate the **glabella**. This structure probably housed intestinal diverticula, stomach, and provided a broad surface to which muscles attached.

4. Lateral to the glabella find the compound eyes. In some species you may be able to see the functional units of the eye, the **ommatidia** (e.g., *Phacops rana*). In many fossils the ommatidia will not be visible.

5. At the posteriolateral corners of the cephalon locate the **genal angles**. If elongated, these are called **genal spines**.

6. The thorax consists of 2–42 unfused, articulating segments each bearing a pair of appendages similar to those on the cephalon. The thorax was capable of flexing so that many trilobites were able to fold over like a pill bug, a process called **enrollment** or **conglobation** (Fig. 15.4). Presumably this protected the vulnerable (thinner) ventral surface called the **doublure**. Do any of the specimens exhibit enrollment?

7. The pygidium, was usually constructed of 2–30 fused segments with the anus located at the end of the last segment. Are the pygidia well defined in the specimens? In some species the axis is ribbed in the pygidium, giving a false impression that it was articulate in this tagma.

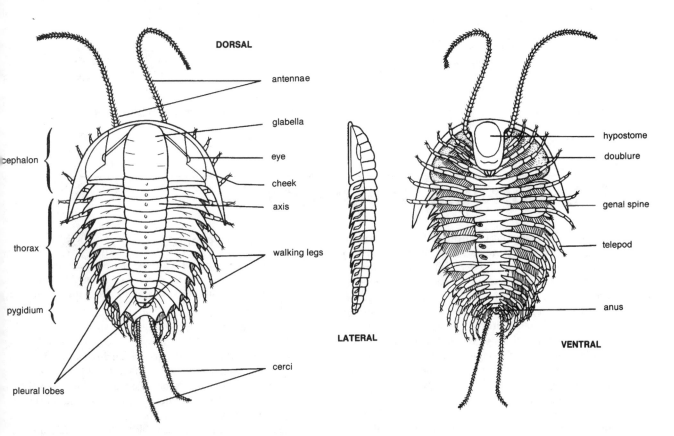

Figure 15.2. Schematic diagram of a generalized trilobite (dorsal, ventral, and lateral views).

LIVERPOOL JOHN MOORES UNIVERSITY
LEARNING SERVICES

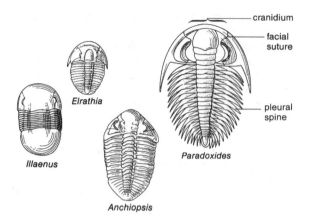

Figure 15.3. Examples of trilobites.

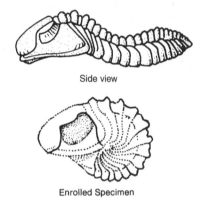

Figure 15.4. Enrollment (conglobation) of *Placops*.

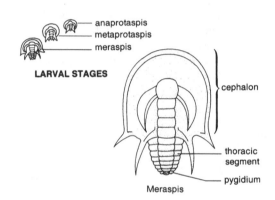

Figure 15.5. Trilobite larval stages (protaspis and meraspis).

8. Although appendages are very infrequently preserved, examine the study specimens for them.
9. The entire **tergal** [L., *tergum*, back] exoskeleton, particularly the cephalon, may be ornamented with bumps, granules, tubercles, lines, and other markings. Using Fig. 15.2 as an aid, identify any other anatomical features exhibited by the specimens.
10. If a whole-mount slide of a trilobite larva from a horseshoe crab is available, compare the larva to the drawings of trilobite larval stages (Fig. 15.5) and to the adult specimens. What similarities are found in these specimens; how do they differ? ■

B. Subphylum Chelicerata

Members of subphylum Chelicerata [ke-LIS-e-ra-tah; G., *chela*, claw + G., *cera*, horn] are distinguished from other arthropods by having two important features. (1) Chelicerates do not possess antennae. (2) The first pair of appendages are modified distally to form opposing clawlike legs called **chelicerae**. Following the chelicerae are the **pedipalps**, and posterior to these are four pairs of walking legs. Mandibles are lacking. The body is divided into two tagmata: anterior **cephalothorax** or **prosoma**, and posterior **abdomen** or **opisthosoma**. All appendages are associated with the prosoma; the opisthosoma lacks appendages except in the Xiphosura. Most of the 75,000 species of chelicerates are predaceous or parasitic on arthropods or other invertebrates.

Class Merostomata

Two subclasses comprise the class Merostomata [mer-o-STO-mah-tah; G., *mero*, part + G., *stoma*, mouth], Xiphosura (including horseshoe crabs) and the extinct Eurypterida.

Subclass Xiphosura. Horseshoe crabs members of the subclass Xiphosura [ZIF-o-SU-rah; G., *xiphos*, sword + G., *oura*, tail] have existed since the **Ordovican Period**, but only four marine species are extant. *Limulus polyphemus* is widely distributed along the Gulf of Mexico and east coast of the United States. The other species occur along coasts in the far east Pacific.

In some ways *Limulus* resembles a trilobite. *Limulus* possesses three indistinct longitudinal lobes and has a convex dorsal shield covering the ventral appendages (Fig. 15.6). However, *Limulus* possesses only two tagmata. From above, *Limulus* resembles an upside-down shovel. Specimens may reach more than 50 cm in length, including the tail spine or **telson**. The sexes are separate and mating pairs may be observed along U.S. coasts in the spring.

■ Observational Procedure: *Limulus*

1. Place a specimen of *Limulus* dorsal side up with the telson to one side. Identify the anterior and posterior ends. Note the shovel-shaped prosoma. How is

Limulus adapted for shallow burrowing? Notice the general resemblance of *Limulus* to that of a trilobite. How does tagmatization in *Limulus* differ from that of the trilobites? Are the ventral sides of the two organisms similar (compare Figs. 15.2 and 15.6)?

2. Return the animal to its dorsal side and examine the **medial** and **compound eyes** with a hand lens. Of what type of eye does this remind you? What defensive structures does *Limulus* possess on its dorsal surface? Examine a molted exoskeleton (**exuvia**) and identify the **suture line** that breaks open when the animal molts.

3. Examine the ventral surface of a preserved specimen. As in the trilobites, the ventral extension of the carapace is called the **doublure**. Locate the seven

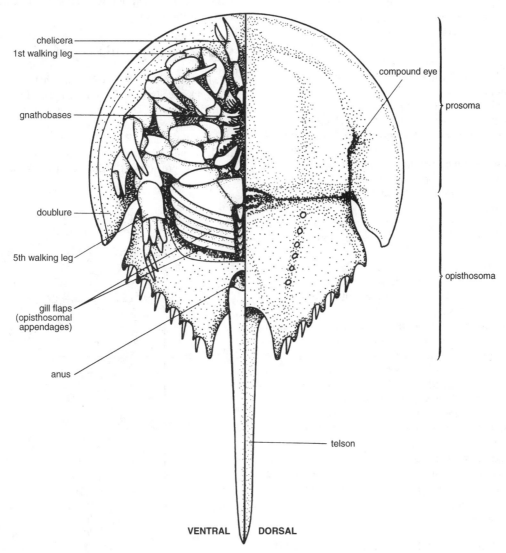

Figure 15.6. Dorsal and ventral views of *Limulus polyphemus.*

pairs of prosomal appendages. The first pair are chelicerae, followed by five pairs of **walking legs**. The first of these being the **pedipalps**. Finally, there is a short, unsegmented pair of appendages called **chilaria**.

4. Trace each leg from its distal end to the point where it attaches to the doublure. How do these seven legs differ?

5. The pedipalps of adult males are different. The chelate ends have a curved claw that is used to hold females while mating. How does this clasping organ function? Is there anything unusual about its size?

6. Observe that the bases of the walking legs, but not the chelicerae, possess spined **coxae** called **gnathobases**. They are grinding organs, processing food to be swallowed by the mouth located posterior to the chelicerae. The chilaria also function as gnathobases.

7. The fifth pair of walking legs are modified in two ways. Observe that they are not chelate, but at their distal end is a structure with four, spatulate blades capable of spreading apart like legs of a tripod. *Limulus* uses these to push against the sediment during locomotion. Also on the lateral side of the same leg, near the doublure, locate the spatulate **flabellum**. This structure is used in cleaning the gills.

8. Observe the opisthosomal region of *Limulus* and examine the six pairs of tough, flaplike appendages that are fused medially. The first pair is thicker and called the **genital opercula**. Move these out of the way and find the next five opercula. These are gill opercula, each possessing many soft, leaflike lamellae known as a **book gill**.

9. In a preserved specimen tease apart some of the gill folds and note their delicate nature. How many gill folds are present?

10. Find the anus at the base of the telson. Carefully move the telson and determine its range of motion. From that observation do you think that the telson is a particularly good defensive structure?

11. Examine a whole-mount slide of a **trilobite larva** of *Limulus*. Note that the larva looks like a minature adult. All the legs are present, but some of the gills are still absent and will be gained in subsequent molts. The medial and compound eyes should be visible through the thin exoskeleton. *Limulus* larvae superficially resemble larval and adult trilobites. Compare the general morphology of a *Limulus* larva to larval and adult trilobites (Figs. 15.2 and 15.3).

12. If live animals are available for study, note how they move, burrow, and right themselves when turned over. Examine adult *Limulus* for the commensal flatworm, *Bdelloura* (Platyhelminthes, Exercise 5), on the ventral side, especially on the legs.

Subclass Eurypterida (Gigantostraca). Eurypterida [U-rip-TER-ee-dah; G., *eury*, wide + G., *pteron*, wing] is a

group of extinct chelicerates that lived during the **Paleozoic Era**. They are sometimes referred to as sea scorpions because the chelicerae and long abdomen resemble that of true scorpions. The etymon, wide wing, refers to the paddle-like swimming legs. Most specimens are less than 30 cm long, but some giant forms reached nearly 3 m, hence the alternate name **Gigantostraca** [JI-gan-tos-tra-kah; G., *gigantos*, giant + G., *ostraca*, shell]. ■

■ Observational Procedure:

If specimens are available for examination, note the general body size and shape of fossil eurypterids (Fig. 15.7). Usually only parts of the organism are present in the sedimentary rock for study. Do eurypterids resemble trilobites or xiphosurans such as *Limulus*? Does the study specimen exhibit any evidence of tagmatization? ■

Class Pycnogonida (Pantopoda)

The Pycnogonida [pik-no-GON-ee-dah; G., *pycno*, thick + G., *gony*, knee or joint] comprises nearly 1,000 species of spiderlike marine chelicerates with very long legs. It is their long-legged condition that provides their other name, **Pantopoda** [pan-to-POD-ah; G., *panto*, all + G., *poda*, feet]. Pycnogonids are common in benthic communities, where they prey on soft-bodied animals such as cnidarians

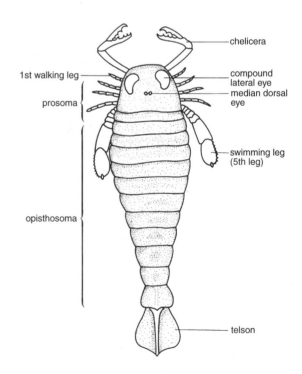

Figure 15.7. A fossil eurypterid (sea scorpion).

(Exercise 3) and bryozoans (Exercise 19) by sucking the body fluids or consuming bits of tissue. Some are not predaceous, but feed on algae accumulated on the surface of sessile organisms. Most pycnogonids are less than 10 mm long, but some have very long legs (ca. 75 cm).

Observational Procedure:

Examine a preserved pycnogonid under a dissection microscope and note the narrow body, composed of a prosoma with cephalon and trunk, and a small, unsegmented abdomen (**opisthosoma**) bearing a terminal anus (Fig. 15.8). Locate the proboscis at the anterior end and, on the anterior dorsal part of the cephalon, find a small tubercle that bears **ocelli**. The number of appendages differs among species; *Nymphon* has seven pairs. Locate the chelicerae that lie alongside the proboscis. Also locate the pedipalps, the second pair of appendages near the chelicerae. The third pair of appendages are the very slender **ovigerous legs** that normally are used by males to carry eggs; females of many species lack these legs. Note that the remaining four pairs of walking legs are long (Fig. 15.8). Does *Nymphon* exhibit any evidence of tagmatization?

Your instructor will assist you in making observations, should live specimens be available for study.

Class Arachnida

Class Arachnida [ah-RAK-nid-ah; G., *arachne*, spider] comprises the vast bulk of chelicerates (more than 60,000 species) and is of great importance because many are parasitic on plants and animals, or prey upon agricultural pests. Arachnids range in size from minute (200 μm) to nearly 20 cm long. The class dates to the **Silurian Period** from which the first aquatic scorpions are known.

The opisthosoma of arachnids usually lacks appendages, while the prosoma possesses six pairs: a pair of chelicerae, a pair of pedipalps, and four pairs of walking legs. Of the 12 orders in this class, seven will be studied here. They represent greater than 99% of arachnids. The remaining five minor orders, Amblypygi, Cyphophthalmi, Palpigradi, Ricinuclei, and Schizomida, will not be studied unless your instructor has specimens for examination.

Order Scorpiones. Order Scorpiones [skor-pi-O-nez; G., *skorpios*, scorpion] is a well-known group of relatively large (2 cm to nearly 18 cm) chelicerates with a

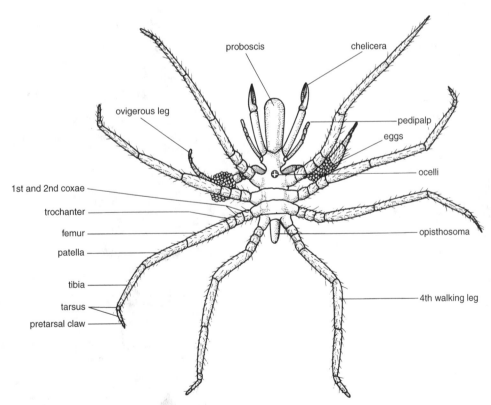

Figure 15.8. Dorsal view of the pycnogonid, *Nymphon*.

long mobile tail that terminates with a poisonous **sting**. Some fossil species were very large (greater than 0.5 m). The prosoma bears chelicerae, pedipalps, and four pairs of walking legs. The chelicerae are small; the large, chelate pedipalps are used to hold prey and are used in courtship. The opisthosoma is elongate and of two parts, a **mesosome** (preabdomen) and a **metasome** (postabdomen). The latter is a segmented tail that possesses the sting. Scorpions are viviparous and live in tropical, subtropical, and south temperate zones where the climate is warm. *Centruroides* is a common genus found in the southwestern United States. The species *C. sculpturatus* can be dangerous to humans, especially children.

■ Observational Procedure: *Centruroides*

1. Examine a specimen of *Centruroides* or of another genus available for study (Fig. 15.9). At the anterior end of the scorpion examine the chelicerae and large pedipalps; behind the pedipalps locate the walking legs.

2. Dorsally the prosoma is covered by a subquadrate plate or **carapace**, that bears, at the anterior end, three groups of simple eyes.

3. Immediately behind the carapace is a short segment (eighth) that begins the opisthosoma. Observe that the opisthosoma is divided into two sections. The first is a broader region called the mesosoma (preabdomen) comprising segments 8–14. The second part (metasoma or postabdomen) is thinner, tail-like and comprises segments 15–19.

4. Locate the **basal bulb** (telson) and sting attached to segment 19. The sting is capable of injecting a neurotoxin into prey from a gland located in the basal bulb.

5. Examine the chelicerae closely. Note that the finger on the outer side has two cusps that overlap the inner finger. The legs end in two pairs of claws on the tarsus.

6. Turn the animal ventral side up (Fig. 15.10). Most of the ventral prosoma is covered by the first joint (**coxae**) of the walking legs. Find the ninth body segment. It may be recognized by the presence of **pectines**, comblike, sensory structures unique to scorpions.

7. Just anterior to the pectines locate the genital operculum of the eighth segment.

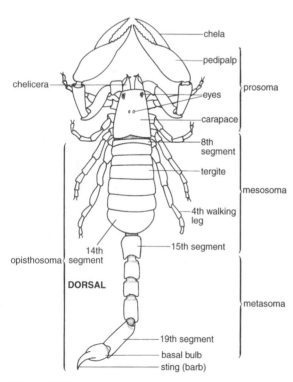

Figure 15.9. Dorsal view of a scorpion.

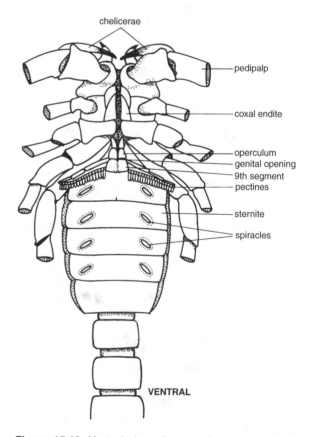

Figure 15.10. Ventral view of a scorpion, prosoma and mesosoma.

8. Find segments 10–13 just posterior to the pectines and, at their lateral margins, locate the spiracular openings to the respiratory organ called the **book lung.** ∎

Order Pseudoscorpiones. Order Pseudoscorpiones [SU-do-SKOR-pi-O-nez; G., *pseudo*, false] comprises about 1,500 species of small (less than 10 mm) chelicerates that resemble true scorpions. Book lungs are also missing; gas exchange takes place through a tracheal system. These animals are common occupants in nests of ground inhabiting animals as well as in organic debris. *Chelifer* is commonly found in houses. Pseudoscorpions feed on other small arthropods by grasping them and injecting a poison from glands in the pedipalps.

∎ Observational Procedure:

Examine a whole-mount slide of a pseudoscorpion, noting their general resemblance to scorpions (Fig. 15.11). What characteristic scorpion features are missing in the pseudoscorpion? Locate the chelicerae, pedipalps, and four pairs of walking legs. The dorsal surface is covered by a carapace or **scutum.** The segmented opisthosoma is broader than the prosoma and rounded at its posterior end. As in scorpions, coxae cover most of the ventral surface of the prosoma. Note that the bulbous pedipalps are chelate and bear several sensory hairs, called **trichobothria,** scattered over their surfaces. Eyes may be present at the anteriolateral regions of the carapace. ∎

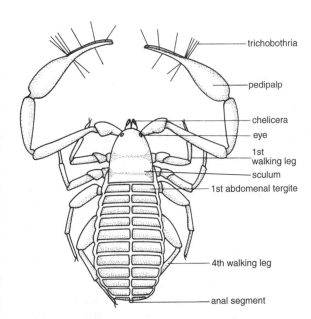

trichobothria

pedipalp

chelicera
eye
1st walking leg
sculum
1st abdomenal tergite

4th walking leg

anal segment

Figure 15.11. Dorsal view of a pseudoscorpion.

Order Araneae. Order Araneae [a-RA-ne-ah; L., *arane*, spider] comprises over 30,000 species of true spiders. Spiders differ from most other chelicerates in that the prosoma and opisthosoma are connected by a short, narrow isthmus called the **pedicel**; both body regions are unsegmented in the adult. The opisthosoma, usually referred to as the abdomen, contains the majority of the viscera, including the silk-spinning organs. Spiders range from less than 1 mm to nearly 10 cm long and are found in most habitats, including the tropics, deserts, arctic and alpine regions, dorm rooms, and at several thousand meters in the air.

∎ Observational Procedure:

1. Obtain several specimens for study (e.g., *Argiope*, orb weaver; *Latrodectus*, black widow; *Tarantula*, tarantula) and examine them, locating the prosoma, opisthosoma, and pedicel (Fig. 15.12).
2. Find the four pairs of walking legs and pedipalps on the prosoma. Note the large, two-jointed chelicerae with **fangs** (Fig. 15.12). The basal joint is the larger and the second joint is composed of a dagger-like claw or fang. In males the terminal joint of the pedipalp is modified as an organ by which sperm are conveyed to the female's genital receptacle. This feature is an important taxonomic character.
3. Note the dorsal shield or carapace that covers the prosoma. Locate two rows, of four eyes each, at the anterior end.
4. Find the pedicel, the first segment of the opisthosoma (Fig. 15.12). Ventrally, segment 2 possesses the genitalia as well as openings to the book lungs. The latter may be found by locating a transverse line called the **epigastric furrow** or fold. The lateral extensions comprise the openings (**spiracies**) to the book lungs.
5. Medial and slightly anterior to the furrow, find the genital opening. In the vicinity of the genital opening of most female spiders is the **epigynum** that contains the opening to the seminal receptacles.
6. Find the **spinnerets** on the posterioventral surface of the opisthosoma near the anal lobe (Fig. 15.12). There are 4–6 spinnerets depending on the species. Spinnerets are silk-producing glands and are considered appendages of the ventral body segments 10 and 11.
7. Anterior to the spinneretes, locate another one or two spiracle opennings. In primitive spiders (mygalomorphs), the posterior spiracles are paired and lead to book lungs. In advanced forms the respiratory organs are trachae, most often with a single spiracle.
8. If live spiders are available, observe web-building and prey-capturing behaviors according to directions provided by your instructor. ∎

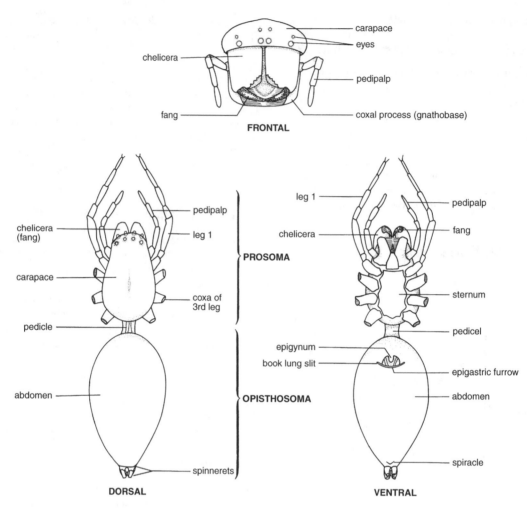

Figure 15.12. Frontal, dorsal, and ventral views of a generalized spider.

Order Opiliones. The familiar daddy longlegs or harvestmen comprise the order Opiliones [o-pil-ee-O-nes; L., *opilio,* shepherd] or Phalangida [fa-LAN-ji-da; G., *phalange,* spider], about 3,000 species of arachnids noted for their long legs. Opiliones have the prosoma and opisthosoma fused into an egg-shaped body (5–20 mm long) that is elevated above the substratum by four pairs of very long legs. They have a cosmopolitan distribution, but are abundant in moist regions.

■ Observational Procedure:

Obtain a daddy longlegs and note that the prosoma and opisthosoma are fused into a single carapace (Fig. 15.13). As in other chelicerates, the prosoma supports the appendages. The opisthosoma may be identified by the presense of 10 weakly lined segments. Examine the anterior end and locate the che-

licerae and leglike pedipalps. Locate the eyes and anteriolaterally the stink gland on the central region of the dorsal carapace. ■

Order Acarina. Acarina [AK-a-RI-na; G., *akari,* mite] comprises about 25,000 species of small (less than 1 cm) chelicerates commonly known as ticks and mites. Acarines are of great medical, veterinary, and economic importance, because many are vectors or intermediate hosts of disease-causing organisms (e.g., Lyme disease and Rocky Mountain spotted fever). The standard tagmatization of chelicerates is not present in acarines. The tagma are fused into a dorsoventrally flattened, ovoid body of several arbitrarily defined regions.

Of the seven suborders, five are parasitic or vectors of other parasites. Anatomical differences that separate ticks and mites are relatively minor. All ticks (suborder Metastigmata) are blood-sucking parasites. Parasitic mites

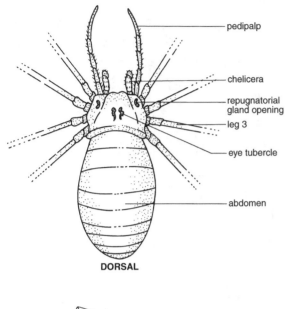

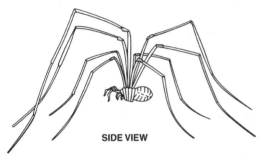

Figure 15.13. Dorsal and lateral views of a daddy longlegs.

feed on blood or burrow through epidermal tissues or internal organs (e.g., respiratory system). The common dust mite (*Dermatophagoides*) is not parasitic, but releases materials that may cause severe allergic reactions in humans. Students will examine several examples of ticks and mites for an understanding of their diversity of form.

■ Observational Procedure: Ticks

Hard Ticks (family Ixodidae): *Dermacentor* spp.
1. Examine male and female specimens of either *Dermacentor andersoni* (Rocky Mountain wood tick) or *D. variabilis* (American dog tick) (Fig. 15.14A). Identify the two body regions: the body or **idiosoma** consists of a region bearing the legs (**podosoma**) and one in which legs are missing (**opisthosoma**).
2. Locate the anterior **capitulum (gnathosoma)** bearing the feeding appendages, including the toothed

hypostome, chelicerae, and pedipalps. Note that the capitulum may be seen in a dorsal view. This feature distinguishes hard from soft ticks.
3. On the ventral side of the opisthosoma, find the genital pore, anus, stigmal plates with spiracle, and marginal **festoons**. The latter have taxonomic importance.
4. Also observe the four pairs of walking legs, each bearing a terminal tarsus with apotele and tarsal claws. At the tip of the tarsus of the first walking leg find **Haller's organ**, which is sensitive to humidity and odors.
5. Compare the dorsal idiosoma of a male and female and note the difference in the size of the **scutum** (Fig. 15.14B).
6. Examine larval and nymph stages of *D. andersoni*. Note that the larval stage is much smaller than the adult and possesses only three pairs of legs. The nymph is larger than the larval stage and resembles the adult, except for lacking the genital opening.
7. For more advanced study, examine other examples of hard ticks: for example, *Haemaphysalis leporispalustris* (rabbit tick); *Amblyomma americanum* (lone star tick); *Rhipicephalus sanguineus* (brown dog tick); *Boophilus annulatus* (American cattle tick). Compare these species to *Dermacentor*.

Soft Ticks (family Argasidae): *Argas persicus*.
Examine a specimen of *Argas persicus*, the fowl tick, a vector of relapsing fever in birds. Observe the sculptured cuticle. Note that the capitulum is obscured by the idiosoma in dorsal view (Fig. 15.15). How would the body regions be designated in this tick? ■

Mites. Representatives of three of the four suborders comprising the mites may be studied depending on the time available.

■ Observational Procedure: Mites

1. *Dermanyssus gallinae* (Chicken Mite): Suborder Mesostigmata. These mites parasitize chickens, hiding in the roosts by day and coming out to feed at night. Large infestations may be sufficient to kill the hosts from loss of blood. Examine a specimen of *Dermanyssus gallinae* (Fig. 15.16). Locate the capitulum with unarmed hypostome, chelicerae, and pedipalps. Note that the chelicerae are needle-like; if you examine them under high magnification you may be able to see their minute chelate ends.
2. *Sarcoptes scabiei* (Human Itch Mite): Suborder Astigmata. These small mites (less then 500 µm long) cause **scabies** in humans. Females burrow through the skin, depositing eggs and excreta that cause itching and may lead to secondary infection by

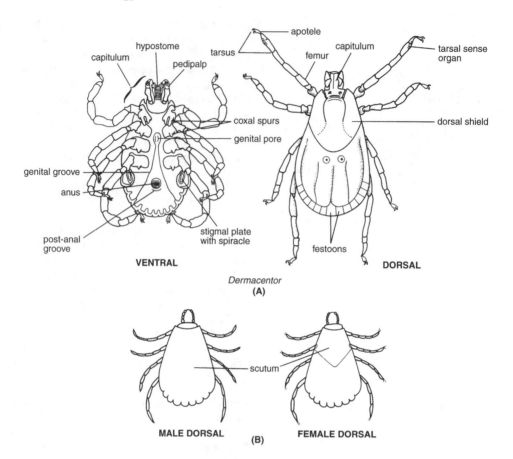

Figure 15.14. The hard tick, *Dermacentor*. (A) Dorsal and ventral views. (B) Dorsal view comparing male and female ticks.

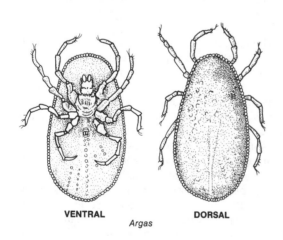

Figure 15.15. The soft tick, *Argas*.

bacteria. Examine a specimen of *S. scabiei* noting the rounded body and fine striations on the cuticle (Fig. 15.17). Also, note the position of the four pair of stubby legs, presence of long setae, and dorsally, a central region of raised scales that extend posterio-laterally. Observe the legs of your specimen. The first and second pairs possess extensions called pulvilli just proximal to the tarsal leg segment. Females lack **pulvilli** on the rear pairs, but males have them on the fourth pair. What might the function be of these structures?

Examine a section of the epidermis from a mammal that had **sarcoptic mange** and attempt to locate adult females in their winding burrows. You may also be able to locate eggs, unhatched embryos, and ecdysed cuticles. What sort of tissue damage do you see in this preparation?

3. *Dermatophagoides* (House Dust Mite): Suborder Astigmata. Most species of this genus are not parasitic; however, certain forms produce powerful aller-

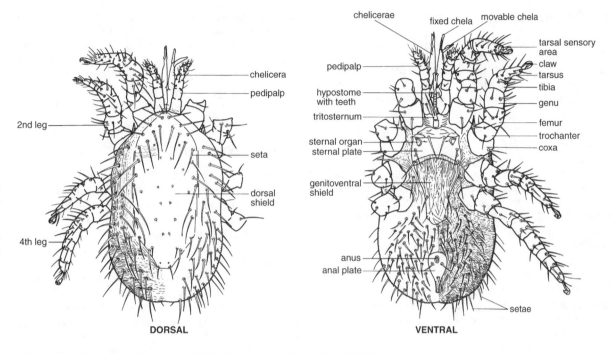

Figure 15.16. Dorsal and ventral views of *Dermanyssus gallinae*. (After T.M. Evans.)

gens causing allergic reactions in sensitive people. Examine a specimen of the house dust mite, *Dermatophagoides* sp., noting the ovoid body, long setae, and fine striations present on the cuticle as in *S. scabiei*. If time permits, you will be instructed on how to collect your own specimens.

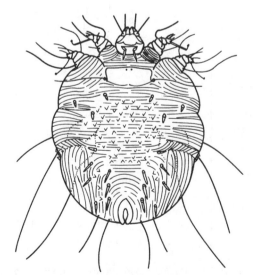

Figure 15.17. Dorsal view of a female *Sarcoptes scabiei*.

4. *Demodex folliculorum* (Follicle Mite): Suborder Prostigmata. The follicle mites, *D. folliculorum* and *D. brevis*, live in human follicles and sebaceous glands, respectively usually without causing serious damage. They are small (less than 500 μm) mites with short legs and an elongated, annulated opisthosoma. Examine a specimen of a follicle mite, noting the four pairs of short legs with claws and the elongate, annulated opisthosoma. What is the significance of the elongated body? Also, examine a cross section of *D. folliculorum* in situ and locate the mites in the hair follicles. Do you notice anything unusual about the orientation of these organisms in the follicle? If time permits, you will be instructed on how to collect specimens from your own eyebrows. Are you parasitized by *Demodex*? ■

Order Solifugae. The Solifugae [so-LIF-u-je; L., *soli*, sun + L., *fug*, flee], also Solpugida [sol-PU-ji-dah; L., *solpug*, venomous spider], is a group of relatively large (1–7 cm), aggressive chelicerates. They are commonly known as sun, camel, soldier, or wind spiders. Solifugids feed on arthropods and vertebrates, tearing prey apart with large, stocky chelicerae. They have a cosmopolitan distribution in tropical, subtropical, and warm, arid temperate regions. They are abundant in desert regions of the southwestern United States. Respiration in the entire order is accomplished by trachea.

■ Observational Procedure:

1. Examine a specimen of a solifugid (Fig. 15.18). Observe that the hirsute (hairy) body appears to be divided into three divisions. Actually the prosoma has a constriction that separates it into two regions of three somites each: head and thorax (or anterior and posterior carapaces, respectively). In nonburrowing species, the head and thorax sections articulate. How can that be advantageous?

2. Note that the thorax has dorsal transverse lines that indicate the position of the three somites. Four to six simple eyes are present on the head, depending on species. Locate the large chelicerae and determine in which direction they articulate. How does this differ from the other chelicerates studied so far?

3. Note the robust nature of the pedipalps and the relatively delicate first legs. The pedipalps are not chelate but they are used in prey capture. The first pair of legs are not used in locomotion but are tactile organs. On the coxae of the posterior legs are the **racquet organs** whose function is unknown. The segmented opisthosoma is broadly joined to the prosoma and comprises 10 segments. ■

Order Uropygi. The Uropygi [U-ro-PI-ge; G., *uro*, tail + G., *pyge*, rump] contains small celicerates (a few mm to about 7 cm) that are flattened both in the prosoma and opistosoma. They are commonly called whip scorpions because of the long **flagellum** or **whip** at the end of the opisthosoma. They are also called **vinegaroons** due to their ability to spray a defensive fluid composed mainly of acetic acid from a pair of glands on either side of the anus. Whip scorpions have a worldwide distribution, but tend to be common in tropical, subtropical, and warm temperate regions. One of the largest forms (ca. 6.5 cm), *Mastigoproctus giganteus*, is found in the southern parts of the United States.

■ Observational Procedure:

Examine a specimen of *Mastigoproctus*, noting the flattened body and long flagellum (Fig. 15.19). At the anterior end distinguish the small, stout chelicerae from the larger, chelate pedipalps used in prey capture. The first pair of legs are thinner and sensory, but the remaining three pairs are used in walking. Note that the opisthosoma is segmented and broadly attached to the prosoma. It is composed of a larger mesosoma of eight somites and a shorter metasoma of three narrow segments before the flagellum. Locate the paired openings to the defensive **anal glands** on the ventral side of the metasoma. Note how the metasoma can be oriented to direct the spray in many directions. ■

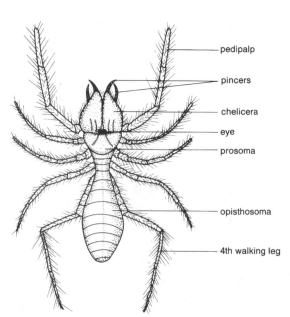

Figure 15.18. Dorsal view of a solifugid. (Total body length ≈2.5 cm.)

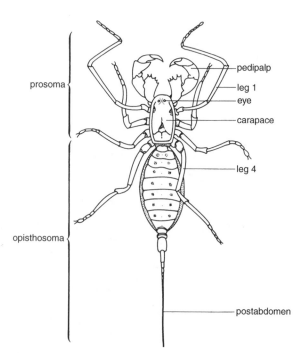

Figure 15.19. The American uropygid, *Mastigoprocetus giganteus.*

C. Subphylum Crustacea

Shrimps, crabs, crayfish, lobsters, barnacles, and pill bugs are members of a large, diverse subphylum of arthropods called Crustacea [krus-TA-shi-ah; L., *crusta*, crust or rind]. Over 40,000 species are recognized, most of which are aquatic. Crustaceans are unique arthropods possessing two pairs of antennae, biramous appendages, and mandibles (jawlike appendages). They range in size from minute (less than 500 µm) to extremely large; the lobster *Homarus americanus* may reach more than 0.5 m long. Many crustaceans are commercially important to the shellfish fishery (e.g., shrimps, crabs, lobsters in marine coastal waters and crayfish in fresh waters). Small free-swimming crustaceans, their larvae, and the larvae of larger forms, collectively called **zooplankton**, are very important ecologically as they constitute a trophic level intermediate between singled-celled organisms and zooplanktivorous fish.

Primitively, the crustacean body is composed of a series of segments each possessing a pair of biramous appendages. In advanced forms there is a strong tendency towards tagmatization, loss and fusion of segments, and loss of appendages. Three tagmata are usually recognized: head, thorax, and abdomen. The head and thorax may be fused into a single cephalic region called the **cephalothorax** (prosoma). Many crustaceans possess a shieldlike extension of the cephalothorax called the **carapace**, which may surround the animal entirely except for a small ventral gap. Most members of the class Malacostraca have a shrimplike body form with obvious tagmatization (Fig. 15.20). This arrangement is called the **caridoid facies** [L., *carid*, shrimp + L., *facies*, appearance].

Typical crustacean appendages are paired and biramous (Fig. 15.21), consisting of a proximal **protopod** (or protopodite) with two distal rami: the **endopod** (inner or medial branch) and **exopod** (outer or lateral branch). The protopod may be composed of three units (precoxa, coxa, and basis) and have other outgrowths on the lateral side (**epipodites** or **exites**) and on the medial side (endites).

Different classification schemes for the crustaceans are found in general and advanced texts, and your instructor may choose to discuss the merits of these various schemes. However, for purposes of laboratory study, taxonomic position is not particularly important, so the following exercises do not emphasize taxonomy. The crustaceans available for study are to be considered only as representatives of this very diverse group. Your instructor may provide additional representatives or substitute others for study.

Class Branchiopoda

Cladocera [kla-DOS-er-ah; G., *cladus*, branch + G., *cera*, horn] is an important suborder of predominantly freshwater crustaceans of class Branchiopoda [brank-ee-o-PO-dah; G., *branchium*, gill + G., *pod*, foot]. Other branchiopods with whom you may be familiar are fairy and brine shrimp. Feeding on algae and bacteria, the cladocerns represent an extremely important link in the food web of freshwater systems. Cladocerans are commonly called water fleas because planktonic forms perform an unusual jump or hop followed by a slow sinking movement. Most individuals are 200–3,000 µm long and possess a transparent, laterally flattened, nonhinged carapace that extends ventrally beyond the body, covering the legs. There is a ventral gap between right and left valves that allows water to pass between them.

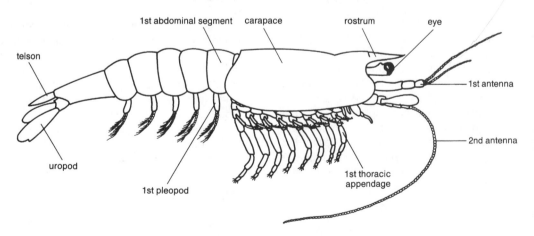

Figure 15.20. Generalized malacostracan crustacean.

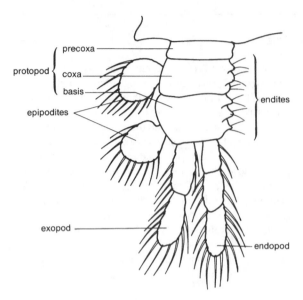

Figure 15.21. Generalized crustacean appendage. (After Schram).

■ Observational Procedure:

Live Specimens

1. Observe the hop-sink motion of live *Daphnia* in a culture vessel under low light. How is the hop motion accomplished? How does drag and the center of gravity of the animal affect the orientation of the animal as it sinks?

2. How does *Daphnia* respond to light shined from above? From the side? From beneath the vessel?

3. Place a single specimen into a drop of pond water on a shallow depression slide and cover it with a coverslip.

4. Observe the beating heart located dorsally near the midpoint of the body. Under high magnification you may be able to see blood cells coursing through the **hemocoel**.

5. Observe feeding activity of *Daphnia* by adding a small drop of a suspension of food particles. Can you see feeding currents? How does *Daphnia* process food particles? How do the mandibles work? Is food present in the food groove? Attempt to observe rejection of food by *Daphnia*. How is this accomplished? Does the postabdomen have any function in food processing? Remember that your observations do not reflect the normal feeding conditions of *Daphnia*. In your preparation the animal is confined in a small amount of water with abnormally high levels of food.

Morphology

The following observations of *Daphnia* morphology may be made using live or preserved specimens. Live specimens will require an anesthetic or methyl cellulose to slow them down.

1. Examine a cladoceran such as *Daphnia* under a compound microscope (Fig. 15.22). Examine the ovoid body and identify the anterior, posterior, dorsal, and ventral ends.

2. Observe the prominent second antennae and the single compound eye enclosed within a head shield at the anterior end. The prominant muscles, that originate on the inner surface of the head shield and insert on the inner surface of the second antennae, are birefringent and may be visualized using crossed polarizing filters. In some populations, the size and shape of the head shield undergoes a dramatic variation (**cyclomorphosis**) depending on season.

3. Find the first antennae, located just beneath the **rostrum** and the small, pigmented **ocellus**, just posterior to the compound eye.

4. Also, locate the mandibles, posterior to the insertion of the second antennae. Using crossed polarized filters may help you locate the mandibles as their striated grinding surfaces are birefringent.

5. Note the spine at the posterior end of the carapace. What function might the spine serve? Does the specimen have an enlarged dorsal **crest** or **helmet** above the compound eye? If so, examine the musculature that originates at the crest and inserts on the second antennae. Compare the development of this musculature with a nonhelmeted form. How do they differ? What function(s) might the helmet serve?

6. Find the gut, which first courses anteriorly and then curves posteriorly, proceeding through the central part of the body. Is food in the gut? The gut exits the body after passing through the flexible **postabdomen**.

7. Locate the thoracic legs within the **branchial chamber**. Note the numerous comblike setae on each leg. In life, the legs are almost constantly in motion producing a water current that flows between the valves of the carapace, bringing food to the animal.

8. Dorsal to the gut is the **brood chamber** that may contain developing embryos. How many embryos are present in your specimens? At what stage of development are the embryos? If sexual eggs have been fertilized, they undergo **diapause** within a thick-walled brood chamber called an **ephippium** [G., saddleshaped]. Are any of the specimens ephippiate? How many diapausing embryos are present in each ephippium? What are the consequences to the female of her ephippium being dark? ■

Class Ostracoda

Ostracoda [os-tra-KO-dah; G., *ostracum*, shell], is a class of small, bean-shaped crustaceans, commonly called

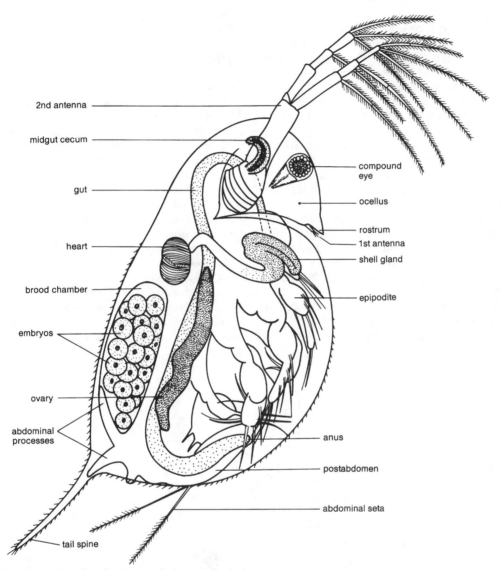

Figure 15.22. Lateral view of the freshwater cladoceran *Daphnia*.

2nd antenna

midgut cecum

gut

heart

brood chamber

embryos

ovary

abdominal
processes

tail spine

compound
eye

ocellus

rostrum

1st antenna

shell gland

epipodite

anus

postabdomen

abdominal seta

seed shrimps. Most are less than 1 to about 3 mm long, although some larger (greater than 2 cm) marine forms are known. Ostracods resemble bivalves in possessing a dorsally hinged carapace of right and left **valves**, with an adductor muscle for closure. Resemblance to bivalves provides another common name, mussel shrimp. The valves are reinforced with calcium carbonate, except at the hinge. Ostracods have a cosmopolitan distribution in benthic freshwater and marine environments; several terrestrial genera are known. As with the cladocerans, ostracods use their antennae for locomotion. Many ostracods are omnivorous.

■ Observational Procedure:

Live Specimens

Observe the locomotory activities of ostracods in a small culture vessel or beaker. How do they move about? What appendages appear to be most important for locomotion? Using a dissection microscope observe the movement of ostracods in a small Petri dish. Compare locomotion in ostracods to that of cladocerans (e.g., *Daphnia*). Do ostracods swim as much of the time as the cladocerans you just studied? If they don't, where do they spend much of their time?

LIVERPOOL
JOHN MOORES UNIVERSITY
AVRIL ROBARTS LRC
TEL. 0151 231 4022

Morphology

The following observations of ostracod morphology may be made using live or preserved specimens. As with cladocerans, live ostracods may require an anesthetic or methyl cellulose to slow them down.

1. Obtain an ostracod and observe it under a compound microscope (Fig. 15.23). Note the ovoid body and determine the anterior, posterior, dorsal, and ventral ends.

2. Observe at the anterior end the prominent, paired first and second antennae. How do these differ from those of *Daphnia*? Note how the valves surround the body and the appendages except for the flagella of the antennae.

3. Unfortunately, much of the internal anatomy of ostracods is obscured by the valves. However, you should be able to observe the single compound eye just posterior to the insertion of the first antennae.

4. In the central region of the valve, locate the adductor muscle.

5. Attempt to locate the gut. Occasionally, portions of the gut may be identified by the presence of food material. The mandibles may been seen about one-third of the way back from the anterior end near the ventral margin of the valves. ■

Class Copepoda

Copepoda [ko-PEP-o-dah; G., *kope*, an oar + G., *podos*, foot] is a large class (more than 8,000 species) of small bullet-shaped crustaceans with prominent first antennae. Like herbivorous cladocerans, copepods represent an extremely important link in aquatic food chains. They range in size from less than 1 mm to several millimeters long. Copepods have three tagmata, including a cephalothorax (head plus 1–2 thoracic segments), the remaining thorax, and an abdomen. Each tagma is comprised of a few to several rigid, cylindrical segments joined by somewhat flexible joints. There is a major junction within the body dividing it into two parts: **metasome** and **urosome**.

Copepods have a cosmopolitan distribution in freshwater and marine environments occurring in both benthic and planktonic habitats. Herbivorous, predaceous, and parasitic forms are known. Ectoparasitic copepods often look very similar to free-living species, but some

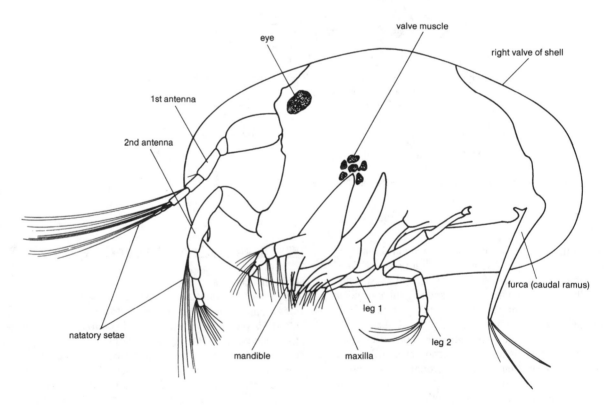

Figure 15.23. Lateral view of a generalized ostracod.

endoparasitic forms are so modified that they hardly resemble other copepods at all.

■ Observational Procedure:

Live Specimens

1. Observe the locomotory activities of copepods in a small culture vessel using a dissection microscope. How do they move about? What appendages appear to be most important for locomotion? Attempt to capture an individual using a wide-mouth Pasteur pipette. Is this an easy task? What does it indicate about copepods? Compare locomotion in copepods to that of cladocerans and ostracods.

2. Place a single specimen into a drop of pond water on a shallow depression slide and cover it with a coverslip. The specimen should be restrained as much as possible, but not crushed.

3. Note that there is no heart as in the cladocerans and ostracods. Are there any colored materials present outside the gut? Colored lipid drops are often found in the hemocoel of copepods (and other microcrustaceans) as forms of stored energy.

4. Occasionally, the exoskeleton will be covered with sessile protozoans or algae. Are any present on your specimen? How might these phoretic [G., *phora*, to bear] organisms influence copepod swimming? ■

Morphology

The following observations of copepod morphology may be made using live or preserved specimens. As with the other microcrustaceans studied, live copepods may require an anesthetic or methyl cellulose to slow them down.

1. Obtain a copepod and observe it under a compound microscope (Fig. 15.24). Examine the bullet-shaped body and identify the anterior, posterior, dorsal, and ventral ends. Also determine the metasome and urosome.

2. At the anterior end observe the long uniramous first antennae and the shorter second antennae. The first antennae are not used in locomotion but are important in sensory reception. Locate the compound **naupilus eye** at the very tip of the head region.

3. Other appendages that should be found include the feeding appendages, the paired swimming legs (in the thorax), and a pair of setaeous **caudal rami** at the posterior end of the urosome.

4. Females may carry a pair of egg sacs near the junction of the metasome and urosome. Are egg sacs present on your specimen?

5. Internally, look for prominent bands of longitudinal and circular muscles. ■

Class Cirripedia

Cirripedia [ser-ree-PED-ee-ah; L., *cirrus*, curl of hair + L., *pedis*, foot] is a small class of about 1,000 species of sessile, marine crustaceans commonly called barnacles. Barnacles hardly resemble crustaceans at all. In fact Carl von Linné (a.k.a. Linnaeus) classified barnacles with the Mollusca where they remained until 1830, when their larvae were discovered and it was deduced that they were crustaceans. Barnacles have tremendous economic importance, chiefly by fouling ocean-going vessels.

Most people are familiar with free-living thoracian barnacles, which are separated into two groups: (1) stalked or goose-neck barnacles (*Lepas*) and (2) stalkless, or sessile barnacles (*Balanus*). Parasitic barnacles, which will not be studied here, have a morphology that is very much different from the thoracicans.

■ Observational Procedure:

Live Specimens

Examine a living barnacle in a small dish filled with seawater and observe movements of the rakelike

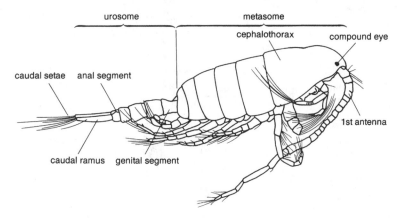

Figure 15.24. Lateral view of typical free-living copepod.

structure that is periodically exposed. These are the legs or **cirri** and they are involved in feeding. The rate of movement of the cirri and the way in which they are used to filter particles from the water depends on factors such as temperature, water movement, and concentration of food. Place a drop of milk near the cirri and observe the movement of milk. How do the cirri respond to the presence of the milk suspension? At the direction of your instructor attempt feeding the barnacles other materials.

Morphology

1. Obtain a specimen of *Lepas*, a stalked barnacle, and place it in a dissection pan with the muscular peduncle towards you. Distinguish the **peduncle** from the laterally compressed capitulum (Fig. 15.25).

2. On one edge of the capitulum, locate the **aperture** and on the opposite edge (dorsal) locate an unpaired, keel-like plate called the **carina**.

3. On either side of the capitulum and near the peduncle, locate a pair of large, broad plates, the **scuta**. Above the scuta is another pair of smaller plates, the **terga**.

4. Remove the mantle and plates from the right side of the animal. Pin the animal to the dissection pan and cover it with water.

5. The most conspicuous structures are six pairs of biramous **cirri** (legs) with numerous setae. The cirri curl towards the mouth and increase in size posteriorly. How is the setation distributed on the cirri? To understand this question you will need to examine the cirri using a dissection microscope.

6. Posterior to the cirri, locate the long, segmented penis and at its base, the posterior anus.

7. At the anterior end, locate the adductor muscle that connects the right and left sides of the mantle. Posterior to this muscle find the mouth and accompanying mouth appendages. The mandibles will be easily recognized by their strong dentition. Other appendages will not be identified.

8. Obtain a specimen of a sessile barnacle such as *Balanus*. Dried specimens will suffice, unless the specimen is to be dissected (Step 10).

9. Determine the anterior-posterior orientation of the barnacle and the various plates (Fig. 15.26). Note that the operculum is composed of two pairs of movable plates (terga and scuta).

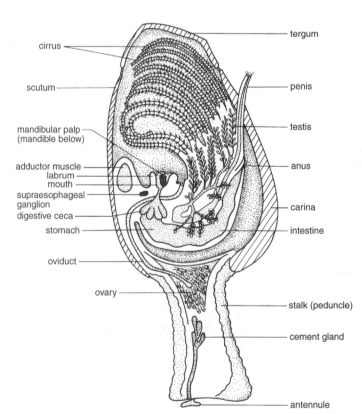

Figure 15.25. Lateral view of the stalked, thoracican barnacle, *Lepas*.

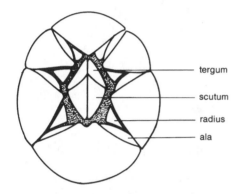

Figure 15.26. External plates of the sessile barnacle, *Balanus*.

10. The internal anatomy of *Balanus* is similar to that of *Lepas*. If advanced dissection is to be done, your instructor will provide additional directions.

11. Obtain a prepared slide of a **naupilus larva** of a barnacle under a compound microscope and examine the shield-shaped body.

12. Determine the anterior, posterior, dorsal, and ventral ends. At the anterior end locate the first antennae, nauplius eye, and frontal horns.

13. Two prominent bulges may be seen on either side of the ventral surface. These are the second antennae, which possess cement glands.

14. Towards the posterior part of the body a metameric tail-like structure may be seen. These metameres will develop into the thoracic legs in subsequent molts. ∎

Class Malacostraca

Class Malacostraca [MAL-ah-co-STRAK-ah; G., *malaco*, soft + G., *ostracum*, shell] encompasses three-quarters of the crustaceans, including those most favored on the dinner table: shrimp, crab, and lobster. The body plan is of two tagma: anteriorly a fusion of the cephalon

and thorax into the cephalothorax and posteriorly an abdomen.

Order Amphipoda. Amphipoda [am-FIP-o-dah; G., *amphi*, double + G., *podus*, foot] is a large order (about 6,000 species) of shrimplike malacostracans that range in size from fewer than 1 mm to nearly 30 cm in a deep-sea benthic form. Most amphipods are marine, but many freshwater species and a single terrestrial group are known. Amphipods are commonly called **scuds**, indicating a swift swimming movement, but benthic burrowing and crawling forms are common. The laterally flattened body is slightly rounded on the dorsal side. Three tagmata are recognized: head, thorax (preaeon), and abdomen with pleon plus urosome. Amphipods lack a carapace.

∎ Observational Procedure:

Live Specimens

If available, observe locomotory activities of amphipods in a small dish with the aid of a dissection microscope. How do the animals move about? What appendages appear to be most important for locomotion? Compare locomotion in amphipods to that of the other crustaceans examined so far.

Morphology

1. Obtain an amphipod such as *Gammarus* and observe it under a compound microscope (Fig. 15.27). Determine the anterior, posterior, dorsal, and ventral ends of the organism.

2. Locate the paired, prominent first and second antennae at the anterior end. Note the well-defined head with paired, sessile compound eyes.

3. Find the feeding appendages in the buccal region. They are usually small and hidden from view. Locate the two pairs of subchelate **gnathopods** that are used in grasping food and, by males, in holding the female during copulation. Gnathopods are the second and third pairs of thoracic legs called **pereiopods** [G., *pereio*, on the other side; also peraeopods]. The third through seventh pereiopods are unspecialized ambulatory limbs.

4. Coxal gills are difficult to locate, but may be seen as ovoid sacs on the inner surface of the coxae of legs 2–6.

5. In the pleon region of the abdomen, locate the three pairs of setous **pleopods** [G., *pleo*, swim + G., *podos* foot] used in swimming.

6. Posterior to the pleon is the urosome of three segments, each with paired uropods.

Order Decapoda. Crayfish, crabs, and lobsters are members of the order Decapoda [DEC-ah-poad-ah; G., *deca*, ten + G., *podos*, foot] by virtue of having five pairs

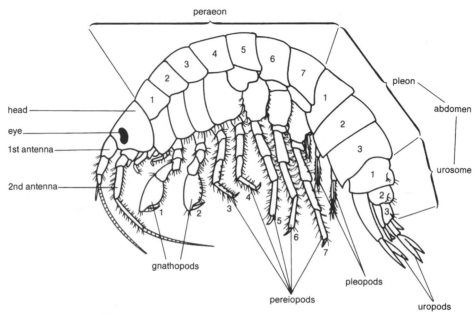

Figure 15.27. Lateral view of the freshwater amphipod, *Gammarus*. (After Pennak.)

of pereiopods. We will study the crayfish as a representative decapod. The study of the blue crab, which then follows, may be substituted for that of the crayfish or these two decapods may be compared. ∎

Crayfish. Crayfish, crawdads, or crawfish are commercially important as food, especially in Louisiana. However, crayfish often are considered to be pests. In some regions they consume young crops and their burrows interfere with cultivation, while in others their feeding habits deteroriate habitat previously suitable for game fish.

∎ Observational Procedure:

General Anatomy

1. Obtain a preserved crayfish and place it in a dissection pan dorsal side up with the anterior end facing away from you (Fig. 15.28A). At first glance it is obvious there has been considerable fusion of segments.

2. The large dorsal, saddle-like structure is the carapace. It represents fused segments of the head and thoracic regions into a single unit, the cephalothorax.

3. Locate the transverse **cervical groove** that divides the cephalothorax into its anterior and posterior regions. Anterior to this groove is the head, and posterior, the thoracic region. In the mid-dorsal region of the thoracic portion of the carapace and bounded by the **branchiocardiac grooves** is the **areola**.

4. The fused terga of the carapace curve ventrolaterally to form a rooflike cover known as the **bran-**

chiostegite. The space underneath it is the branchial chamber.

5. Turn the specimen over and reflect to one side one of the branchiostegites (Fig. 15.28B). Observe the branchial chambers containing the gills.

6. Return the specimen to its dorsal side and locate the **rostrum**. To each side of the rostrum are located the stalked compound eyes that are set in ocular depressions (Fig. 15.28A).

7. Turn the specimen ventral surface up and find the biramous **antennules** (first antennae). These are the anteriormost appendages.

8. Just behind them are the elongate antennae (second antennae).

9. The next six pairs of appendages surround the mouth. These feeding structures are, from front to rear, the **mandibles**, first and second **maxillae**, and the first, second, and third **maxillipeds**.

10. The next five pairs of appendages are termed pereiopods. The first pair of pereiopods (**chelipeds**) are large and stout, distally bearing massive chelae. They are used in prey capture, defense, and aggression, and in manipulation of food.

11. The next two pairs of appendages posterior to the chelipeds are also chelate, but are much smaller. These and the following two nonchelate pairs are the walking legs.

12. Turn your attention to the abdomen. There are six abdominal segments (**tergites**). The first five are somewhat bandlike and extended laterally over the sternal plates. The sixth segment is flattened and

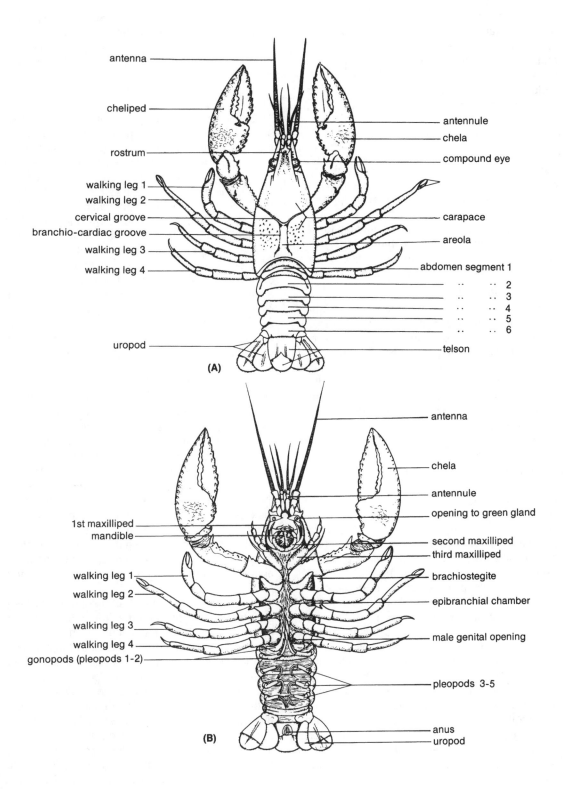

antenna

cheliped

rostrum

walking leg 1
walking leg 2
cervical groove
branchio-cardiac groove
walking leg 3
walking leg 4

uropod

antennule
chela
compound eye

carapace

areola

abdomen segment 1
··   ·· 2
··   ·· 3
··   ·· 4
··   ·· 5
··   ·· 6

telson

(A)

antenna

chela

antennule
opening to green gland

1st maxilliped
mandible

walking leg 1
walking leg 2

walking leg 3
walking leg 4
gonopods (pleopods 1-2)

second maxilliped
third maxilliped
brachiostegite

epibranchial chamber

male genital opening

pleopods 3-5

anus
uropod

(B)

Figure 15.28. Male crayfish. (A) Dorsal view. (B) Ventral view.

broader than the others, and bears the tail, consisting of two fanlike biramous appendages called **uropods** and a central plate, the **telson**.

13. Ventrally, the abdomen is somewhat flattened and possesses six pairs of lateral appendages called **pleopods**. In the female the first pair of appendages is diminutive, while in male the first two pairs are modified for reproductive purposes.

14. In the central part of the telson find the anus.

Appendages

This section gives direction for a detailed study of crayfish appendages. In each instance, observe the appendages in place before they are removed, noting their position, movements, and relationship to other appendages. Each appendage will be compared to the hypothetical biramous crustacean limb (Fig. 15.21). As the appendages are removed from one side of the body, lay them out in their proper sequence.

Sensory Appendages

1. Manipulate the first antennae (antennule) and determine what its range of motion is. Two **flagella** of unequal length and a **protopod** of three segments compose the first antennae (Fig. 15.29). The proximal, single segment of the protopod is the coxa. The next two distal segments are a basis of two parts. The flagella attach to the distal joint of the basis. The shorter, inner flagellum is the **endopod** and the longer, outer one is the **exopod**.

2. The dorsal surface of the coxa of the first antennae when viewed under magnification will show a depressed area. This is the external opening to the **statocyst**.

3. Observe the second antenna, manipulating it to determine its range of motion.

4. Using a scalpel, carefully remove the second antenna, making sure that the coxa remains attached. The large leaflike structure is the exopod, also termed the **squame** or **antennal scale** (Fig. 15.29).

5. Locate the slightly elevated structure on the coxa and note the opening at the apex. This is the excretory organ commonly called the **green gland**.

6. Anterior and lateral to the coxa there is the broad second joint of the protopod, the basis. Inward to the basis is the first joint of the endopod, called the **ischiopodite**. The endopod consists of three joints. First is the ischiopodite, next the **meropodite**, and then the **carpopodite**. The latter supports the many-jointed flagellum. The straight margin of the squame is the outer edge; the inner margin is clothed on its margin with a fringe of long setae. Examine the distribution and kinds of setae on all parts of the second antenna.

7. Remove one eye, including the **peduncle** (stalk) and examine it under a dissection microscope.

8. Scrape a portion of the eye's surface and place the scraping on a slide. Add a drop of water and a coverslip; examine under low power of a compound microscope. What pattern do you see? What does this represent?

Feeding Appendages

1. Observe the appendages in the region of the mouth and move them about using a probe. Observe how they articulate and nest together. It is easier to study the crayfish mouthparts by removing the third maxilliped and then proceed anteriorly, appendage by appendage, until the mandibles are reached. The following procedures use this approach.

2. Locate the three maxillipeds. The third maxilliped is the largest and most posterior feeding appendage. It is located anterior to the base of the chelipeds. Lying just anterior to the third maxilliped is the second and ahead of it the first. Carefully remove all three maxillipeds by grasping and pulling each frimly at its base with forceps. Observe that the second and third maxillipeds are similar, being composed of an outer exopod with an extension called the flagellum and an inner endopod of several joints (Figs. 15.29 and 15.30).

3. Examine the first maxilliped. The elongate, basal rectangular piece of the first maxilliped is the epipodite, which fits into the gill chamber and assists in water movement (Fig. 15.29). Anterior to the epipodite is the long, narrow, and medially grooved exopod. Distally the exopod has a many-jointed filament called the flagellum. The large, leaflike inner structure is divided into two parts, endite 1 proximally and endite 2 distally. At the base of and in between endite 2 and the exopod is the relatively diminutive endopod. As with the second maxilla, the coxa and basis are not distinctly separated. The central complex of chitinized structures represents a fusion of parts and constitutes the protopod. A large epipodite is attached to a coxa, as are the endites. The endopod and exopod are attached to the basis.

4. Locate the second maxilla. The largest part is the **scaphognathite** or **gill bailer** (Fig. 15.29). The narrower part lies next to the mandible, while the broader portion extends into the branchial chamber and assists water movement in the chamber. The lateral (inner) part of the second maxilla is divided into four narrowly foliaceous subdivisions. The anterior pair is considered to be endite 2 and the posterior endite 1. The slender, elongate appendage between endite 2 and the scaphognathite is the endopod. The basis and coxa are not distinguishable as separate components.

5. The first maxilla is leaflike (Fig. 15.29). The smaller, pointed part is the endopod. The two parts lateral to the endopod are endites; the broadest one is endite 1 and the longest is endite 2.

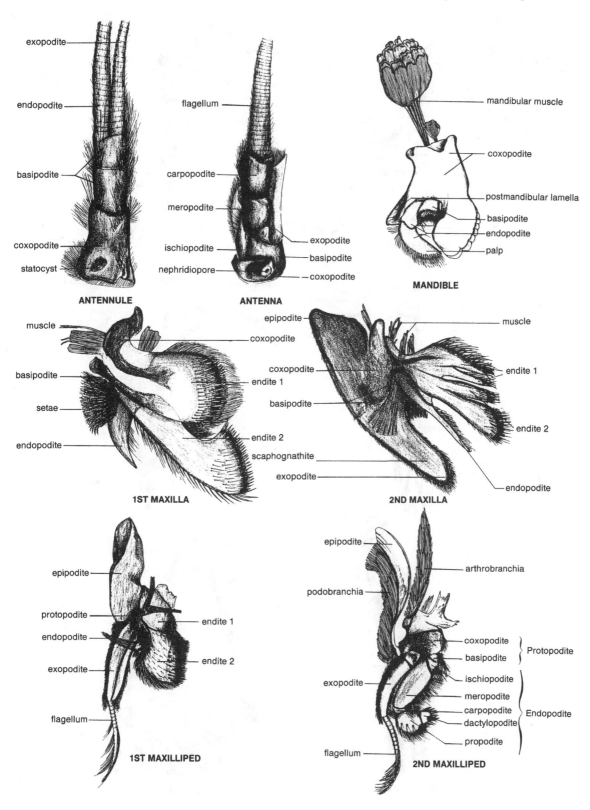

Figure 15.29. Crayfish appendages: first antennae through second maxilliped.

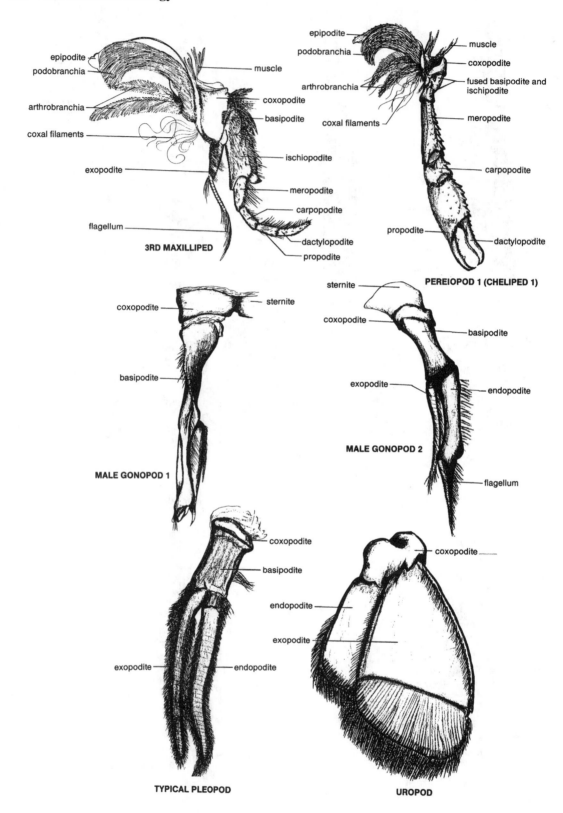

Figure 15.30. Crayfish appendages: third maxilliped through uropod.

6. Locate the anteriormost feeding appendages, the mandibles, composed of broad, smooth, convex surfaces, with an inner sawlike edge and a blunt finger-like process (Fig. 15.29). The basal part of the mandible is the coxa and the finger-like structure the **palp**, comprised of three joints. The first joint of the palp is considered the basis and the outer two segments compose the endopod. The exopod is not present on the mandible.

Walking Appendages The first, second, and third maxillipeds are considered the first appendages of the thorax. The next five to be studied are the pereiopods.

1. Locate the first pair of pereiopods, the chelipeds (Figs. 15.28 and 15.30). They are the largest of the appendages and are used to capture prey and to tear food apart. The remaining four pairs of pereiopods are the walking legs. Of these, only the first two pairs are chelate.

2. Carefully remove all the pereiopods from one side of the body, severing them from the body at the coxa. Note that the exopod is not present in these appendages.

3. On the posteriormost part of the cheliped is a bilamellate, folded **podobranch gill** (arthrobranchia). The lamellated epipodite is membranous and folded as in the second and third maxilliped. The upper lamella is covered with the gill filaments. Near the point where the long, threadlike coxal filaments are located, the plumose arthrobranchia are attached. Even with the greatest of care, the arthrobranchia may be difficult to remove intact.

4. Observe that the chela is made up of the propodite and dactylopodite (Fig. 15.30). The propodite is extended finger-like at its outer edge, while the dactylopodite is moveable and opposes the propodite, forming a pincer-like structure.

5. Remove a chela from one of the large chelipeds. This is best done by severing the appendage at the carpopodite joint (Fig. 15.30).

6. Remove the exoskeletal covering from one side of the chela. Judicious use of a nut cracker will be helpful. Note the two sets of muscles. The larger is the adductor muscle; it inserts on the inner edge of the dactylopodite and is instrumental in pulling the dactylopodite toward the propodite, thus providing the pincer action. The smaller abductor muscle reverses the movement.

7. Pereiopods 2 and 3 are chelate and other than in size are very similar to the cheliped. The female crayfish has a **genital operculum** on the inside surface of the coxa of the third pereiopod. The male genital opening is located on the inner surface of the coxa of the fifth pereiopod (Fig. 15.28).

8. Identify the sex of your specimen.

9. The fourth and fifth pereiopods are strictly ambulatory appendages and are nonchelate. Between the coxopodites of the fifth pereiopods, the sternum in some female crayfish is modified as a **seminal receptacle**. This structure is termed the **annulus ventralis**. Sperm released from the receptacle at egg-laying time fertilize the eggs, which are then retained on the under surface of the tail region.

10. The remainder of the crayfish appendages are located on the abdominal region and are called pleopods or **swimmerettes** (Figs. 15.28 and 15.30). Some of them are similar in structure, but others differ considerably, depending on the sex.

11. Compare the position and morphology of the pleopods on the first and second abdominal segments in male and female crayfish.

12. In the male, the first pair of pleopods are closely associated, directed forward, and modified for sperm transport to the female during copulation. They are called **gonopods** (Fig. 15.30).

13. In the female the first pleopod is small and is composed of a single stalk (of two joints) with a flattened, distal flagellum. The second female pleopod is the typical biramous appendage characteristic of most abdominal segments (Fig. 15.30). The sternite to which the pleopod is attached is narrow and ringlike.

14. The remaining pleopods are the same in both sexes (Fig. 15.30). The coxa is a small ringlike structure and may be lost unless the limb is carefully removed. The endopods and exopods are flattened, setose, and flagellate. They arise from the stocky basis.

15. The sixth pair of pleopods is the uropods. They are similar in both sexes, but radically different in shape compared to the other pleopods. The exopod and endopod are large paddle-like flaps (Fig. 15.30). The exopod is divided into two sections along a broad joint. The coxa and basis are fused into a large, angular segment from which the exo- and endopods arise.

16. The telson is not considered to be an appendage in the same sense as the others studied here. Note that the anus opens through the telson's ventral surface (Fig. 15.28). The uropods and the telson comprise the tail fan. This structure is used for rapid backward movement.

Gill Chamber Cut away the carapace which covers the gills on the side where appendages are still intact. Be careful to cut only the branchiostegite, not the gills within the chamber. The entire gill system should be exposed undisturbed. Starting at the anteriormost gill, note how the gills are arranged. Using dissection needles, tease the gills apart to understand their places of attachment and serial arrangement.

Internal Anatomy

1. Make a very shallow, midsagittal incision through the carapace, beginning at the posterior dorsal margin, and extend it to the rostrum. Avoid damaging the muscles and other internal organs just beneath the carapace. Remove the carapace anterior to the **cervical groove** (Fig. 15.28A).

2. With the carapace removed, several muscle groups and other internal organs are exposed. Extending to the rostrum from the stomach are the **anterior gastric muscles** (Fig. 15.31). The two large muscles at the lateral edge of the stomach are the **mandibular adductors**, and those attached to the posterior dorsal margin of the stomach are the **posterior gastric muscles**. One of the mandibular muscles may have been removed previously when the appendages were being studied, but there should be one left in position. The anterior and posterior set of muscles are important in the mechanical operation of the stomach. Dissection a little later will show an intricate set of **ossicles** in the stomach that are moved by muscles assisting in grinding food that enters the stomach.

3. Gentle movement of the mandibular muscles to one side will reveal the dorsal anterior portion of the digestive gland (Fig. 15.31).

4. Cut away the carapace posterior to the cervical groove. This will expose the heart and the principal arterial vessels, reproductive bodies, and the remainder of the hepatic bodies. On each side of the heart are the elongate extensor muscles that manipulate the abdomen. Closely examine the heart and note the **ostia** [L., opening]. If the specimen is a female, the ovary will be found directly in front of and slightly below the heart. If the sex is male, the testes will be located in about the same position.

5. Remove the right lateral wall of the cephalothorax to the base of the legs. The expansive liver is best shown by this dissection. Identify the green gland (Fig. 15.31).

6. Remove the stomach by severing the esophagus from the intestine. Place it in a dissection pan, partially filled with water. Turn the stomach ventral surface upward. Examine the organ with the dissection microscope. Make a longitudinal midsagittal incision and then pin the stomach membrane aside. Identify the several sets of ossicles (Fig. 15.32) that give the stomach the name **gastric mill**.

7. Also note the unusual lining of portions of the stomach. This serves as a filtering mechanism. In the anterior, lower portion of the stomach, there may appear rounded, flattened structures on each side. These are **gastroliths**. They occur seasonally and will not be present in recently molted individuals.

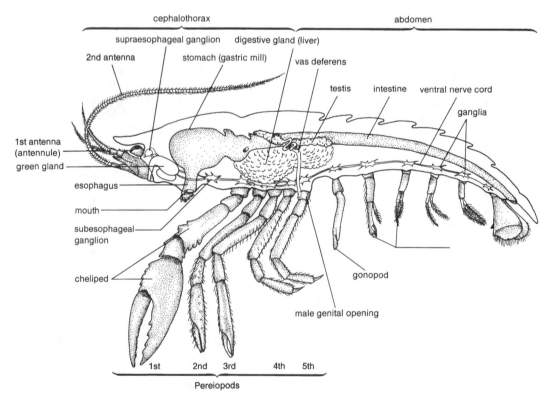

Figure 15.31. Internal anatomy of the crayfish (diagrammatic).

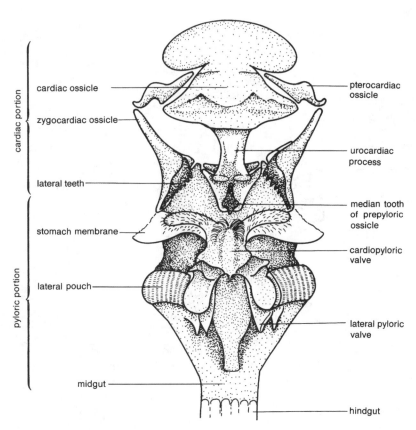

cardiac portion
- cardiac ossicle
- zygocardiac ossicle
- lateral teeth

pyloric portion
- stomach membrane
- lateral pouch
- midgut

- pterocardiac ossicle
- urocardiac process
- median tooth of prepyloric ossicle
- cardiopyloric valve
- lateral pyloric valve
- hindgut

Figure 15.32. An internal, ventral view of the gastric mill.

8. Now that the stomach and other viscera in the cephalothorax have been removed, direct your attention to the nervous system (Fig. 15.31). Locate the stub of the esophagus and posterior to it find the **subesophageal ganglion**. Extending around the esophagus are single nerve connectives that communicate with the **supraesophageal ganglion**.

9. Follow the nerve tract posteriorly to the abdomen. This will necessitate careful dissection as the nerve trunk is closely associated with the inner surface of the exoskeleton.

10. Remove the upper exoskeletal surface of the abdomen to the telson (Fig. 15.31). Locate the hingut and the superior abdominal artery with its lateral branches. Note the arrangement and distribution of the abdominal flexor muscles.

Live Specimens
Observe the live crayfish on display. How do they move about the aquarium? How does the crayfish react when threatened? As your instructor directs, observe the feeding behavior of crayfish. ■

Blue Crab. The blue crab, *Callinectes sapidus*, is a common Atlantic and Gulf coast inhabitant of the United States, ranging from Massachusetts Bay to the northern shores of South America. It is commercially important to the fishing industry in the United States. This crab lives in waters ranging in salinity from brackish to normal sea water.

■ Observational Procedure:

External Morphology
Obtain a specimen and compare its morphology with that of the crayfish (see Figs. 15.28 and 15.33A,B).
1. Place the crab dorsal side up in a dissection pan. Note that, unlike the crayfish, the blue crab is compressed dorsoventrally and is greatly expanded laterally. The crab is very broad and the legs are spread comparatively farther apart than in the crayfish.
2. Turn the crab ventral side up and find the abdomen, which is folded under the thorax (Fig. 15.33B). Note that the abdomen is proportionally smaller than that of the crayfish. The abdomen consists of six segments and is narrow and T-shaped in males, triangular in immature females, and broad in mature females.
3. What is the sex of your specimen? Observe the abdomen of a crab of the opposite sex. Find the tuber-

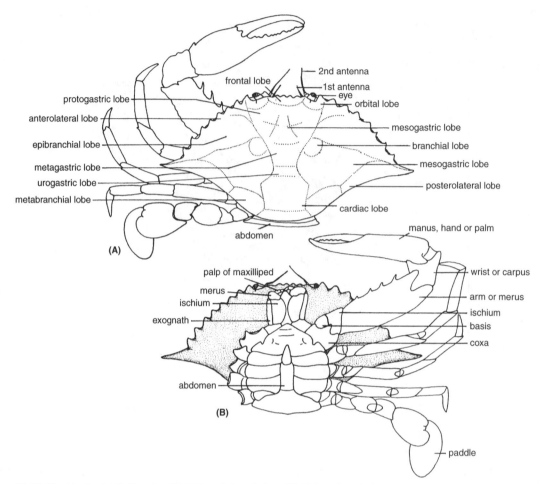

Figure 15.33. The blue crab, *Callinectes*. (A) External dorsal view. (B) External ventral view.

cles on the fifth thoracic sternum of males and immature females that lock the abdomen to the thorax.

4. Locate the anus on the ventral side of the abdomen at its distal end.

5. The ventral side is divided into several regions (Fig. 15.33B). The area reflected around from the spines to the legs is called the **sub-branchial carapace**. Most of the ventral surface is composed of the thoracic sterna, separated from one another by distinct suture lines. The thoracic sterna are more or less covered by the abdomen, depending on the sex.

6. Although the position and anatomy of the gills will be considered later, locate the right and left **inhalant** and **exhalant apertures**, through which water enters and leaves the branchial chambers. The inhalant apertures are located anterior to the coxae of the chelipeds; the exhalant apertures are on either side of the mouth (Fig. 15.33B).

Appendages

1. Identify the first antennae (antennules), second antennae, and eyes noting their size, shape, and relative position (Fig. 15.33A and B). Compare these structures to those of the crayfish. Determine the movements of which each are capable. Locate the depression into which the first antennae fold. What is the significance of the ability to protect the antennae in this way?

2. The feeding appendages are similar to those of the crayfish. Remove a set from one side of the crab and compare them with those of the crayfish. As in the crayfish, it is best to locate the three pairs of maxillipeds in reverse order, followed by the two pairs of maxillae, and then the mandibles.

3. The third maxillipeds are the largest feeding appendages. Progressing towards the mouth, locate the second and third maxillipeds.

4. On each maxilliped, locate the long, setous epipodite called the **flabellum**, that enters the branchial chamber. The three flabella are fringed with setae and function to clean debris from the surface of the gills.

5. Find the second and first maxillae. The second maxillae has a large, platelike exopodite called the **scaphognathite** or **gill bailer**. The scaphognathite beats back and forth to circulate water through the branchial chamber.

6. Locate the mandibles surrounding the mouth. Note their broad, smooth, heavily calcified surfaces. These appendages are used to cut, crush, and hold food, while other appendages process it.

7. The pereiopods are similar in crabs and crayfish, although the basis and ischium are fused into a single unit called the basi-ischiopodite. As in the crayfish, the first pair of pereiopods (chelipeds) are chelate and are used to obtain food and for defense.

8. Note how the articulation of this pereiopod permits it to be folded closely against the body. The next three pairs of pereiopods are not chelate and are used for walking. If a leg is caught or damaged, it is cast off or **automized** by the crab at the basi-ischiopodite and is later to be regenerated.

9. Find the last pair of pereiopods (Fig. 15.33). How do they differ from the other pereiopods? These paddle-shaped legs are used in a sculling fashion for swimming.

10. On the ventral surface of the abdomen, locate the pleopods. Only the first two pairs of pleopods are present in the male. They are highly modified and used in copulation. The first pair is larger and possesses an elongate, hollow spine (endopod). The second pair is similar to the first, but its spine is shortened and not hollow.

11. Observe that the spine of the second pleopod enters the first pleopod through a hole (anterior foramen) at its base. Spermatophores are pushed through the hollow spine of the first pair by the spine of the second.

12. Examine the pleopods of a female crab. In the female the first pleopods are absent, but the second through the fifth pairs are present and similar. The long endopod and exopod of these appendages possess many setae. These setae help hold the numerous embryos while they develop.

13. Examine a female specimen in **sponge**. The spongy material, composed of hundreds of small balls, is the egg mass. Ocassionally a specimen will be found with 2–3 large roundish masses where the eggs should be. These are the reproductive organs of *Sacculina*, an unusual parasitic barnacle.

14. Note that uropods are absent in both sexes.

Internal Anatomy

1. Carefully remove the top of the carapace by making an incision a short distance inward from the dorsal border of the carapace and separate the underlying tissues from the exoskeleton.

2. Locate the heart enclosed within the pericardial cavity by the translucent pericardium and the vessels leading from the heart (Fig. 15.34). The circulatory system is almost identical to that of the crayfish. The most noticeable organ exposed at this point is the large digestive gland (**hepatopancreas**) with left and right lobes.

3. Carefully determine what the extent of the digestive gland is without damaging other tissues (Fig. 15.34).

4. Find the gonads on the surface of the digestive gland and trace their ducts ventrally just anterior to the heart. Ovarian size may vary considerably in different specimens.

5. Paired **antennary glands** that are excretory in function lie just lateral to the cardiac stomach; they usually are difficult to find.

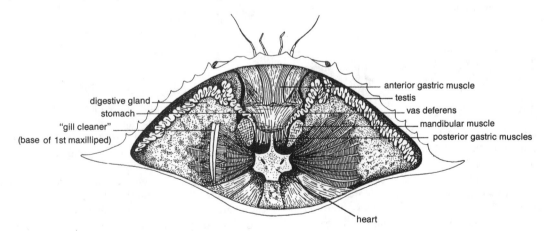

Figure 15.34. Dorsal internal view of *Callinectes*.

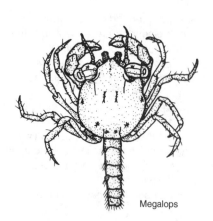
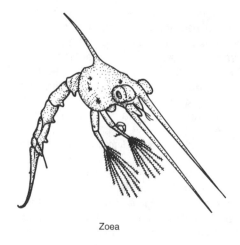

Megalops

Zoea

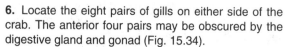

Figure 15.35. Zoea and megalops larvae of a crab.

6. Locate the eight pairs of gills on either side of the crab. The anterior four pairs may be obscured by the digestive gland and gonad (Fig. 15.34).

7. Remove a gill and examine it using a dissection microscope. Each gill has a central axis and consists of an anterior and posterior row of lamellae.

8. The gut is similar to that of the crayfish. Locate the esophagus, cardiac stomach, pyloric stomach, and intestine. Find the long, narrow midgut cecum extending from the pyloric stomach.

9. Remove the stomach region and identify the nerve ring around the esophagus and the nerves that radiate from it.

10. Locate the cerebral ganglion anteriorly and a large, posterior concentration of ganglia around the sternal artery. From this conspicuous mass radiate separate nerves to all mouthpart appendages, to the pereiopods, to the abdomen, and to a pair of circumesophageal commissures that connect with the cerebral ganglion.

11. If advanced study is to be done of blue crab anatomy, your instructor will provide the necessary directions.

Crab Larval Stages

1. Examine prepared slides of **zoea** and **megalops** larvae (Fig. 15.35). In each locate the carapace, abdomen, rostrum, and stalked compound eyes.

2. Find the biramous thoracic appendages in the zoea. These will develop into the maxillipeds.

3. Locate the chelipeds and walking legs in the megalops. This is the last stage before final metamorphosis into the adult crab.

Live Specimens

Observe the live specimens of crabs on display. How do they move about the aquarium? How do the various species react when threatened? Can you observe the bailing activity of the scaphognathites? As your instructor directs, place a small piece of shrimp to one side of the crab. How does the crab respond? Observe the elaborate movements of the mouthparts during feeding. ■

D. Subphylum Uniramia

Class Insecta

All insects belong to the class Insecta (in-SEC-tah; L., *insectum*, insect). These arthropods are the dominant and most successful group of all invertebrates, especially in the invasion of terrestrial habitats. Insects are the only invertebrates capable of flight. There are probably not less than 750,000 described insect species; however, some recent estimates of the actual number of extant insect species may reach 10 million. The ecological and economical impacts insects have on local and worldwide populations of plants and animals are tremendous. The study of insects is called **entomology** [G., *entomon*, insect + G., *logy*, study], often with emphasis in medical entomology and economic entomology.

The bodies of adults are divisible into three **tagmata**: an anterior **head**, middle **thorax**, and posterior **abdomen**. The head typically bears two antennae, five eyes (two compound and three ocelli), and mouthparts. Six legs and usually two pairs of wings in the adults are attached ventrolaterally and dorsolaterally to the thorax, respectively. The thorax is always composed of three segments.

Classification

Class Insecta is often divided into two large subclasses based on the presence or absence of wings. Insects lacking wings are believed to be the most primitive. Each subclass is further divided into orders based on features including morphology of mouthparts, modes of development, and wing venation. Familiarity with the principal characteristics of the orders is a necessity for the beginning student of insect biology. The more common orders are characterized below.

1. **Subclass Aperygota**. Wingless and most primitive of living insects with no metamorphosis.

   a. **Order Thysanura.** Silverfish and bristletails. These fast-moving insects possess two or three styliform appendages on the abdomen and chewing mouthparts. They are destructive to clothing and books.

   b. **Order Collembola.** Springtails. These small, insects possess an abdominal organ for jumping and chewing mouthparts; They are found in leaf litter.

2. **Subclass Pterygota**. Primarily winged insects; the wingless condition, if present, has been secondarily acquired. Metamorphosis present or absent depending on the group.

   a. **Order Ephemeroptera.** Mayflies. The first pair of net-veined wings are larger than second pair. Animals possess two or three caudal appendages. Mouthparts of short-lived adults are a vestigial chewing type. Metamorphosis is gradual.

   b. **Order Odonata.** Dragonflies and damselflies. Wings are long and net-veined. Mouthparts are chewing type. Dragonflies possess a stout-body and are stronger fliers than are damselflies. Metamorphosis is gradual.

   c. **Order Orthoptera.** Grasshoppers, katydids, crickets, roaches, and allies. The winged forms of this large order generally have the membranous hind wings folded beneath the leathery forewings (**tegmina**). The femur of hind legs is enlarged in many species. Mouthparts are chewing type. Metamorphosis is gradual.

   d. **Order Isoptera.** Termites. The abdomen is broadly joined to the thorax. Bodies are soft and often whitish. These colonial, social insects have winged and wingless individuals. Wings are equal in length. Mouthparts are chewing. Metamorphosis is gradual.

   e. **Order Hemiptera.** True bugs. The basal part of the forewings is thickened, while the distal part is membranous, overlaping at rest. The hind wings are membranous. Piercing-sucking mouthparts form a beak arising at the front of the head. Metamorphosis is gradual.

   f. **Order Homoptera.** Aphids, cicadas, and leafhoppers. Both winged and wingless individuals are known. The forewings are membranous unlike those of Hemiptera. (Many entomologists make this order a suborder of Hemiptera.) Piercing-sucking mouthparts from a beak arising form back of the head. Metamorphosis is gradual.

   g. **Order Neuroptera.** Lacewings, ant lions, and dobsonflies. Both pairs of wings are veined and membranous. Mouthparts are a chewing type. Metamorphosis complete. Larva of ant lions commonly called doodlebug; larva of dobsonfly is a hellgrammite; larvae of one family (Sisyridae) live in and eat freshwater sponges.

   h. **Order Coleoptera.** Beetles. Anterior wings are thick and hard and called **elytra**. Membranous hind wings, when not in use, are folded under the elytra. Metamorphosis is complete. Mouthparts are a chewing type. This is the largest insect order, with over 300,000 species. The larva stage commonly called a grub.

   i. **Order Lepidoptera.** Butterflies and moths. Body and wings in the adult are covered with minute scales. A coiled proboscis is used for sucking nectar from flowers. Metamorphosis is complete. The larva is commonly called a caterpillar.

   j. **Order Diptera.** Flies and mosquitoes. Hind-wings reduced to knoblike **halteres**. Mouthparts are piercing-sucking and lapping types. Metamorphosis is complete. The larva commonly called a maggot.

   k. **Order Hymenoptera.** Ants, bees, wasps. Winged and wingless species occur in this large order. The wings are transparent few veins. Both solitary or social species; are known. Mouthparts are modified for chewing, lapping, or sucking. Metamorphosis is complete.

   l. **Order Siphonaptera.** Fleas. The bodies of small, wingless insects are compressed laterally. The legs are modified for jumping. Adults possess piercing-sucking mouthparts. Metamorphosis is complete.

■ Observational Procedure: *Romalea*

The grasshopper or locust (order Orthoptera) will be used to illustrate basic insect anatomy. Obtain a large specimen such as the lubber grasshopper (*Romalea*), a common species found in the southern United States.

1. Observe the body markings, coloration, and reduced wings that are characteristic of the species (Fig. 15.36). Note that the grasshopper is segmented. Is there evidence of tagmatization?

2. As in all insects, the external, protective covering is a tough, nonliving **cuticle** or **exoskeleton** containing **chitin**. The hardened sclerotized plates of the cuticle are called **sclerites**. The major sclerites are typically separated by membranous areas called sutures that permit movement of body parts. A dorsal sclerite is the **tergum**, a ventral one the **sternum**, and a lateral one the **pleuron**. In areas (e.g., head) where individual segments are fused, the terga (or tergites) are fused as a solid hardened mass. The sclerites may also be divided into subplates.

3. Locate the three tagmata composing the grasshopper's body: (1) anterior head, (2) middle thorax, and (3) posterior abdomen (Fig. 15.36).

Head

1. The head is in form of a capsule (Figs. 15.36 and 15.37A). The upper part is heavily sclerotized and the ventral area contains the mouth surrounded by several mouthparts used in feeding. Exact agreement by entomologists does not exist as to the number of segments in the insect head. Note the shape of the head and its position in relation to the long axis of the body. The mouthparts of the grasshopper are directed downward, a position called **hypognathous**.

2. The head capsule is divided by sutures into several regions. It is often difficult to discern where one region begins and the other ends. The dorsum of the head between and behind the eyes is the **vertex** (Fig. 15.37A). The **frons** is the anteriofrontal region. Note on the frons the two arms or forks of the Y-shaped **epicranial suture** whose stem begins on the back of the head. The forks are called the **frontal suture** and the stem the **coronal suture**. The epicranial suture represents a line of weakness that splits the head capsule at molting.

3. Find the **gena** or cheek, which is located at the lateral lower part of the head posterior to the frons and below the eyes. The back of the head is the **occipital** that contains the **occipital foramen** (or foramen magnum).

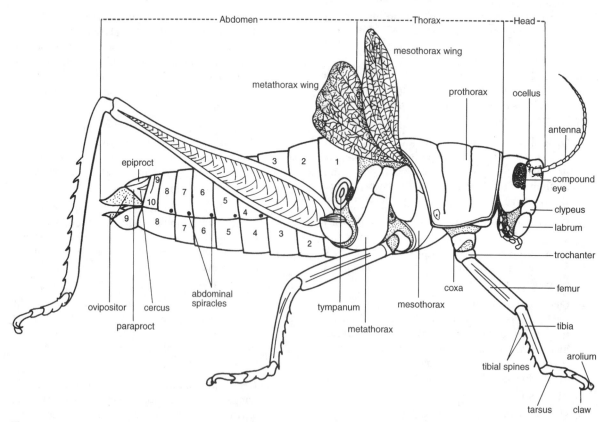

Figure 15.36. Lateral view of a female lubber grasshopper.

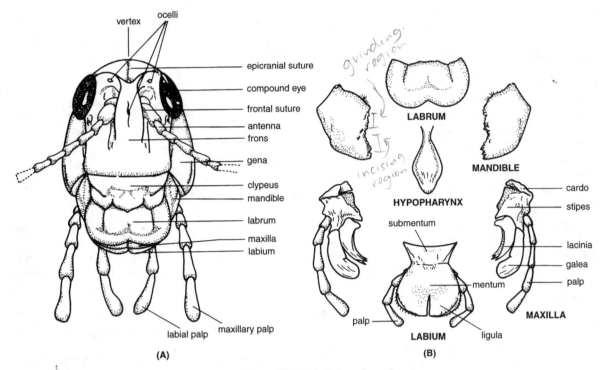

Figure 15.37. Head of the grasshopper. (A) Frontal view. (B) Detailed view of mouthparts.

4. Examine the anterior face of the head (Fig. 15.31). Locate the liplike sclerite, the **clypeus**, below the frons to which it is attached. Hanging down from the clypeus is the **labrum** or upper lip.

5. Lift the clypeus with forceps to fully expose the bilobed labrum (Fig. 15.37B). Does the labrum articulate directly to the clypeus? A swollen area, the **epipharynx**, is prominent in many insects on the ventral or posterior side of the labrum. Does the grasshopper have an epipharynx?

6. Behind the labrum are paired, unsegmented **mandibles**. Lift up the labrum with your forceps to expose these paired, hard mouthparts with teeth used to cut, chew, and grind food. The mandibulate insects include other orthopterans, odonates, isopterans, and coleopterans, so-named because of the type of mandibles present.

7. Behind the mandibles are the paired maxillae. These masticate and manipulate food. Note the antenna-like, segmented **palps** attached on top of the maxillae. How many segments compose a palp? The palps are sensory structures.

8. Other parts of a maxilla are as follows: the basal **cardo**, the **stipes** with a palp, the outer **galea**, and the inner jawlike **lacinia**. These areas are best observed when the maxillae are dissected out (Fig. 15.37B).

Although the **labium** is a single structure, it represents fused paired second maxillae. The labium also manipulates food and has paired sensory palps. Note the three regions on the labium: basal **postmentum** (divided into a basal submentum and distal mentum), middle **prementum** with palps, and distal **liguia** with apical lobes.

9. The labium should be dissected from the grasshopper and observed with the dissection microscope (Fig. 15.37B).

10. Locate the median, unpaired tongue-like **hypopharynx** that protrudes in the preoral cavity. The hypopharynx is closely associated with the base of the labium. Your instructor may have you dissect out all of the mouthparts and examine them under the dissection microscope (Fig. 15.37B).

11. Note the shape and position of the two large immovable **compound eyes** (Fig. 15.37A). Examine an eye under the dissection microscope. How does it compare to the eyes of the other arthiopods that you have examined? Note the many hexagonal **facets**. Each facet is a lens to an **ommatidium**, the basic structure of the eye. The number of ommatidia varies; some dragonflies have up to 30,000.

12. Between the compound eyes and on the frons are three single-faceted simple eyes or **ocelli**. Note that the arrangement of the ocelli forms a triangle. The ocelli apparently perceive changes in light intensity, whereas the compound eyes form images.

13. Below the two lateral ocelli are two antennae (Fig. 15.37A). These movable sensory appendages articulate in antennal sockets. A variety of shapes and sizes of antennae occurs in insects. They function in touch, smell, and hearing. The grasshopper's antennae are threadlike or **filiform**. Remove an antenna from its socket and examine it under the dissection microscope. Note the tiny sensory hairs on the antenna. Locate the basal **scape**, the middle **pedicel**, and the long **flagellum** consisting of many segments (Fig. 15.38).

Cervix or Neck

Connecting the head and thorax is a membranous region called the **cervix** or **neck**. It is not a separate body segment, but a contribution from the labial segment and the prothoracic segment. The cervix allows flexibility of movement between the head and thorax regions.

Thorax

1. All insects have three segments to the thorax: anterior **prothorax**, middle **mesothorax**, and posterior **metathorax** (Fig. 15.36). Each bears a pair of legs. The parchment-like forewings (or **tegmina**) are attached to the mesothorax and the hind wings are attached to the metathorax.

2. Observe the wing coloration and the folded shape of the hind wings (Fig. 15.36). Flight is primarily by the hind wings. Note the many veins in the wings. What function might they perform? The venation pattern is important in insect taxonomy. Both wings are outgrowths of the body wall and lack internal muscles.

3. Examine the sclerites of the thorax. Each tergum or dorsal plate on the thorax is called a **notum**. The wings are attached to the lateral margins of the nota.

4. Note the greatly sclerotized sterum. The different specialized plates and sutures of the notum and sternum will not be studied.

5. Each pleuron is divided by a suture forming two sclerites. The anterior sclerite is the **episternum** and the posterior one the **epimeron**. Observe these with the dissection microscope.

6. Two pairs of **spiracles**, external openings to the tracheal respiratory system, occur on the thorax above the attachments of the second and third pairs of legs. You will observe that these locations are between the

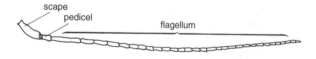

Figure 15.38. Insect antenna.

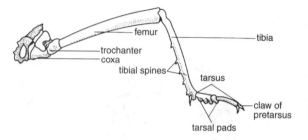

Figure 15.39. Insect leg.

prothorax and mesothorax and between the mesothorax and metathorax, respectively.

7. Observe the legs. Each leg consists of six parts: **coxa, trochanter, femur, tibia, tarsus**, and **pretarsus** (Fig. 15.39). The short basal coxa articulates with the body.

8. Note the shape of the greatly enlarged femur of the last pair of legs. What adaptation might this enlarged part serve?

9. Note the two rows of spines on the tibia. Are the spines of both rows equal in size and number? Why are the spines located on the dorsal surface of the tibia instead of the ventral surface? What function do they serve?

10. Count the segments of the tarsus. Note the tarsal pads or **pulvilli**. How many are there and what function might the pads serve? The short pretarsus also has an adhesive pad, the **arolium**, and a pair of **claws**.

Abdomen

1. Observe that the abdomen is distinctly segmented and lacks jointed appendages (Figs. 15.36 and 15.40). Count the abdominal segments. The first segment bears a **tympanum**, an oval membrane covering the organ of hearing. The last pair of segments is reduced and limited to a tergite, the **epiproct**, below which lies the **anus**.

2. Observe the tergite and anus under the dissection microscope. Note the small spine or **cercus** projecting behind the tenth segment. The plate below the cercus is the **paraproct**. Although reduced in size, the cercus is one of the few appendages found on the abdomen. The other appendages on the abdomen are reproductive in function.

3. At the end of the abdomen in female grasshoppers is a pincer-shaped **ovipositor** for digging a hole in the ground for deposition of eggs. Observe the bladelike structures (**valvulae**) that compose the ovipositor (Fig. 15.40).

4. Male grasshoppers lack the ovipositor. Note that the expanded sternum of segment 9 (the **subgenital plate**) encloses the male copulatory organs.

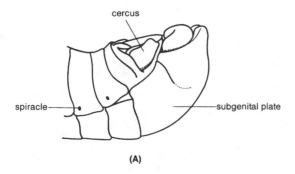

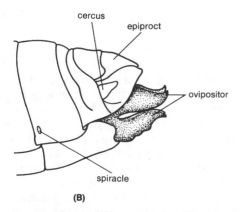

Figure 15.40. Posterior end of the grasshopper. (A) Male. (B) Female.

5. Observe the terga and sternal plates of the abdominal segments. Note their arrangement. Are pleural sclerites present?

6. On the lower anterior edge of each abdominal tergite of segments 1 to 8 is a small oval spiracle that leads to an internal trachea of the respiratory system (Fig. 15.40). Observe the spiracles with the dissection microscope. Some may be situated in a pigmented area and be difficult to see. Counting those on the thorax and those on the abdomen, the grasshopper has 10 pairs of spiracles. The first four pairs are used in inspiration and the remaining pairs function in expiration.

Internal Anatomy

Much of the internal anatomy of the grasshopper will be difficult to observe. Therefore, exercise care in dissection and observation.

1. With a pair of fine-pointed scissors, clip off the wings close to the body. Make a very shallow midsagittal cut through the dorsal body wall beginning at the head and ending at the posterior end. Pin down the two halves of the body wall to the dissection tray.

2. The large cavity exposed and containing the internal organs is the **hemocoel** (Fig. 15.41). This is not a true body cavity or coelom but part of the **open (lacu-**

nar) circulatory system. The hemocoel is filled with blood or **hemolymph**. The **heart** is a elongated slender tube in the dorsal part of the abdomen. It extends anteriorly as an aorta. Although the heart is difficult to see, make an attempt to locate it.

3. Note the muscles along the body wall, especially those in the thorax associate with the legs and wings. The muscles of the grasshopper are in bands.

4. Note the whitish, delicate tubes extending from the body wall to the viscera. These are **tracheae**. Prepare a wet mount of a piece of a trachea and examine it under low and high powers of a compound microscope. Describe the appearance of the tracheae. How are they constructed?

5. Components of the digestive system are the most conspicuous of the internal organs. Much of the hemocoel is filled with the digestive system (Fig. 15.41). Locate the tube-shaped **esophagus** leading from a short **pharynx** that connects to the mouth. The esophagus enlarges posteriorly as a thin-walled **crop** in the thorax. The crop leads to a narrow **proventriculus** (or **gizzard**) with internal gastric teeth for shredding food.

6. You may see salivary or labial glands below the crop, with a duct opening in the preoral cavity between the labium and hypopharynx. The components above compose the foregut (or **stomodeum**) of the grasshopper's digestive system.

7. Posteriorly the proventriculus leads to an elongated **stomach** or **ventriculus**. Much of the stomach is concealed by finger-shaped **gastric ceca**. These are involved in nutrient absorption. How many are present? The stomach and gastric ceca constitute the midgut (**mesenteron**).

8. Connecting to the stomach posteriorly is an **intestine**, with an enlarged anterior component and a slender posterior component, an enlarged **rectum** and the **anus**. The intestine, rectum, and anus constitute the hindgut (**proctodeum**).

9. The many threadlike excretory organs called **Malphigian tubules** and other organs such as those originating from the reproductive system may conceal parts of the hindgut. Care must be exercised not to destroy these structures.

10. Grasshoppers are dioecious. Females possess the egg-laying ovipositor, whereas males lack this structure and thus have a rounded posterior end (Fig. 15.40). Females lay their eggs in the ground in late summer or early fall. In the following spring, the eggs hatch into immature grasshoppers (or **nymphs**). They undergo several molts before attaining the adult form.

11. Examine a male grasshopper and locate as many of the following parts as possible (Fig. 15.42). The two fused **testes** of the male lie dorsal to the intestine. A slender tube, the **vas deferens**, extends from each testis. The tube passes posterioventrally and becomes

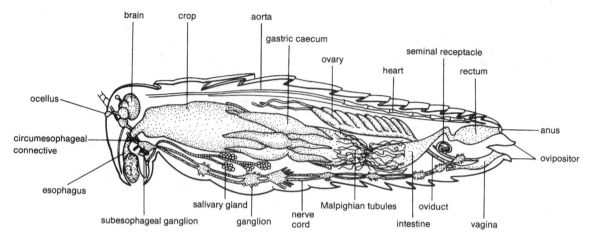

Figure 15.41. Internal anatomy of a female grasshopper.

enlarged as a reservoir, the **seminal vesicle**, before uniting to form a common **ejaculatory duct** that enters the **penis**. Two **accessory glands** that secrete a fluid join the ejaculatory duct.

12. Examine a female grasshopper. Several tapering **ovarioles** or egg tubes where eggs are produced from an **ovary**. Extending from each ovary is an **oviduct**. The two oviducts join to form the **vagina**, an egg-holding chamber located under the intestine. Two **accessory glands** and a **seminal receptacle** (or **spermatheca**) enter the vagina separately. Sperm received from the male during copulation are stored in the receptacle.

13. Insects possess a highly developed nervous system (Fig. 15.41). It is similar to that of the crayfish and, for the most part, the system lies below the digestive system. Attempt to observe the following components of the grasshopper's nervous system. The central nervous system consists of a **trilobed brain** located in the head above the esophagus. Extending off the brain is a pair of **circumesophageal connectives** that connect to a **subesophageal ganglion**. The ganglion innervates the mandibles, maxillae, and labium. Extending posteriorly in the insect is the paired **ventral nerve cord** with one pair of ganglia per segment.

Adaptability of Structure

Insects provide excellent opportunities for students to study specializations and adaptability of structures. Morphological modifications in insects can be seen in most structures, but especially in the mouthparts, legs, antennae, and wings. Selected examples of insects showing adaptations and specializations of these structures will be observed.

Mouthparts Insect mouthparts are adapted for feeding on different types of food. As a result, different methods are employed for food procurement. Some of the major types are biting and chewing, chewing and sucking, lapping (sponging), piercing and sucking, and shiphoning. The grasshopper and cockroach (order Orthoptera) are good examples of insects with biting and chewing mouthparts. Many zoologists believe the other specialized mouthparts observed in many insects have evolved from the biting-chewing type. As you study the following examples, it is important that you keep in mind the mouthparts of the grasshopper previously examined as a basis for making comparisons.

1. Chewing and sucking: honeybee (*Apis*). The mouthparts of the honeybee (order Hymenoptera) are used for biting and chewing wax and sucking or lapping nectar from flowers. Examine a preserved honeybee or prepared slide of the insect's mouthparts (Fig. 15.43). Note the arrangement of hairs on the head and mouthparts. Below the clypeus is the narrow **labrum** and a smaller fleshy **epipharynx**. Note that the **mandibles** project laterally alongside the labrum. Are teeth similar to those observed on the grasshopper's mandibles present? Locate the medial, hairy **labium**. The specialized labium, also called the **glossa** or **alaglossa**, forms an extensible organ for probing into flowers. **Labial palps** and **maxillae** occur on each side of the tongue-like labium. These fit against the labium, forming channels through which saliva is discharged and food drawn up to the mouth.

2. Lapping (sponging): housefly (*Musca*). The non-biting fly (order Diptera) extrudes saliva onto food and then sponges or laps the dissolved food-saliva mixture into the mouth. Examine the mouthparts of a housefly from a preserved specimen or prepared slide. The feeding part is a highly modified **proboscis** with a basal rostrum and a distal labium which is modified at the extreme end as a sponge-like, bilobed **labellum**

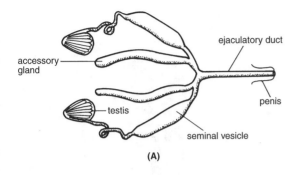

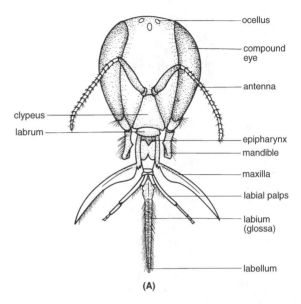

Figure 15.43. Mouthparts of the honeybee.

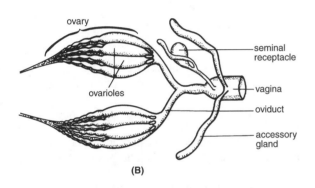

Figure 15.42. Reproductive systems of the grasshopper. (A) Male. (B) Female.

is the hairy **labium**, with a flexible tip called the **labellum**. Note the central groove of the labium, into which the other needle-like mouthparts lie when not in use. Alongside the labium are two hairy, segmented sensory **maxillary palps**. These are as long as the labium in *Anopheles*, but generally shorter in *Culex*. The non-hairy **labrium-epipharynx** and narrow ribbon-like **hypopharynx** form a tube for drawing blood into the mouth. The hypopharynx contains a small tube for

(Fig. 15.44). On the anterior face of the proboscis is the **labrum-epipharynx**, which covers a labial groove in which lies the bladelike **hypopharynx**. The epipharynx and hypopharynx interlock, forming a food tube to the esophagus. **Mandibles** are lacking as are most of the **maxillae** except for two maxillary palps. Unlike the housefly, other dipterans, such as horse-flies and deer flies, have sharp, bladelike mandibles to cut flesh and maxillae serve as long probing styes. These flies are blood feeders. In contrast to the mandibulate insects with mandibles designed for cutting, grinding, and chewing, the housefly has **haustellate** mouthparts, in which the mandibles are stylelike for sucking or are completely absent.

3. Piercing and sucking: mosquito (*Anopheles*). The haustellate mouthparts of the female mosquito (order Diptera) are elongated for piercing and sucking. Except for the paired maxillary palps, the mouthparts collectively form a **proboscis**. Examine a prepared slide of a female mosquito's mouthparts that have been teased apart (Fig. 15.45). The largest mouthpart

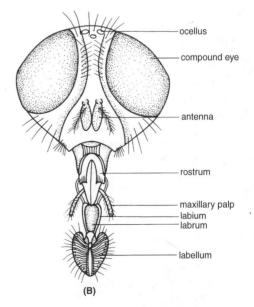

Figure 15.44. Mouthparts of the housefly.

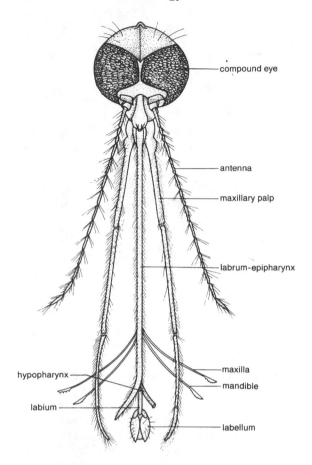

Figure 15.45. Mouthparts of the mosquito.

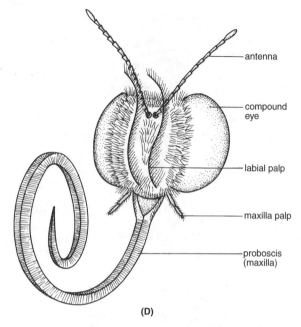

(D)

Figure 15.46. Mouthparts of the butterfly.

ejecting saliva into the puncture and thus prevents clotting of the blood. The flexible labellum of the labium guides the piercing mouthparts. Observe the thin, flat paired **maxillae** and paired **mandibles** also used in piercing. The maxillae bear teeth on one edge of a narrow fin-shaped tip. The mandibles are similar but are more delicate, with finer teeth.

4. Siphoning: butterfly. Moths and butterflies (order Lepidoptera) have mouthparts designed for siphoning nectar and juices from flowers. Examine a preserved specimen or prepared slide of the mouthparts of a butterfly (Fig. 15.46). The long siphoning tube is the **proboscis**, fused galeae of the maxillae. When not in use, the proboscis is held coiled between the sensory labial palps.

Legs Although the insect legs are composed of six basic parts (Fig. 15.39), many modifications and specializations of these appendages can be observed. Certain insects possess legs designed for performing more than one type of locomotion. For example, the first and second pairs of legs of the grasshopper are used in walking or crawling, whereas the third pair is designed for leaping. Selected examples of insect legs that illustrate adaptations to different types of locomotion will be available for your study. Your instructor will direct your study.

1. Ambulatory (e.g., walking, running, crawling): cockroach, walkingstick (order Orthoptera), butterfly (order Lepidoptera), stink bug (order Hemiptera), ant (order Hymenoptera), and termite (order Isoptera). Describe the function of these legs in terms of a tripod.

2. Grasping and holding prey: first pair of legs of a praying mantis (order Orthoptera), ambush bug, and giant water bug (order Hemiptera). What crustacean has similar grasping legs?

3. Swimming (flattened parts with long brushes of hair): third pair of legs of a backswimmer, water boatman (order Hemiptera), and diving beetle (order Coleoptera). How are these legs like paddles?

4. Digging: first pair of legs of the mole cricket (order Orthoptera) and digger wasp (order Hymenoptera). Are these legs like shovels or rakes?

5. Clinging or holding by suction: housefly (order Diptera). How are connections to surfaces maintained by housefly legs?

6. Hopping, leaping, or jumping: third pair of legs of a cricket, grasshopper, katydid (order Orthoptera), flea (order Siphonaptera), and leafhopper (order Homoptera). Describe the springlike mechanisms of these legs.

7. Floating or skimming on water surface: water strider (order Hemiptera). How are these legs like oars?

8. Gathering and carrying pollen: honeybee (order Hymenoptera). Where is the pollen held?

9. Sound production: file on inner surface of hind femur of an acrid grasshopper (order Orthoptera). Are the spines the noise-making devices (i.e., do they vibrate) or do they pluck something that vibrates?

Antennae Other extremely varied appendages of insects are the antennae. These delicate structures range in size from very short (e.g., housefly, order Diptera; dragonfly, order Odonata; giant water bug, order Hemiptera) to being very long (e.g., camel cricket and true katydid, order Orthoptera; longhorn beetle, order Coleoptera). In the latter, the antennae may be longer than the insect's entire body. Antennae even vary in shape between the male and female of the same species. For example, female mosquitoes have beadlike antennae, whereas the male's antennae are plumose. The antennae of males in many of the giant silkworm moths (order Lepidoptera) are feathery and larger than those of the females. Within large orders of insects such as Hymenoptera and Coleoptera, great variations in the shape, number of segments, and size occur in the antennae. Some of the more common and perhaps striking types of antennae are shown in Fig. 15.47 and listed below along with representative adult insects. Observe these and other examples with the dissection microscope.

1. Filiform (threadlike). The segments have a uniform size. Examples: silverfish (order Thysanura), dytiscid and ground beetles (order Coleoptera), vespid wasp (order Hymenoptera), mole cricket (order Orthoptera).

2. Setaceous (tapering). The segments become more slender distally. Examples: dragonfly and damselfly (order Odonata), lacewing (order Neuroptera).

3. Serrate (sawlike). The segments are triangular in shape. Examples: click beetle and firefly (order Coleoptera).

4. Clavate (clubbed). The segments increase gradually in diameter distally. Examples: ladybird and June beetles (order Coleoptera).

5. Capitate (having a head). The terminal segments enlarge suddenly. Example: histerid beetle (order Coleoptera).

6. Lamellate (leaflike). The terminal segments form oval lobes. Examples: June, dung, and unicorn beetles (order Coleoptera).

7. Pectinate (comblike). Most segments have long lateral processes. Examples: male European sawfly (Hymenoptera), fire-colored beetle (order Coleoptera).

8. Plumose (feathery). Most segments have many hairs. Examples: males of chironomids, mosquitoes (order Diptera); giant silkworm moths (order Lepidoptera).

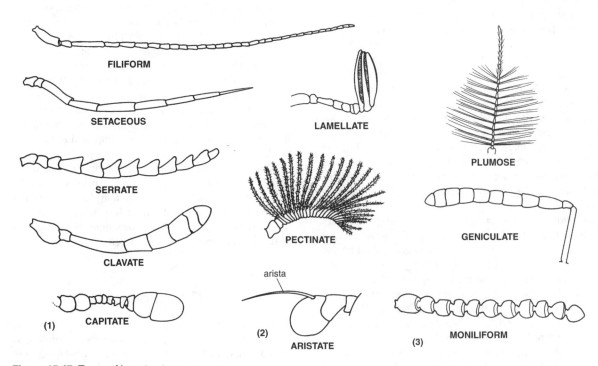

Figure 15.47. Types of insect antennae.

9. **Aristate.** The last segment with a dorsal bristle (**arista**). Examples: housefly and syrphid fly (order Diptera).

10. **Geniculate (elbowed).** The first segment is long and the following smaller segments are at an angle. Examples: ant (order Hymenoptera), stag and histerid beetles (order Coleoptera).

11. **Moniliform (beadlike).** The segments resemble a string of similar-shaped beads. Examples: wringled bark and tenebrionid beetles (order Coleoptera).

Wings The presence of wings undoubtedly is a major feature that has contributed to the success of insects. Most insects have two pairs of wings. Insect wings are outgrowths of the body wall and structurally unlike those of bats and birds, which are modified forelimbs. Considerable variation in the size, texture, shape, number, venation, and position held at rest exists in insect wings. Examples of these variations will be observed in selected representative insects.

1. **Size.** The more primitive winged insects have both pairs of wings equal or nearly equal in length. Observe a dragonfly or damselfly (order Odonata). Note the size of the wings in relation to the body length. Unlike many insects, mayflies (order Ephemeroptera) and dragonflies and damselflies cannot fold their wings over the abdomen when at rest. Examine the wing lengths of other insects from different orders. How do they compare?

2. **Texture.** Many insects, such as the dragonfly, have all four wings of the same membranous texture. Members of the orders Orthoptera and Coleoptera have the first pair of wings parchment-like or hard, respectively. The grasshopper's forewings are called **tegmina** because of their leathery texture. Examine other orthopterans, such as the field cricket and American cockroach. Note the texture of the forewings compared to that of the hindwings. The hard forewings of beetles are called **elytra**. Observe examples of beetles showing variations of the elytra; note how the membranous hindwings are folded under the forewings. True bugs of the order Hemiptera have the proximal part of the forewings harden and the distal part membranous. This half-wing condition (**hemilytron**) is reflected in the ordinal name, Hemiptera. Butterflies and moths (order Lepidoptera) have wings covered with numerous **scales**, hence the name Lepidoptera. The scales easily rub off when the wings of these insects are held by the fingers. The different types and arrangement of scales give the different patterns observed in these insects' wings. Examine several scales from different moths or butterflies under the compound microscope. Describe the patterns you observe. Are all of the scales of the same type on a given individual?

3. **Number.** Most adult insects have four wings. Adult members of the order Diptera (two wing) are unique in having only one pair of well-developed wings. The second pair, called **halteres**, is reduced and knoblike. Examine representative flies or mosquitoes with a dissection microscope to see the halteres. What function might these reduced wings serve?

4. **Venation.** One of the most important features of insect wings used by the taxonomist is the venation pattern. An extensive terminology, which will not be elaborated on, has been developed to describe the number and arrangement of the veins in insect wings. The more primitive insects have a large number of veins than do the advanced forms. Compare the venation of a dragonfly or damselfly (order Odonata) with that of a wasp or bee (order Hymenoptera). Examine the venation in the wings of insects representing different orders. Can you determine whether a basic pattern exists in the venation? The veins are trachae that extend into the wing. The areas between the veins are called **cells**. ■

Myriapodous Arthropods

The myriapodous (G., *myrias*, number 10,000 + G., *podos*, foot) are wormlike, uniramous arthropods that live in moist dark places such as beneath wood, rocks, leaf litter, and in unused parts of buildings. The 10,500 species are divided into four classes: Chilopoda (ki-LOP-o-dah; G., *cheilos*, lip + G., *podos*, foot) the centipedes; Diplopoda (dip-LOP-o-dah; G., *diplos*, double + G., *podos*, foot) the millipedes; Pauropoda (paur-OP-o-dah; G., *pauros*, small + G., *podos*, foot) the pauropods; and Symphyla (sim-PHY-la; G., *symphylos*, of the same race + G., *phylon*, a tribe) the symphylans.

Two tagmata, the head and trunk, compose the bodies of these arthropods. The head bears a pair of segmented antennae, a pair of mandibles, an upper lip (labrum), a tongue-like hypopharynx, and one (in millipedes and pauropods) or two (in centipedes and symphylans) pairs of maxillae. Some species of millipedes and centipedes have nonimage-forming ocelli or compound eyes on the head. Pauropods and symphylans lack eyes.

Most people are familiar with centipedes and millipedes, that comprise the vast majority of myriapods. However, the pauropods and symphylans are seldom observed; the former at 2 mm or less in length, and the latter are no more than 8 mm in length. Here you will study only centipedes and millipedes.

Classification

The four classes of myriapods are probably polyphyletic and may be related to insects. Pauropods and diplopods are considered closely related, whereas chilopods and symphylans form another group of similarity.

(1) **Class Chilopods.** Centipedes. Myriapods with flattened bodies. Trunk with many segments each possessing one pair of legs. The first pair of trunk appendages (maxillipeds) are poisonous. Head with two pairs of maxillae. A genital opening is located at the posterior end. Most centipedes are swift-moving, carnivorous predators. Examples: *Scolopendra* and *Scutigera*.

(2) **Class Symphyla.** Symphylans. Myriapods with a small, flattened body. The trunk is composed of 12 segments, with one pair of legs per segment. Head with two pairs of maxillae. A genital opening is located on the third trunk segment. Symphylans are plant feeders and are often found in greenhouses. Examples: *Scuitgerella* and *Scolopendrella*.

(3) **Class Diplopoda.** Millipedes. Myriapods with a cylindrical body. The trunk is composed of diplosegments each with two pairs of legs. Head with one pair of maxillae (gnthochilarium). A genital opening is located anteriorly on trunk. Most millipedes are slow-moving herbivores. Examples: *Julus* and *Spirobolus*.

(4) **Class Pauropoda.** Pauropods. Small, soft-bodied myriapods with biramous antennae and one pair of maxillae (gnathochilarium). The trunk is composed of 11 segments, but legs are absent on segments 1 and 11. Pauropods are vegetarians and inhabitants of leaf litter and soil. Examples: *Pauropus* and *Allopauropus*.

■ Observational Procedure:

1. Obtain a specimen each of a centipede and a millipede (Figs. 15.48 and 15.49). Compare the shapes of their bodies, the number of segments, and number and arrangement of legs attached to the segments.

2. Note the less flexible and harder body of the millipede compared to that of the centipede. Calcium salts in the millipede's procuticle harden the sclerites, especially the tergites.

3. What type of segmentation occurs in the two animals? Are all segments the same size? Compare the tergites, pleurites, and sternites of the two animals. Are interterites present? What adaptations might be correlated with a flattened body as in the centipede and the cylindrical body of the millipede?

4. Observe the specimens under the dissection microscope. Locate the anterior **head** and the posterior **trunk** (Figs. 15.48 and 15.49). The greater part of the body of both animals is the trunk. The first four trunk segments of millipedes are often called collectively a **thorax**. The first thoracic segment is called the

collum and is legless (Fig. 15.48). Segments 2 to 4 bear one pair of legs. Locate these segments. Note the flattened head of the centipede and the round head of the millipede. The individual segment composing the heads of both arthropods is not readily discernible.

5. Examine with the dissection microscope the legs of both animals. Are all of the legs on each animal of the same size? Note the **diplocondition** in the millipede; this results from fusion of two originally separate segments.

6. Are claws or pads present similar to those on the grasshopper's legs? Are the number of leg segments for both animals the same? In centipedes the last pair of legs are longer than the others and are sensory or defensive in function.

7. Locate the **spiracles**. Look for the small openings in the pleural membrane dorsal to each leg (except the first and last three segments) of the centipede (Fig. 15.49). In the house centipede (*Scutigera coleoptrata*) each spiracle is centrally placed on a tergum and opens to a saclike tracheal lung. Millipedes have two spiracles per segment on the sternal plate near the base of the legs. How does this relate to the diplocondition? It may be necessary to remove carefully two or three pairs of legs to clearly reveal the spiracles. In some millipedes there are paired lateral openings on the terga that communicate with internal **repugnatorial glands** that secrete hydrogen cyanide, iodine, and phenol.

8. Now locate the **gonopores** behind the second pair of legs in the millipede. The gonopore of centipedes is on the last trunk segment, that also bears the anus (Fig. 15.49). In some millipedes the legs of segment 7 or 8 may be modified for reproductive purposes.

9. Examine the centipede's head with the dissection microscope (Fig. 15.49). Observe the two antennae and paired, nonimage-forming eyes. How many pairs of eyes are on your specimen? Not all centipedes have eyes, and some centipedes (e.g., order Scutigeromorpha) have clusters of ocelli that form compound eyes.

10. Observe the ventral area of the head and locate the paired **forcipules** (Fig. 15.49). Each is composed of an enlarged basal coxa and poison claw or fang; internally, a poison gland and duct are present. The forcipules represent the first trunk appendages or maxillipeds. A centipede's bite, inflicted by the fang, can be painful.

11. Carefully remove both forcipules with your scissors. This will expose the leglike second maxillae (**palpognaths**) with paired palps. Lift up these maxillae with a probe to see the smaller first maxillae. Below the first maxillae are paired mandibles with distal teeth,

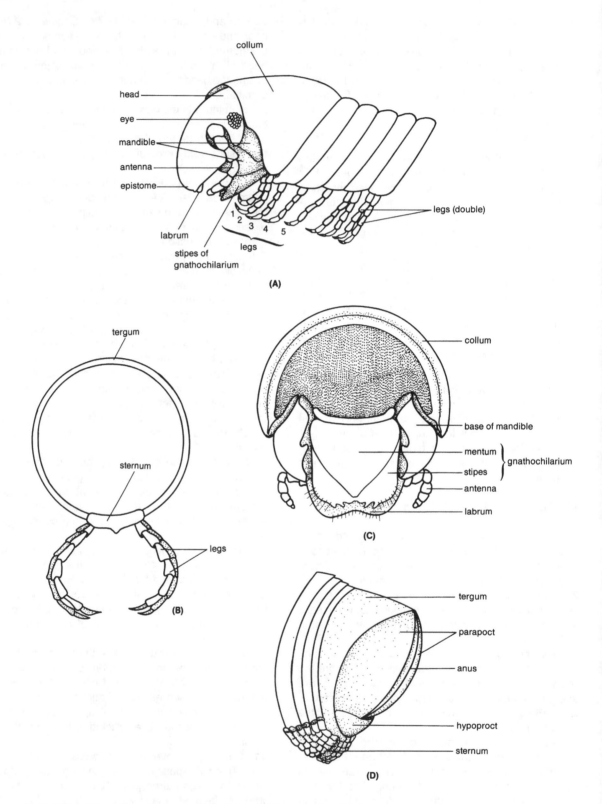

Figure 15.48. Millipede body form. (A) Anterior lateral view. (B) Cross section. (C) Posterior view of head. (D) Posterior lateral view.

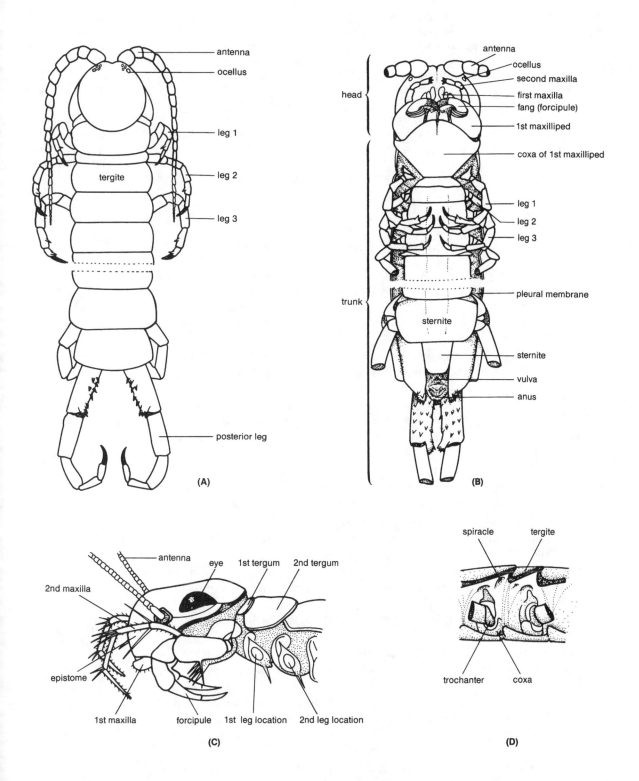

Figure 15.49. Centipede body form. (A) Dorsal view. (B) Ventral view. (C) Anterior lateral view. (D) Lateral view of trunk, showing location of a spiracle.

between which is located the mouth. Anterior to the mouth and mouthparts is a hard plate, the **clypeus**, bordered anteriorly by the **epistome**.

12. Observe with the dissection microscope the round head of your millipede. Locate the broad epistome, that at its ventral margin is inseparably connected to the labrum.

13. Find the paired antennae, segmented, teeth-bearing mandibles, and the broad ventral plate called the **ganthochilarium** formed by fusion of the first maxillae (Fig. 15.48). Note that the lateral mandibular bases are large and immovable. The flexible cutting edges of the

mandibles are covered by the gnathochilarium. The gnathochilarium in some species is divided into a medial **mentum** and lateral **stipes** (Fig. 15.48). Unlike centipedes, millipedes lack second maxillae.

14. Are eyes present on your specimen? Some species of millipedes have ocelli that are grouped together and give the impression of being compound.

15. Locate the anus of the millipede at the posterior end between two **paraproct plates** (Fig. 15.48). Ventral to these is a single plate, the **hypoproct**. Find the anus of your centipede on the ventral area of the last trunk segment (Fig. 15.49).

Supplemental Readings*

Abele, G. (ed.). 1982. Biology of Crustacea, Vols. 1 and 2. Academic Press, New York. (CR)

Akre, R.D., L.D. Hansen, and E.A. Myhre. 1995. The collection and maintenance of ants to use for teaching. The Kansas School Naturalist 41(1): 3–8 (I)

Alcock, J. 1995–continuing. Various titles. The Cornell Series in Arthropod Biology. Cornell University Press, Ithaca, NY. (G, I)

Anderson, D.T. 1973. Embryology and Phylogeny in Annelids and Arthropods. Pergamon, Press, Oxford. (G)

Anderson, D.T. 1980. Cirral activity and feeding in the lepadomorph barnacle *Lepas pectinata*. Proc. Linn. Soc. N.S.W. 104: 147–159. (CR)

Anderson, D.T. 1981. Cirral activity and feeding in the barnacle *Balanus perforatus*, with comments on the evolution of feeding mechanisms in thoracican cirripedes. Philos. Trans. R. Soc. Lond. B 291: 411–449. (CR)

Arnette, R.H. 1985. American Insects: A Handbook of the Insects of North America. Van Nostrand Reinhold, New York. (I)

Atkins, M.D. 1978. Insects in Perspective. Macmillan, New York. (I)

Bergström, J. 1973. Organization, life, and systematics of trilobites. Fossils Strata 2: 1–60. (T)

Bland, R.G., and H.E. Jaques. 1978. How to Know the Insects, 3rd ed. Wm. C. Brown, Dubuque, IA. (I)

Bliss, D.E. (ed.). 1982–1985. The Biology of Crustacea, Vols. 1–10. Academic Press, New York. (CR)

Blower, J.G. (ed.). 1974. Myriapoda. Symposia of the Zoological Society of London, 32. Academic Press, New York. (M)

Blum, M.S. (ed.). 1985. Fundamentals of Insect Physiology. Wiley-Interscience, New York. (I)

Borro, D.J., and R.E. White. 1974. A Field Guide to the Insects of America North of Mexico. Houghton Mifflin, Boston. (I)

Borror, D.J., D.M. DeLong, and C.A. Triplehorn. 1981. An Introduction to Insect Biology and Diversity. W.B. Saunders, Philadelphia. (I)

Boudreaux, H.B. 1979. Arthropod Phylogeny with Special Reference to Insects. Wiley, New York. (I)

Briggs, D.E.G. 1989. The early radiation and relationships of the major arthropods groups. Science 246: 241–243. (G)

Brodsky, A.K. 1994. The Evolution of Insect Flight. Oxford University Press, New York. (I)

Brownell, P.H. 1984. Prey detection by the sand scorpion. Sci. Am. 251(6): 86–97. (CH)

Budd, T.W., J.C. Lewis, and M.L. Tracy. 1978. The filter feeding apparatus in crayfish. Can. J. Zool. 56: 695–707. (CR)

Burgess, J.W. 1976. Social spiders. Sci. Am. 234: 101–106. (CH)

*The supplemental literature for Arthropoda is coded to indicate the following groups: G = General; T = Trilobites; CH = Chelicerates; CR = Crustaceans; I = Insects; M = Myriapods.

Camatini, M. (ed.). 1979. Myriapod Biology. Academic Press, New York. (M)

Cameron, J.N. 1985. Molting in the blue crab. Sci. Am. 252: 102–109. (CR)

Chapman, R.F. 1982. The Insects: Structure and Function, 3rd ed. Harvard University Press, Cambridge, MA. (I)

Chen, J.-y., G.D. Edgecombe, L. Ramsköld, and G.-g. Zhou. 1995. Head segmentation in early Cambrian Fuxianhuia: implications for arthropod evolution. Science 268: 1339–1343. (G, T)

Christiansen, K. 1992. Springtails. The Kansas School Naturalist 39: 3–16. (I)

Cisne, J.L. 1973. Life history of an Ordovician trilobite *Triarthrus eatoni*, Ecology 54: 135–142. (T)

Cisne, J.L. 1974. Trilobites and the origin of arthropods. Science 186: 13–18. (T)

Cisne, J.L., G.O. Chandless, B.D. Rabe, and J.A. Cohen. 1980. Geographic variation and episodic evolution in a trilobite. Science 209: 925–927. (T)

Clarke, U. 1973. The Biology of the Arthropoda. Elsevier, New York. (G)

Cloudsley-Thompson, J.L. 1968. Spiders, Scorpions, Centipedes, and Mites. Pergamon Press, Elmsford, NY. (CH, M)

Cohen, J.A. and H.J. Brockmann. 1983. Breeding activity and mate selection in the horseshoe crab, *Limulus polyphemus*. Bull. Mar. Sci. 33: 274–281. (CH)

Connell, J.H. 1961. The influence of interspecific competition and other factors on the distribution of the barnacle *Chthamalus stellatus*. Ecology 42: 710–723. (CR)

Crompton, J. 1987. The Spider. (Reprint of 1950 ed.). Nick Lyons, New York. (CH)

Cronin, T.W., N.J. Marshall, and M.F. Land. 1994. The unique visual system of the mantis shrimp. Am. Sci. 82: 356–365. (CR)

Denno, R.F., and H. Dingle. 1981. Insect Life History Patterns. Springer-Verlag, New York. (I)

Downer, R.G.H. (ed.). 1981. Energy Metabolism in Insects. Plenum Press, New York. (I)

DuBois, M.B. 1995. Studying ants: a beginning. The Kansas School Naturalist 41(1): 9–16. (I)

Dunkel, F.V. (ed.). 1988–present. The Food Insects Newsletter. Montany State University, Bozeman, MT. (I)

Ehrlich, P.R., and A.H. Ehrlich. 1961. How to Know the Butterflies. Wm. C. Brown, Dubuque, IA. (I)

Elzinger, R.J. 1981. Fundamentals of Entomology, 2nd ed. Prentice-Hall, Englewood Cliffs, NJ. (I)

Emerson, M.J., and F.R. Schram. 1990. The origin of crustacean biramous appendages and the evolution of Arthropoda. Science 250: 667–669. (G)

Evans, H.E. 1984. Insect Biology. Addison-Wesley, Reading, MA. (I)

Factor, J.R. 1995. Biology of the Lobster. Academic Press, Orlando, FL. (CR)

Foelix, R.F. 1996. Biology of Spiders. 2nd ed., Harvard University Press, Cambridge, MA. (CH)

Fortey, R.A., and S.F. Morris. 1978. Discovery of nauplius-like trilobite larvae. Paleontology 21: 823–833. (T)

Govind, C.K. 1989. Asymmetry in lobster claws. Am. Sci. 77: 468–474. (CR)

Grimaldi, D.A. 1996. Captured in amber. Sci. Am. 274(4): 84–91. (I)

Hadley, N.F. 1986. The arthropod cuticle. Sci. Am 255(1): 104–112. (G)

Halder, G., P. Callaerts, and W.J. Gehring. 1995. Induction of ectopic eyes by target expression of the eyeless gene in Drosophila. Science 267: 1788–1792. (I)

Hamner, W.M., P.O. Hamner, S.W. Strand, and R.W. Gilmer. 1983. Behavior of Antarctic krill, *Euphausia superba*: chemoreception, feeding, schooling, and molting. Science 220: 433–435. (CR)

Hessler, R.R., and W.A. Newman. 1975. A trilobite origin for the Crustacea. Fossils Strata 4: 437–459. (T, CR)

Hoffman, R.L., and J.A. Payne. 1969. Diplopods as carnivores. Ecology 50: 1096–1098. (M)

Hölldobler, B., and E.O. Wilson. 1990. The Ants. Harvard University Press, Cambridge, MA. (I)

Hopkin, S.P., and H.J. Read. 1992. The Biology of Millipedes. Oxford University Press, New York. (M)

Kaston, B.J. 1978. How to Know the Spiders, 3rd ed. Wm. C. Brown, Dubuque, IA. (CH)

Kerfoot, W.C. (ed.). 1980. Evolution and Ecology of Zooplankton Communities. University Press of New England, Hanover, NH. (CR)

Kettle, D.S. 1984. Medical and Veterinary Entomology. Wiley, New York. (I)

Kim, K.C. (ed.). 1985. Coevolution of Parasitic Arthropods and Mammals. Wiley, New York. (CH, I)

King, P.E. 1973. Pycnogonids. St. Martin's Press, New York. (CH)

Labandeira, C.C., B.S. Beal, and F.M. Hueber. 1988. Early insect diversification: evidence from a Lower Devonian bristletail from Québec. Science 242: 913–916. (I)

Levi-Setti, R. 1993. Trilobites, 2nd ed. University of Chicago Press, Chicago. (T)

Lewis, J.G.E. 1981. The Biology of Centipedes. Cambridge University Press, Cambridge. (M)

Lynch, M. 1980. The evolution of cladoceran life histories. Rev. Biol. 55: 23–42. (CR)

Manton, S.M. 1977. The Arthropoda: Habits, Functional Morphology and Evolution. Clarendon Press, Oxford. (G)

Matthews, R.W., and J.R. Matthews. 1978. Insect Behavior. Wiley, New York. (I)

Mauchline, J., and L.R. Fischer. 1969. The biology of the euphausiides. Adv. Mar. Biol. 7: 1–454. (CR)

McCafferty, W.P. 1981. Aquatic Entomology. Jones & Bartlett, Boston. (I)

McDaniel, B. 1979. How to Know the Mites and Ticks. Wm. C. Brown, Dubuque, IA. (CH)

McLaughlin, P.A. 1980. Comparative Morphology of Recent Crustacea. W. H. Freeman, San Francisco. (CR)

Moore, R.C. 1959. Treatise on Invertebrate Paleontology. Part O: Arthropoda 1. Geol. Soc. Am., University of Kansas Press, Lawrence, KS. (T)

Neville, C. 1975. Biology of the Arthropod Cuticle. Springer-Verlag, New York. (G)

Novak, V.J.A. 1975. Insect Hormones, 2nd ed. Wiley, New York. (I)

Paffenhöfer, G.A., J.R. Stickler, and M. Alcaraz. 1982. Suspension feeding by herbivorous calanoid copepods: a cinematrographic study. Mar. Biol. 67: 193–199. (CR)

Panganiban, G., A. Sebring, L. Nagy, S. Carroll. 1995. The development of crustacean limbs and the evolution of arthropods. Science 270: 1363–1366. (G, CR)

Polis, G.A., and S.J. McCormick. 1987. Intraguild predation and competition among desert scorpions. Ecology 68: 332–343. (CH)

Richards, O.W., and R.G. Davies. 1978. Imm's Outline of Entomology, 6th ed. Chapman & Hall, London. (I)

Rinderer, T.E., B.P. Oldroyd, and W.S. Sheppard. 1993. Africanized bees in the U.S. Sci. Am. 269(6): 84–90. (I)

Schram, F.R. 1986. Crustacea. Oxford University Press, New York. (CR)

Seeley, T.D. 1989. The honey bee colony as a superorganism. Am. Sci. 77: 546–553. (I)

Shear, W.A. (ed.). 1986. Spiders: Webs, Behavior and Evolution. Stanford University Press, Stanford, CA. (CH)

Shear, W.A. 1994. Untangling the evolution of the web. Am. Sci. 82: 256–266. (CH)

Sonenshine, D.E. 1992/1993. Biology of Ticks, Vols. 1 and 2. Oxford University Press, New York. (CH)

Vollrath, F. 1992. Spiders webs and silks. Sci. Am. 266(3): 70–76. (CH)

Weygoldt, P. 1969. The Biology of Pseudoscorpions. Harvard University Press, Cambridge, MA. (CH)

Wootton, R.J. 1990. The mechanism of insect wings. Sci. Am. 263(5): 114–120. (I)

OTHER PROTOSTOME PHYLA

Three small, schizocoelomate, protostome phyla are listed here as a matter of convenience: Echiura, Priapulida, and Sipuncula. Priapulida is believed by some zoologists to be a member of the pseudocoelomate complex. The specific relationship of these phyla to the protostome line is uncertain. Only the Sipuncula are studied here due to lack of readily available study material.

Kohn, A.J., and M.E. Rice. 1971. Biology of the Sipuncula and Echiura. BioScience 21: 583–584.

Rice, M.E. 1985. Sipuncula: developmental evidence for phylogenetic inference. *In*: The Origins and Relationships of Lower Invertebrates. S. Conway Morris, J.D. George, R. Gibson, and H.M. Platt (eds.). Clarendon Press, Oxford, pp. 274–296.

Storch, U. 1991. Pripulida. *In*: Microscopic Anatomy of Invertebrates, Vol. 4. Harrison, F.W. and E.E. Ruppert (eds.). Wiley-Liss, New York, pp. 333–350.

■

Phylum Tardigrada

Phylum Tardigrada [TAR-di-grad-a; L., *tardi*, slow + L., *grada*, walk] comprises a small group of bilaterally symmetrical, arthropod-like animals. About 400 species of tardigrades, or water bears as they are commonly known, are recognized. These minute (50–1,000 μm) Protostomes possess short cylindrical bodies with four pairs of short, stubby, nonjointed legs extending ventrolaterally as if they were attached to a fat little bear (Fig. 16.1). Each leg has two or more toes, claws, or a combination of both at the distal end for grasping the substrate. Many semiterrestrial species have eyespots anteriorly. The bearlike image in tardigrades is reinforced when one observes the slow pawing motion that they make with their legs when moving.

Tardigrades are principally freshwater, but marine species have been found in the littoral area and at considerable depths (5,000 m). Semiterrestrial forms often are found in the thin water films in soils or covering mosses, lichens, and liverworts. In both freshwater and marine habitats, they may be found on surfaces of aquatic plants and in interstitial spaces among sedimentary particles. Many species have a cosmopolitan distribution.

Classification

Two classes are recognized based on presence or absence of cephalic appendages. Malpighian tubules, morphology of solid supports in the pharynx, and types and number of claws and toes.

1. **Class Heterotardigrada.** Armored tardigrades. Example: *Echiniscus.*

2. **Class Eutardigrada.** Unarmored tardigrades. Example: *Macrobiotus.*

A third class Mesotardigrada consisting of a single, poorly described species collected from a hot spring near Nagasaki (Japan) should be dismissed; neither the original specimen, nor the location of the collect still exist.

■ Observational Procedure:

Observe the movements of a live tardigrade under a compound microscope. Note the four pairs of legs, each of which possesses a set of claws. Describe the locomotory activity of your specimen. How fast does the animal move? Does the specimen stop to feed? How is feeding accomplished?

Using live or preserved specimens, attempt to locate the following internal structures: (1) muscular sucking pharynx, (2) salivary glands, (3) esophagus, (4) midgut, (5) Malpighian tubules, (6) rectum, (7) anus, and (8) unpaired gonad (Figs. 16.1 and 16.2). Eggs may be present in the ovaries of females. Locate several of the body muscle fibers. Trace a muscle from its attachment to the body wall at one point to another. What is the net result of the contraction of this single muscle fiber?

If time and specimens are available for experimental work, follow the directions given by your instructor to initiate and revive animals from the anabiotic resting state. One way to undertake this study is to rehydrate small samples of dried moss using aged tap water. Then, over the next few days, periodically examine the moss for active tardigrades using a dissection microscope. ■

241

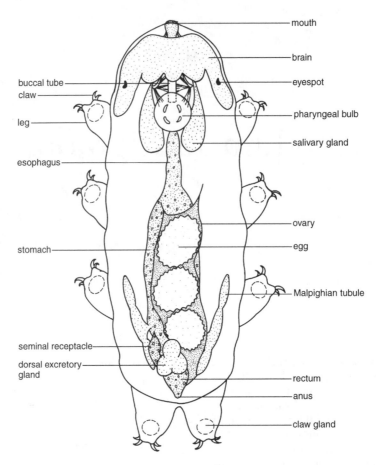

Figure 16.1. Dorsal internal view of a female tardigrade of the genus *Macrobiotus*.

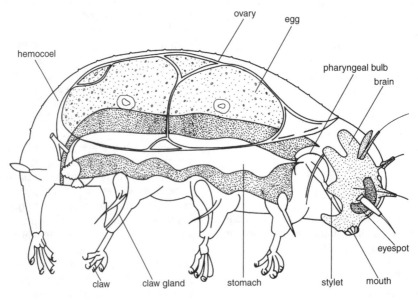

Figure 16.2. Lateral internal view of a female marine tardigrade of the genus *Styraconyx*. (After Kristensen and Higgins.)

Supplemental Readings

Bertolani, R. (ed.). 1987. Biology of Tardigrades. Selected Symposia and Monographs, I. Collana U.Z.I., Modena, Mucchi Editore.

Crowe, J.H. 1975. The physiology of cryptobiosis in tardigrades. Mem. Ist. Ital. Idrabiol. Suppl. 32: 37–59.

Crowe, J.H., and A.F. Cooper, Jr. 1971. Cryptobiosis. Sci. Am. 225(12): 30–36.

Crowe, J.H., and R.P. Higgins. 1967. The revival of *Macrobiotus areoatus* Murry (Tardigrada) from the cryptobiotic state. Trans. Am. Microsc. Soc. 86: 286–294.

Everitt, D.A. 1981. An ecological study of an Antartic freshwater pool with particular reference to Tardigrada and Rotifera. Hydrobiologia 83: 225–237.

Higgins, R.P. (ed.). 1975. International symposium on tardigrades. Mem. Ist. Ital. Idrobiol. 32 (Suppl.): 1–469.

Kinchin, I.M. 1994. The Biology of Tardigrades. Portland Press, London.

Nelson, D.R. (ed.). 1982. Proceedings of the 3rd International Symposium on the Tardigrada. East Tennessee State University Press, Johnson City, TN.

Pollock, L.W. 1970. Distribution and dynamics of interstitial Tardigrada at Woods Hole, Massachusetts, USA. Ophelia 7: 145–166.

Pollock, L.W. 1995. New marine tardigrades from Hawaiian beach sand and phylogeny of the family Halechiniscidae. Invertebr. Biol. 114: 220–235.

Schuster, T.R., and A.A. Grigarick. 1965. Tardigrada from western North America with emphasis on the fauna of California. Univ. Calif. Publ. Zool. 76: 1–67.

Wright, J.C., P. Westh, and H. Ramløv. 1992. Cryptobiosis in Tardigrada. Biol. Rev. Camb. Phil. Soc. 67: 1–29.

THE LOPHOPHORATES

Lophophorates constitute a group of coelomates (Phoronida, Bryozoa, Brachiopoda) each possessing an anterior crown of tentacles called a **lophophore**. This structure is responsible for feeding and gaseous exchange. Traditionally a lophophore is defined as a set of tentacles encircling the mouth, but not the anus, and into which a portion of the coelom (mesocoel) extends (Hyman 1959: 229). Although there are exceptions, typical lophophorates are sessile, vermiform or clamlike organisms with a reduced head, U-shaped digestive tract, and a leathery to hard case.

Based on Hyman's defination of the lophophore, phylum Entoprocta does not possess a lophophore. Nevertheless some argue for an affinity between Entoprocta and Bryozoa (Nielsen 1977b) and we cover them here (cf. Mackey et al. 1996).

The phylogenetic position of the lophophorates is as controversial as ever, and several competing phylogenies have been proposed over the years. They may be a closely related group (clade) with affinities to the deuterostomes (Zimmer 1973, Farmer 1977; Valentine 1981; Willmer 1990). Valentine (1973) suggests placing them in the superphylum Lophophorata and Emig (1984) recommends that that group receive the status of phylum. However, an opposing view is offered by Nielsen (1977a, 1985, 1995) who considers lophophorates to be an unnatural assemblage of protostomes (Bryozoa) and deuterostomes (Phoronida and Brachiopoda).

The latest addition to this group is Phylum **Cycliophora**. It is believed to have affinities to the Bryozoa and Entoprocta (Funch and Kristensen 1995).

Emig, C. 1984. On the origin of the Lophophorata. Z. Zool. Syst. Evolut.-Forsch. 22: 91–94.

Farmer, J.D. 1977. An adaptive model for the evolution of the Ectoproct life cycle. R.M. Woolacott and R.L. Zimmer (eds.), Biology of Bryozoans. Academic Press, New York, pp. 487–517.

Funch, P., and R.M. Kristensen. 1995. Cycliophora is a new phylum with affinities to Entoprocta and Ectoprocta. Nature 378: 711–714.

Halanych, K.M., J.D. Bacheller, A.M.A. Aguinaldo, S.M. Liva, D.M. Hillis, and J.A. Lake. 1995. Evidence from 18S ribosomal DNA that the lophophorates are protostome animals. Science 267: 1641–1643.

Hyman, L.H. 1959. The Invertebrates: Smaller Coelomate Groups, Vol. 5. McGraw-Hill New York.

Mackey, L.Y., B. Winnepenninckx, R. De Wachter, T. Backeljau, P. Emschermann, and J.R. Garey. 1996. 18S rRNA suggests that the Entoprocta are protostomes, unrelated to Ectoprocta. J. Mol. Evol. (in press).

Nielsen, G. 1977a. The relationship of Entoprocta, Ectoprocta and Phoronida. Am. Zool. 17: 149–150.

Nielsen, C. 1977b. Phylogenetic considerations: the Protostomian relationships. R.M. Woolacott and R.L. Zimmer (ed.), Biology of Bryozoans. Academic Press, New York, pp. 519–534.

Nielsen, C. 1985. Animal phylogeny in the light of the trochaea theory. Biol. J. Linn. Soc. 25: 243–299.

Nielsen, C. 1995. Animal Evolution. Oxford University Press, New York.

Valentine, J.W. 1973. Coelomate Superphyla. Syst. Zool. 22: 97–102.

Valentine, J.W. 1981. The Lophophorate Condition. Lophophorates, Notes for a Short Course. University of Tennessee, Knoxville, TN.

Willmer, P. 1990. Invertebrate Relationships: Patterns in Animal Evolution. Cambridge University Press, Cambridge, UK.

Zimmer, R.L. 1973. Morphological and developmental affinities of the Lophophorates. G.P. Larwood (ed.), Living and Fossil Bryozoa. Academic Press, New York, pp. 593–599.

■

Phylum Sipuncula (Sipunculida)

Sipuncula [sigh-PUN-ku-la; L., *sipunculus*, little siphon] is a small phylum of about 320 species of vase-shaped, wormlike, marine organisms. They are commonly called peanut worms because their contracted bodies resemble a peanut. When extended, the **trunk** is usually vaselike or plump (Fig. 17.1). The anterior part of the body or the neck region, called the **introvert**, constitutes at least one-half to two-thirds the length of the worm, and is capable of being rapidly everted and retracted. Segmentation and setae are lacking. At the end of the introvert is the mouth, surrounded by modified, lobate, ciliated tentacles. The tentacles are hollow, but do not communicate with the body cavity. Most sipunculans (sipunculids) live a sedentary life in burrows or tubes. Internally, sipunculans have a spacious coelom and an elongated, U-shaped, digestive system (Fig. 17.1). The mouth is located at the anterior part of the body from which the gut is directed posteriorly only to be recurved, opening at the anus, which is located near the junction of the introvert and trunk.

Classification

Two classes are recognized.

1. **Class Phascolosomatidea.** Sipunculans with tentacles confined to an arc surrounding the nuchal organ.

Peripheral, circumoral tentacles are absent. Examples: *Aspidosiphon*, and *Phascolosoma*.

2. **Class Sipunculidea.** Sipunculans with peripheral, circumoral tentacles. Examples: *Phascolopsis*, *Golfingia*, *Phascolosoma*, and *Sipunculus*.

■ Observational Procedure:

Examine the external anatomy of several peanut worms (Fig. 17.1). Locate the anterior and posterior ends, trunk, introvert, mouth with tentacles, and anus. Note the variations in size and shape of tentacles among species available for study.

If specimens are available for dissection, make a shallow, midsagittal, dorsal incision with scissors through the body wall slightly to one side of the anus. Extend the cut forward to the tip of the introvert and backwards as far as possible. Pin the body wall open and locate the internal organs (Fig. 17.1). Observe the introvert under a dissection microscope to see the many minute spines and sensory papillae. Examine the inner side of the tentacles surrounding the mouth and find the groove that conveys food to the mouth. ■

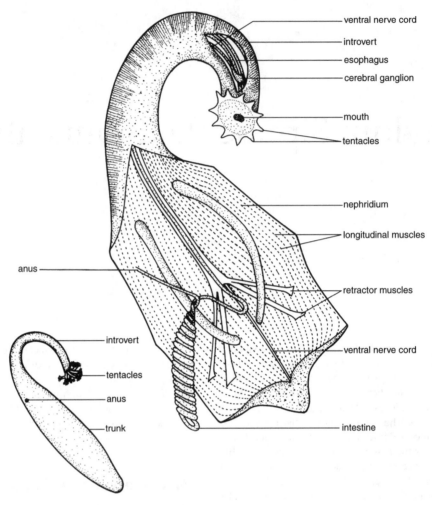

ventral nerve cord
introvert
esophagus
cerebral ganglion
mouth
tentacles
nephridium
longitudinal muscles
retractor muscles
ventral nerve cord
intestine
anus
introvert
tentacles
anus
trunk

Figure 17.1. External and internal anatomy of the sipunculan *Phascolosoma agassizi*.

Supplemental Readings

Clark, R.B. 1969. Systematics and Phylogeny; Annelida, Echiura, Sipuncula. *In*: M. Florkin and B.T. Scheer (eds.). Chemical Zoology, Vol. 4. Academic Press, New York, pp. 1–68.

Cutler, E.B. 1973. Sipuncula of the western North Atlantic. Bull. Am. Mus. Nat. Hist. 152: 103–204.

Cutler, E.B. 1994. The Sipuncula: Their Systematics, Biology, and Evolution. Comstock Publishing Associates, Ithaca, NY.

Cutler, E.B., and P.E. Gibbs. 1985. A phylogenetic analysis of higher taxa in the phylum Sipuncula. Syst. Zool. 34: 162–173.

Dybas, L.K. 1981. Cellular defense reactions of *Phascolosoma agassizii*, a sipunculan worm: phagocytosis by granulocytes. Biol. Bull. 161: 104–114.

Murina, V.V. 1984. Ecology of Sipuncula. Mar. Eco. Prog. Ser. 17: 1–7.

Pilger, J.F. 1982. Ultrastructure of the tentacles of *Themiste lageniformis* (Sipuncula). Zoomorphology 100: 143–156.

Rice, M.E. 1981. Larvae adrift: patterns and problems in life history of sipunculans. Am. Zool. 21: 605–619.

Rice, M.E. 1985. Sipuncula: Developmental Evidence for Phylogenetic Inference. *In*: S. Conway Morris, J.D. Morris, R. Gibson, and H.M. Platt (eds.). The Origins and Relationships of Lower Invertebrates. Systematics Association Spec. Vol. 28, Clarendon Press, Oxford, pp. 274–296.

Rice, M.E. 1993. Sipuncula. *In*: F.W. Harrison and M.E. Rice (eds.). Microscopic Anatomy of Invertebrates. Wiley-Liss, New York, pp. 238–325.

EXERCISE 18

■

Phylum Phoronida

Phylum Phoronida [fo-RON-i-da; G., *Phoronis*, from Greek mythology] contains only about 10 species of non-segmented, tube-dwelling, marine worms. Ranging in length from a few millimeters to over 30 cm, these sedentary animals secrete a chitinous, parchment-like tube to which sand, small stones, shells, and other debris may be attached (Fig. 18.1). Although usually not common, they may be found occasionally carpeting shallow to moderately deep (ca. 400 m) tropical and temperate waters (Emig 1979, 1982; Johnson 1959). Phoronids usually live as solitary individuals buried in soft sediments (mud or sand), or as aggregates (Fig. 18.2) enhabiting burrows among rocks and mollusk shells, or attached to pilings and other organisms. Some species penetrate calcareous shells or rocks. Phoronids move freely in their tubes, but they do not leave them. The horseshoe-shaped **lophophore** is located at the anterior end of the animal. Normally this structure is extended from the tube for feeding, but upon disturbance it is quickly retracted. The gut is elongate, U-shaped, and extends from the lophophore to the posterior end and back toward the anterior end. Phoronids have a complex larval stage known as an **actinotroch** (Fig. 18.3), sometimes interpreted as a modified **trochophore** larva which it is not (Zimmer 1973). [NB: the concocted etymon (L., *nidus*, nest + G., *phora*, to bear) should be rejected.]

Classification

No taxonomic division higher than genus (*Phoronis* and *Phoronopsis*) is recognized.

■ Observational Procedure:

Observe a phoronid specimen under a dissection microscope and identify the structures discussed above and shown in Fig. 18.1. Note the lophophore at the anterior end with its tentacles surrounding the mouth. A small flap of tissue, the **epistome**, may be seen above the mouth. The anus is external to the lophophore and may be located by carefully rotating the animal about its longitudinal axis. The paired **nephridiopores**, which are located on either side of the anus, will probably not be seen without considerable effort. Given adequate lighting and a sufficiently transparent specimen you may be able to trace the gut from mouth to anus. Make of list of the characteristics you believe are the most important ones in defining the phoronids. This compilation will be useful in comparing phoronids with the other lophophorate phyla. ■

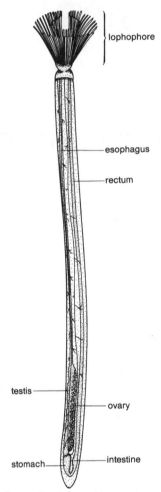

Figure 18.1. Generalized phoronid.

Figure 18.2. Aggregate of the phoronid *Phoronis hippocrepis*.

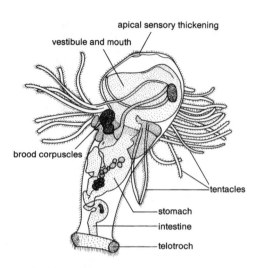
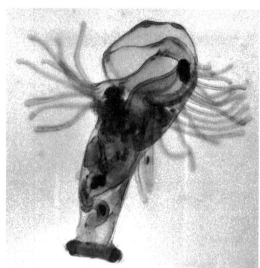

Figure 18.3. Photomicrograph of an actinotroch larva of the phoronid *Phoronis muelleri*, with a line drawing interpretation. (Photomicrograph courtesy of K. Herrmann.)

Supplemental Readings

Abele, L.G., T. Gilmour, and S. Gilchrist. 1983. Size and shape in the phylum Phoronida. J. Zool. Lond. 200: 317–323.

Emig, C.C. 1979. A synopsis of British and other Phoronids. Academic Press, London.

Emig, C.C. 1982. The biology of Phoronida. Adv. Mar. Biol. 19: 1–89.

Harrison, F.W., and R.W. Woollacott. 1996. Microscopic Anatomy of Invertebrates, Vol. 13. Lophophorates and Entoprotca. Wiley-Liss, New York.

Johnson, R.C. 1959. Spatial distribution of *Phoronopsis viridis* Hilton. Science 129: 1221.

Marsden, R.C. 1957. Regeneration in *Phoronis vancouverensis*. J. Morphology 101: 307–324.

Nielsen, C. 1977. The relationship of Entoprocta, Ectoprocta, and Phoronida. Am Zool. 17: 149–150.

Strathmann, R. 1973. Function of lateral cilia in suspension feeding of lophophorates (Brachiopoda, Phoronida, Ectoprocta). Mar. Biol. 23: 129–136.

Zimmer, R.L. 1973. Morphological and development affinities of the lophophorates. *In*: G.P. Larwood (ed.), Living and Fossil Bryozoa, Academic Press, New York, pp. 593–599.

Zimmer, R.L. 1978. The comparative structure of the preoral hood coelom in Phoronida and the fate of this cavity after metamorphosis. *In*: F.-S. Chia and M.E. Rice (eds.), Settlement and Metamorphosis of Marine Invertebrate Larvae. Elsevier, New York, pp. 23–40.

EXERCISE 19

■

Phylum Bryozoa (Ectoprocta)

Phylum Bryozoa [BRI-o-ZO-a; G., *bryo*, moss + G., *zoa*, animals] comprises a group of sessile, predominantly colonial lophophorates, some of which superficially resemble hydroids. First appearing in the early **Ordovician Period** nearly 500 million years ago (McLeod 1978), bryozoans have a rich geologic history (Fig. 19.1) with more than 16,000 described species. Many of the 4,000 living species of bryozoans are found in the marine intertidal zone, but specimens have been recovered from depths greater than 8,000 m. There are a few common freshwater genera. The common name "moss animal" is fitting, because bryozoans grow in water as colonies, sometimes forming mosslike blankets of growth over the object to which they are attached. Two alternate names are used for this phylum. **Ectoprocta** emphasizes that the anus is located outside of the lophophore in contrast to phylum Entoprocta (Exercise 20), while **Polyzoa** emphasizes the colonial aspect of members of the phylum.

Although the individual organism, or **zooid**, is microscopic in size (usually less than 1 mm), the colony or **zoarium** may attain many centimeters in length and contain thousands of zooids. Zooids consist of two basic parts: a protective exoskeleton called the **zoecium** (zooecium) and the softer, internal viscera (Figs. 19.2 and 19.3). Zoecia may be gelatinous, chitinous, or calcareous and are secreted by the epidermis of the body wall. The body wall and the zoecium together are referred to as the **cystid**, while the rest of the animal, including the lophophore, digestive apparatus, and associated viscera, constitutes the **polypide**. In forms with calcareous zoecia, an orifice permitting eversion and retraction of the lophophore, may be covered by a lid or **operculum**.

Some bryozoans show considerable polymorphism in zooid structure. **Autozooids** are responsible for feeding, while heterozooids have other functions. **Avicularia** (Fig. 19.4) and **vibracula** (Fig. 19.5) are two kinds of

TIME (IN MILLIONS OF YEARS)

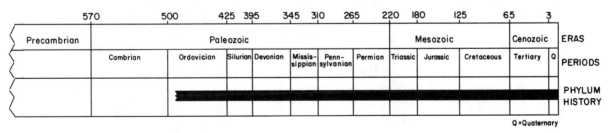

570		500	425	395	345	310	265	220	180	125	65	3	

Precambrian	Paleozoic							Mesozoic			Cenozoic		ERAS
	Cambrian	Ordovician	Silurian	Devonian	Missis-sippian	Penn-sylvanian	Permian	Triassic	Jurassic	Cretaceous	Tertiary	Q	PERIODS
													PHYLUM HISTORY

Q =Quaternary

Figure 19.1. Geologic history of the phylum Bryozoa.

253

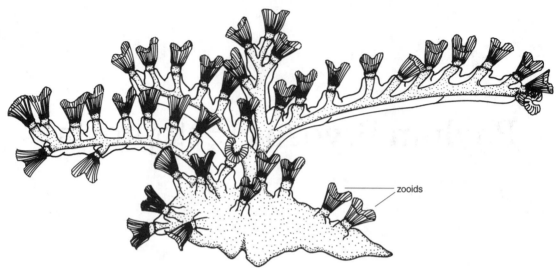

Figure 19.2. Freshwater bryozoan colony, *Hyalinella*.

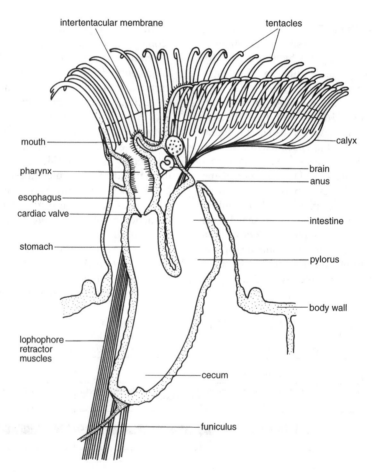

Figure 19.3. A freshwater bryozoan zooid showing internal morphology.

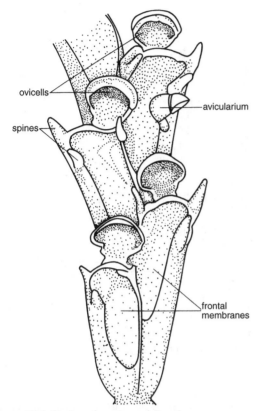

Figure 19.4. Portion of a colony of *Bugula*.

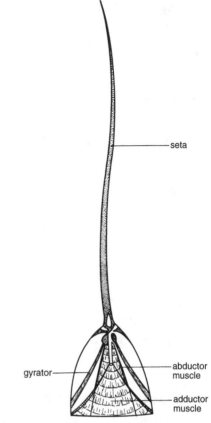

Figure 19.5. Vibraculum.

heterozooids which function to keep the colony surface free from debris and encrusting organisms.

There are two types of bryozoan larvae. One is produced by nonbrooding species and is named a **cyphonautes larva**. This is a flattened, bivalve-like larva which feeds in the plankton for up to several months. The other larva is more spherical and nonfeeding (lecithotrophic); it is produced by brooding species and is known as a **coronate larva** (Fig. 19.6). After attachment and metamorphosis, the resulting zooid, known as the **ancestrula**, grows and produces additional zooids by budding.

Classification

Three classes of bryozoans are recognized.

1. **Class Phylactolaemata.** Freshwater bryozoans comprising about 50 species usually with a horseshoe-shaped lophophore and statoblasts. Calcified zoecium is lacking. Polymorphism. Examples: *Fredericella, Pectinatella, Plumatella,* and an unusual form, *Cristatella,* where the entire colony slowly glides on a muscular sole.

2. **Class Stenolaemata.** Marine bryozoans with tubular, calcified zoecia. One extant order (Cyclostomata) and several orders that became extinct during the Paleozoic. Example: *Crisia.*

3. **Class Gymnolaemata.** Mostly marine bryozoans with a circular lophophore and polymorphic zooids.

   a. **Order Ctenostomata.** Mostly marine bryozoans with noncalcified zoecia and a stoloniferous to somewhat flattened growth form. Zooids without an operculum. Examples: *Zoobotryon* and *Paludicella* (freshwater genus).

   b. **Order Cheilostomata.** A large group of widely-distributed marine bryozoans having a box-like, variably calcified zoecium, with an operculum. Zooids are polymorphic with avicularia and/or vibracula present. Examples: *Bugula Electra, Membranipora,* and *Schizoporella* (all with encrusting growth).

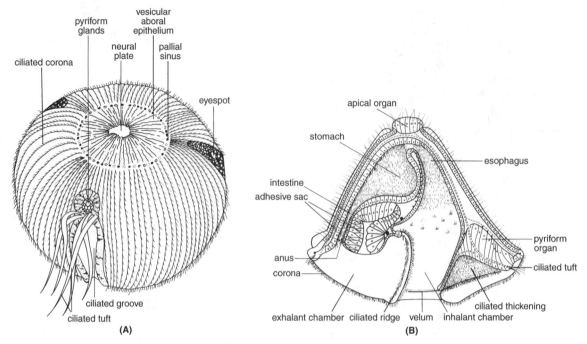

Figure 19.6. Generalized bryozoan larval types. (A) Coronate. (B) Cyphonautes. (After Woolacott & Zimmer, and Nielsen, respectively.)

■ Observational Procedure:

We will begin our observations of bryozoans with a prepared slide of the common, marine cheilostome *Bugula* because it is easy to see the general features of the phylum. Specimens from other genera will then be compared to *Bugula*.

Bugula

1. Although the zooids of *Bugula* are small (ca. 500 μm long), colonies may attain heights of a few centimeters (Fig. 19.4). Examine a prepared slide or preserved specimen under a dissection microscope or low power of a compound microscope. Start by focusing on a zooid at the base of the colony and scan the entire colony, taking note of the branching pattern formed by asexual budding. What type of branching pattern does *Bugula* exhibit? Occasionally one may find a specimen with a developing bud.

2. As you scan the colony you should note the predominance of the autozooid. However, the avicularia may be seen attached to the autozooid by a short **peduncle**. Upon examination of an avicularium in profile under higher magnification you should observe the rostrum (beak), mandible, and powerful musculature that closes the mandible. A sharp tooth may be seen on the mandible.

3. Now observe the lophophores of several autozooids. How many tentacles does an autozooid have?

If the bases of the tentacles were connected by a line, what would be the general shape of the figure thus created?

4. Try to locate the mouth which is centrally located at the base of the tentacles. Inside the zooid lies the gut and associated viscera.

5. Scan the colony and find an individual in which you can trace the entire digestive tract. Beginning with the mouth, locate the esophagus, cardiac valve, stomach with caecum, pyloric valve, rectum, and anus. Note in particular that the anus empties outside the lophophore. Trace the guts of other zooids and determine what food this specimen of *Bugula* ate.

6. If the specimen were part of a fully grown colony it may have **ovicells**. These globular or hemispherical structures will be situated at the distal (upper) end of some of the zoecia. Inside you may be able to see a developing embryo.

7. Not all of the autozooids are still functional. At the base of the colony several almost empty zoecia may be seen. In some of these you may see a small, darkly stained structure. This is the **brown-body**; it consists of the cellular remnants of a degenerated zooid.

8. Compare *Bugula* and a cnidarian hydroid such as *Obelia*. To do this make a list of their similarities and differences; you may wish to organize this list in some sort of table form. For example, compare *Bugula* and *Obelia* in size and overall body shape. Are their tentacles grouped the same way about the mouth? How are the zooids located on each specimen? Is polymorphism displayed to the same extent in each phyla? If not, how is it different? Trace the gut in each specimen; how do they differ? Finally, devise a simple key, table, and/or word description that will permit someone who has never seen these organisms to distinguish them.

Encrusting Marine Bryozoans

Observe a specimen of an encrusting bryozoan (Fig. 19.7). Note how different the growth form is from than that of *Bugula*. The autozooids are found in calcareous, boxlike zoecia so that the entire colony resembles a brick wall. At one end of the zooid an orifice can be seen which may or may not have a lidlike operculum. Ovicells, avicularia, vibracula, and spines may be present depending on the species.

Freshwater Bryozoans

In contrast to the two marine gymnolaemates just studied, we will now observe a freshwater phylactolaemate such as *Pectinatella* and/or *Plumatella*. Is the shape of their lophophores different from that of *Bugula*? How would you describe its shape? Running from the lophophore to the base of the animal you may see the lophophore retractor muscle; it should not be confused with the funiculus. Note the numerous tentacles in this species. How many are there? Trace the gut of your specimen. How is it similar to that of *Bugula*? The funiculus, if present, may be seen attached to the base of the stomach and connected to the body wall. Scattered along the length of the funiculus may be several statoblasts in various stages of development (Fig. 19.8).

Live Specimens

1. Observe the manner of rapid retraction and gradual extension of the lophophore. Describe how the lophophore is retracted and extended. (NB: This is a good example of a muscular/fluid hydraulic mechanism.)

2. Note the movement of the cilia on the tentacles in expanded individuals. One also may observe flicking of the whole tentacles. This process, which varies among species, is a mode of transporting food particles toward the mouth and rejecting unwanted materials.

3. Using a Pasteur pipette, gently add a few drops of water containing some yeast, carbon black, or carmine red near the lophophore and observe the feeding currents. Note that the feeding currents run down into the center of the lophophore crown and then out the sides between the tentacles.

4. In active colonies of some encrusting cheilostomates, such as *Membranipora*, the autozooids lean away from a particular area thereby leaving a blank site called a chimney. Chimneys represent areas where water, filtered by the lophophores, exits the colony surface. These transient structures cannot be seen when the lophophores are retracted but appear to be important in preventing interference between neigh-

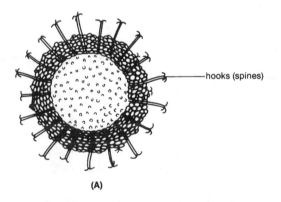

(A)

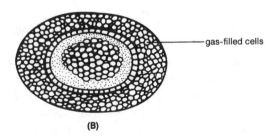

(B)

Figure 19.8. Phylactolaemate statoblasts. (From Barnes, after Allman.)

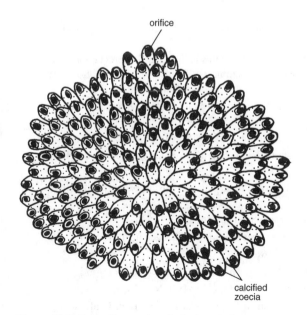

Figure 19.7. Encrusting marine cheilostomate bryozoans.

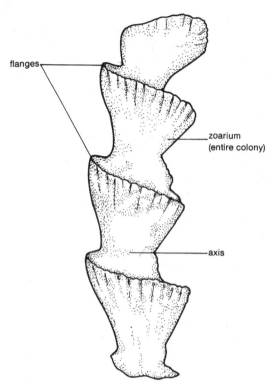

flanges

zoarium
(entire colony)

axis

Figure 19.9. The fossil bryozoan *Archimedes*.

boring lophophores. Chimneys apparently have a constant location on the surface of the colony. To verify this, observe the pattern of chimneys across the colony's surface before you disturb the zooids and after the lophophores re-emerge. You should also be able to see periodic ingestion of particles by the pharynx and the rotating food cord in the stomach.

Fossil Forms

Unfortunately, a formidable nomenclature has been developed within the literature on fossil bryozoans. We will not take time to present the details of this discipline. Identification and detailed study are done from longitudinal and tangential thin sections which have been prepared from fossils (Moore et al. 1952).

Observe the fossil specimens available for study using a hand lens or dissection microscope. Students should be able to observe the orifice of each individual zooid. One interesting genus, *Archimedes*, which occurred from the **Mississippian** through **Permian Periods**, resembles a water-screw and was named after the inventor of that device (Fig. 19.9). Examine several fossil bryozoans for the presence of numerous small bumps scattered fairly uniformly over the surface. Many fossil stenolaemates have these bumps. They are known as a **monticules** and are believed to be the Paleozoic equivalent of the chimneys discussed above. ■

Supplemental Readings

Banta, W.C. 1977. Body wall morphology of the sertellid cheilostome bryozoan, *Reteporellina evelinae*. Am. Zool. 17: 75–91.

Banta, W.C., F.K. McKinney, and R.L. Zimmer. 1974. Bryozoan monticules: excurrent water outlets? Science 185: 783–784.

Bushnell, J.H., and K.S. Rao. 1974. Dormant or quiescent stages and structures among the Ectoprocta; physical and chemical factors affecting viability and germination of statoblasts. Trans. Am. Micros. Soc. 93: 524–543.

Buss, L.W. 1981. Group living, competition, and the evolution of cooperation in a sessile invertebrate. Science 213: 1012–1014.

Cheetham, A.H. 1986. Branching, biomechanics and bryozoan evolution. Proc. R. Soc. Lond. B 228: 151–171.

Farmer, J.D., J.W. Valentine, and R. Cowen. 1973. Adaptive strategies leading to the ectoproct groundplan. Syst. Zool 22: 233–239.

Harvell, C.D. 1984. Why nudibranchs are partial predators: intracolonial variation in bryozoan palatability. Ecology 65: 716–724.

Harvell, C.D. 1984. Predator-induced defense in a marine bryozoan. Science 224: 1357–1359.

Larwood, G.P. (ed.). 1973. Living and Fossil Bryozoa. Recent Advances in Research. Academic Press, New York.

Mayr, E. 1968. Bryozoa versus Ectoprocta. Syst. Zool. 17: 213–216.

McLeod, J.D. 1978. The oldest bryozoans: new evidence from the early Ordovician. Science 200: 771–773.

Moore, R.C., C.G. Lalicker, and A.G. Fischer. 1952. Invertebrate Fossils. McGraw-Hill, New York.

Nielsen, C., and G.P. Larwood (eds.). 1985. Bryozoa: Ordovician to Recent. Olsen and Olsen, Fredensborg, Denmark.

Ryland, J.S. 1970. Bryozoans. Hutchinson University Library, Hutchinson and Co., London, UK.

Ryland, J.S., and A.R.D. Stebbing. 1971. Settlement and orientated growth in epiphytic and epizooic bryozoans. *In*: D.J. Crisp (ed.), Proc. 4th European Mar. Biol. Symp., Cambridge University Press, Cambridge, UK, pp. 283–300.

Seed, R., and R.J. O'Connor. 1981. Community organization in marine algal epifaunas. Ann. Rev. Ecol. Syst. 12: 49–74.

Sutherland, J.P. 1978. Functional roles of *Schizoporella* and *Styela* in the fouling community at Beaufort, North Carolina. Ecology 59: 257–264.

Wendt, D.E., and R.M. Woollacott. 1995. Induction of larval settlement by KCl in three species of *Bugula* (Bryozoa). Invertebr. Biol. 114: 345–351.

Wood, T.S. 1971. Laboratory culture of freshwater Ectoprocta. Trans. Am. Microsc. Soc. 90: 229–231.

Wood, T.S. 1991. Bryozoans. *In*: J.H. Thorpe and A.P. Covich (eds.), Ecology and Classification of North American Freshwater Invertebrates. Academic Press, New York, pp. 481–499.

Woollacott, R.M., and R.L. Zimmer (ed.). 1977. Biology of Bryozoans. Academic Press, New York.

Yoshioka, P.M. 1982. Role of planktonic and benthic factors in the population dynamics of the bryozoan *Membranipora membranacea*. Ecology 63: 457–468.

■

Phylum Entoprocta (Kamptozoa)

Entoprocta [EN-to-PROK-ta; G., *ento*, within + G., *proct*, anus] or Kamptozoa [KAMP-toe-zoo-ah; G., *campto*, bending + G., *zoa*, animal] is a little known phylum of mostly sessile invertebrates numbering about 100 species. Although considered by some to be pseudo-coelomates, their relationship to other organisms is uncertain. Currently, many invertebrate zoologists place entoprocts as a separate phylum related to the Bryozoa or a subphylum of Bryozoa (see Nielsen 1971, 1977 and Farmer 1977 for an in-depth discussion).

Except for a single, freshwater genus (*Urnatella*, Fig. 20.1) entoprocts are marine, growing in short tufts or mats on mollusk shells, pilings, rocks, and on living animals such as polychaetes and sponges. The small, somewhat transparent zooids are composed of two main parts: a slender **stalk** and an enlarged body or **calyx** with a crown of tentacles (Figs. 20.2 and 20.3). The stalk elevates the calyx a few millimeters above the substrate and may be smooth, spined, or have swellings that resemble beads on a string. The swellings are muscular joints that permit bending movements in the animal. The viscera are completely enclosed in the calyx which typically is vase- to boat-shaped. Enclosed by the crown of tentacles is a flattened to slightly concave region termed the **vestibule**. The mouth is at the anterior end of the vestibule and the anus is at the posterior end. The term entoproct is derived from the fact that the anus is located inside the crown of tentacles.

Classification

Three families occur in the phylum.

1. **Family Loxosomatidae.** Solitary entoprocts. Examples: *Loxosoma* and *Loxosomella*.

2. **Family Pedicellinidae.** Colonial marine and brackish water entoprocts. Examples: *Pedicellina*, *Myosoma*, and *Barentsia*.

3. **Family Urnatellidae.** Composed of only one freshwater genus, *Urnatella*.

■ Observational Procedure:

Examine a whole-mount slide of an entoproct and identify the calyx, with its crown of tentacles, and stalk (Figs. 20.1–20.3). Internally the body is filled with a gelatinous material containing mesenchymatous cells, a U-shaped, ciliated digestive tract, and reproductive and excretory systems. There is a cuticular covering of variable thickness over the entire animal, except for the tentacles and vestibule.

Compare the entoproct's external and internal structures to those of a hydrozoan (e.g., *Campanularia*) and a bryozoan (e.g., *Bugula*). Take particular note of (1) the striking differences in overall body architecture and reproductive structures, (2) the lack of tentacular ciliation in hydrozoans, and (3) the lack of nematocysts in entoprocts and bryozoans. ■

LIVERPOOL JOHN MOORES UNIVERSITY
LEARNING SERVICES

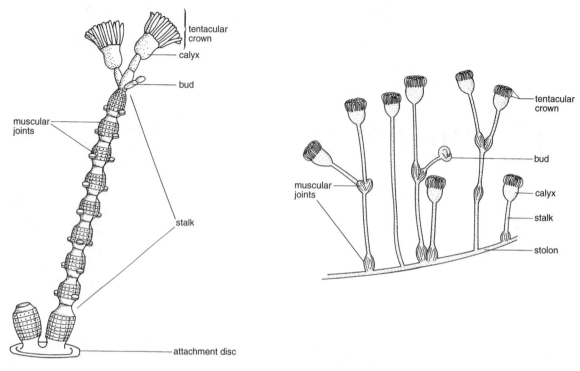

Figure 20.1. Colony of *Urnatella gracilis*. (After Leidy, from Pennak.)

Figure 20.2. Colony of *Barentsia*. (After Robertson, from Hyman.)

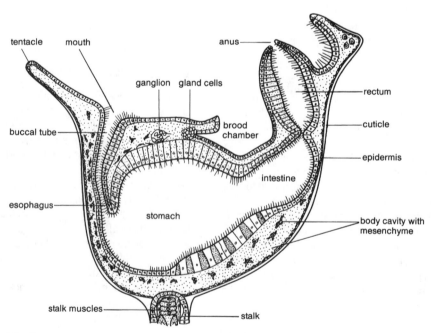

Figure 20.3. Calyx of *Pedicellina* (median sagittal section). (After Becker, from Hyman.)

Supplemental Readings

Cusak, T.M., and J.D. McCullough. 1985. *Urnatella gracilis* (Entoprocta) from Caddo Lake, Texas and Louisiana. Tx. J. Sci. 37: 141–142.

Eng, L.L. 1977. The freshwater entoproct, *Urnatella gracilis* Leidy, in the Delta-Mendota Canal, California. Wassman J. Bio. 35: 196–202.

Farmer, J.D. 1977. An adaptive model for the evolution of the Ectoproct life cycle. *In*: R.M. Woolacott and R.L. Zimmer (eds.), Biology of Bryozoans. Academic Press, New York, pp. 487–517.

King, D.K., R.H. King, and A.C. Miller. 1988. Morphology and ecology of *Urnatella gracilis* Leidy, (Entoprocta), a freshwatter macroinvertebrate from artificial riffles of the Tombigbee River, Mississippi. J. Freshwater Ecol. 4: 351–359.

Leidy, J. 1884. *Urnatella gracilis* a freshwater Polyzoan. J. Acad. Nat. Sci. 9: 5–16.

Nielsen, C. 1971. Entoproct life cycles and the entoproct/ectoproct relationship. Ophelia 9: 209–341.

Nielsen, C. 1977. The relationship of Entoprocta, Ectoprocta, and Phoronida. Am. Zool. 17: 149–150.

Nielsen, C., and J. Rostgaard. 1976. Structure and function of an entoproct tentacle with a discussion of ciliary feeding types. Ophelia 15: 115–140.

Weise, J.G. 1961. The ecology of *Urnatella gracilis* Leidy: Phylum Endoprocta. Limnol. Oceanogr. 6: 228–230.

Wood, T.S. 1991. Bryozoans. *In*: J.H. Thorp and A.P. Covich (eds.), Ecology and Classification of North American Freshwater Invertebrates. Academic Press, New York, pp. 481–499. [pp. 490–491 on Entoprocta.]

EXERCISE 21

■

Phylum Brachiopoda

Brachiopoda [brak-e-OP-o-da; G., brachio, arm, G., pod, foot] is a small group (ca. 400 species) of lophophorates found in marine intertidal areas to depths up to 7,600 m. In comparative terms, fossil brachiopods (ca. 40,000 species) far outnumber present-day forms. They are found in **Cambrian** deposits, but achieved their greatest diversity during the **Ordovician** and **Devonian Periods** and their greatest specialization in the **Permian Period** (Fig. 21.1).

These animals are less than 10 cm long and are enclosed in two shells. At first glance brachiopods resemble bivalve mollusks. However, they are different in two major ways: (1) brachiopods possess a lophophore and (2) the shell is very different from that of bivalves. The shells of most bivalves are arranged laterally, consisting of right and left valves, the plane of symmetry passing between the valves. Valve orientation in brachiopods is such that the two valves assume a dorsoventral position and are normally bilaterally symmetrical; the plane of symmetry passes through the valves (Fig. 21.2). Furthermore, brachiopod valves are often dissimilar in size and shape (Fig. 21.3); in most bivalves the shells are nearly mirror images of one another. In most brachiopods the larger pedicle valve is ventral and the smaller brachial valve dorsal. In epifaunal inarticulate brachiopods, this size relationship is reversed. Another feature of most modern brachiopod shells is the posterior extension of the ventral valve to form a break that bears a foramen for exit of an attachment stalk or pedicle. This makes the shell resemble a Roman oil lamp and it is for this reason that brachiopods are called lamp shells (Fig. 21.3). The two valves may articulate by teeth and sockets at the posterior margin; if teeth and sockets are absent, a complex musculature permits sliding and rotation of the valves.

Classification

The phylum is divided into two classes.

1. **Class Articulata.** Brachiopods with articulating teeth and sockets at the posterior margin of the shell. Valves possess a foramen in the beak of the ventral valve through which the pedicle passes, permitting attachment of the animal to hard substrates. The digestive tract is incomplete, ending as a blind intestine. The shell is composed of $CaCO_3$ (calcite). Examples: *Laqueus, Terebratalia,* and *Terebratulina.*

2. **Class Inarticulata.** Brachiopods that lack articulating teeth and sockets. A large pedicle serves as an anchor in sediments or for attachment to hard substrates. In some the ventral valve is cemented firmly to the substrate. Digestive tract is complete. Most shells are composed of $Ca_3(PO_4)_2$. Examples: *Crania, Discinisca, Glottidia,* and *Lingula* (a genus in existence since the **Silurian Period**).

265

LIVERPOOL JOHN MOORES UNIVERSITY
LEARNING SERVICES

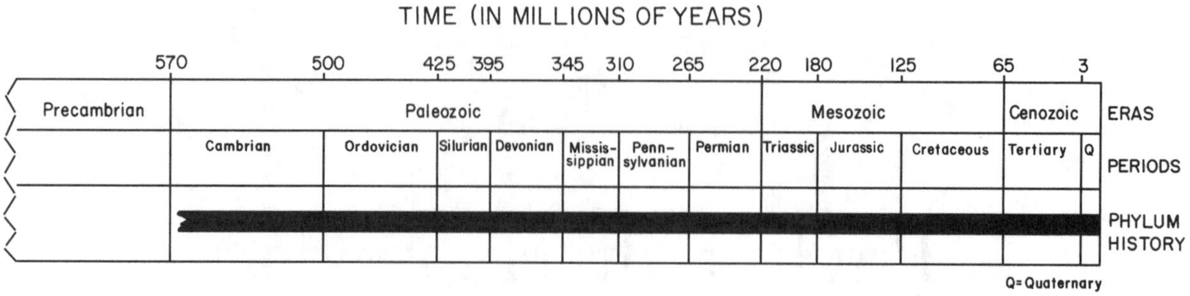

TIME (IN MILLIONS OF YEARS)

570   500   425 395   345 310   265   220 180   125   65   3

Precambrian	Paleozoic							Mesozoic			Cenozoic		ERAS
	Cambrian	Ordovician	Silurian	Devonian	Missis-sippian	Penn-sylvanian	Permian	Triassic	Jurassic	Cretaceous	Tertiary	Q	PERIODS
													PHYLUM HISTORY

Q= Quaternary

Figure 21.1. Geologic history of the phylum Brachiopoda.

■ Observational Procedure:

1. Compare the general shell morphology of an articulate brachiopod (*Terebratalia*), an inarticulate brachiopod (*Lingula* or *Glottidia*), and a bivalve mollusk (*Mercenaria*) (Figs. 21.3, 21.4, and 12.11, respectively). Identify dorsal and ventral valves of each specimen. In particular, be sure to contrast the symmetry of the brachiopods with that of the bivalve. Add to this suite of specimens another bivalve, the oyster *Crassosterea*. How does the shell of the oyster resemble the two brachiopods? What are the differences?

2. Identify anterior and posterior ends of each brachiopod specimen. Setae will be visible along the lateral shell margins of *Lingula* (Fig. 21.4). The brachiopods should have at least a portion of the pedicle visible. How might each of these specimens be oriented in its natural habitat? Compare these positions to the two bivalves and/or other specimens provided by your instructor (see Exercise 12.1).

3. Examine the internal anatomy of an articulate brachiopod by separating the two valves (Figs. 21.3 and 21.5). To do this gently pry apart the valves at the anterior until the hinge breaks. During this procedure, one

or more sets of muscles may be pulled off their attachments. Note where they were attached. Sometimes it may be necessary to cut the adductor and/or diductor muscles using a fine dissection blade in order to separate the valves completely. Take care not to destroy the lophophore and other internal organs.

4. Observe the looped lateral arms and central coiled portion of the lophophore. How does the lophorphorate of brachiopods compare to that of other lophorphorates? How are they similar; how do they differ?

5. The digestive glands should be visible at the posterior end of the shell (Fig. 21.3). They are composed of many spherical structures called **acini** [L., *acinus*, a berry] and resemble bunches of grapes.

6. The gonads are often best seen in the posterior parts of the branches of the coelom that extend into the mantle as structures called mantle canals.

7. Your instructor will provide additional directions if live brachiopods are to be examined (e.g., a study of feeding currents).

Fossil Forms

Some fossil specimens were much larger than the modern species, measuring greater than 35 cm wide. As in other animal groups, that possess a rich fossil

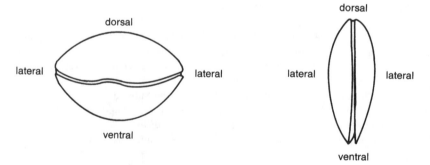

Figure 21.2. Comparison of the shell morphology of a brachiopod and a bivalve mollusk. Both are shown in an anterior view, thus right- and left-hand sides of the organisms may be determined.

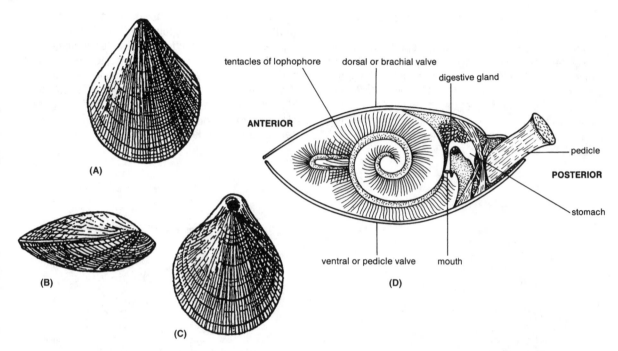

Figure 21.3. *Terebratulina*, an articulate brachiopod. (A) Ventral view of pedicle valve. (B) Side view. (C) Dorsal view of brachial valve. (D) Sagittal section showing the internal anatomy.

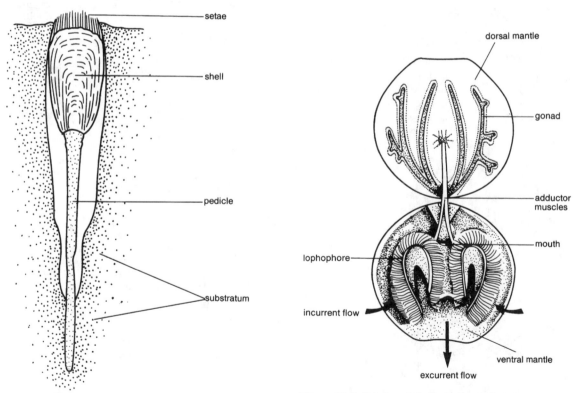

Figure 21.4. *In situ* view of the inarticulate brachiopod *Lingula*.

Figure 21.5. Interior of the articulate brachiopod *Laqueus californianus*. Arrows indicate the course of water currents.

history, a complex nomenclature has been developed by paleontologists.

Observe specimens of fossil brachiopods comparing their general shell morphology with specimens of living species (Fig. 21.6). Attempt to find the following structures: dorsal and ventral valves; beak with foramen; presence of any external ornamentation (spines, plication furrows and crests, fold and sulcus); growth lines; commissure line between the valves.

Occasionally, specimens are found with the shell cavity exposed. Such specimens should be examined for the presence of the following features: hinge dentition, lophophore skeletal supports, pallial markings (mantle), platform (muscle attachment), and scars on the inside of the shell indicating muscle attachments. How do fossil specimens compare in their general structure to extant specimens? Are there any significant differences? ■

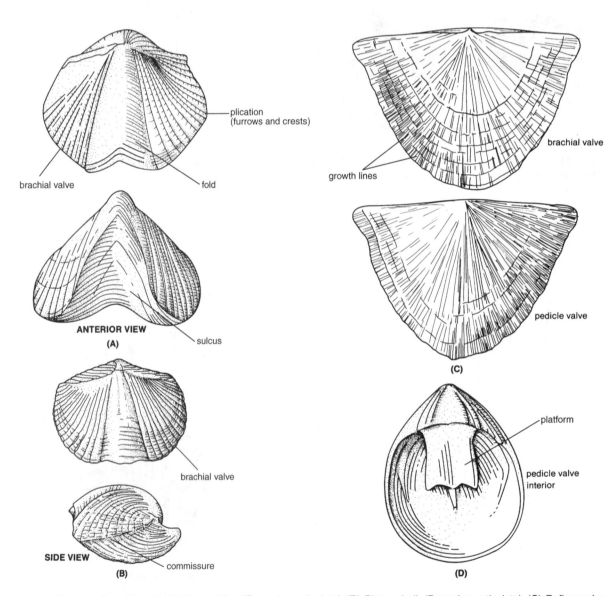

Figure 21.6. Fossil brachiopods. (A) *Paraspirifera* (Devonian, articulate). (B) *Platyrachella* (Devonian, articulate). (C) *Rafinesquina* (Ordovician, articulate). (D) *Trimerella* (Silurian, inarticulate).

Supplemental Readings

Broadhead, T.W. 1981. Lophophorates, Notes for a short course. University of Tennessee, Department of Geological Sciences, Studies in Geology 5, Knoxville, TN.

Chuang, S.H. 1956. The ciliary feeding mechanism of *Lingula unguis* (L.) (Brachiopoda). Proc. Zool. Soc. Lond. 127: 167–189.

Cowen, R. 1971. The food of articulate brachiopods—a discussion. J. Paleontol. 45: 137–139.

Craig, G.Y. 1952. A comparative study of the ecology and paleoecology of *Lingula*. Trans. Edin. Geol. Soc. 15: 110–120.

Emig, C.C 1981. Observations on the ecology of *Lingula reevei*. J. Exp. Mar. Biol. Ecol. 52: 47–62.

Gould, S.J., and C.B. Calloway. 1980. Clams and brachiopods—ships that pass in the night. Paleobiology 6: 383–396.

Hammen, C.S. 1977. Brachiopod metabolism and enzymes. Amer. Zool. 17: 141–147.

LaBarbera, M. 1977. Brachiopod orientation to water movement. I. Theory, laboratory behavior, and field orientations. Paleobiology 3: 270–287.

McCammon, H.M., and W.A. Reynolds. 1976. Experimental evidence for direct nutrient assimilation by the lophophore of articulate brachiopods. Mar. Biol. 34: 41–51.

Paine, R.T. 1963. Ecology of the brachiopod *Glottidia pyramidata*. Ecol. Monogr. 33: 255–280.

Richardson, J.R. 1986. Brachiopods. Sci. Am. 255(3): 100–106.

Rudwick, M.J. 1970. Living and Fossil Brachiopods. Hutchinson University Library, London.

Smith, S.A., C.W. Thayer, and C.E. Brett. 1985. Predation in the Paleozoic: gastropod-like drillholes in Devonian brachiopods. Science 230: 1033–1035.

Suchanek, T.H., and J. Levinton, 1974. Articulate brachiopod food. J. Paleont. 48: 1–5.

Thayer, C., and M. Steele-Petrovic. 1975. Burrowing of the lingulid brachiopod *Glottidia pyramidata*: its ecological and paleoecologic significance. Lethaia 8: 209–221.

Thayer, C.W. 1985. Brachiopods versus mussels: competition, predation, and palatability. Science 228: 1527–1528.

Wright, A.D. 1979. Brachiopod radiation. *In*: The Origin Of Major Invertebrate Groups. M.R. House (ed.), Academic Press, New York, pp. 235–252.

DEUTEROSTOME PHYLA

All deuterostomes are enterocoelomates with radial, indeterminate cleavage. If we ignore for the moment the putative relationship of certain lophophorates (see Nielsen 1977, 1985, 1995), only four phyla make up the deuterostome line: Echinodermata, Hemichordata, Chaetognatha, and Chordata. These four phyla are very heterogeneous in structure, function, and development. Echinodermata is a large phylum exhibiting the greatest diversity among the deuterostomes, including forms familiar to people visiting the seashore (e.g., sea stars, brittle stars, sand dollars, sea urchins, and sea cucumbers). Hemichordata and Chaetognatha are small phyla of marine worms, together containing fewer than 200 species. Placement of the Chaetognaths remains controversial. Most members of the phylum Chordata are vertebrates and possess a vertebral column.

Ghirardelli, E. 1995. Chaetognaths: two unsolved problems: the coelom and their affinities. *In*: Body Cavities: Function and Phylogeny. G. Lanzavecchia, R. Valvassori, and M.D. Candia Carnevali (eds.). Mucchi, Modena, pp. 167–185.

Nielsen, C. 1977. The relationship of Entoprocta, Ectoprocta and Phoronida. Am. Zool. 17: 149–150.

Nielsen, C. 1985. Animal phylogeny in the light of the trochaea theory. Biol. J. Linn. Soc. 25: 243–299.

Nielsen, C. 1995. Animal Evolution. Oxford University Press, New York.

Telford, M.J., and P.W.H. Holland. 1993. The phylogenetic affinities of the chaetognaths: a molecular analysis. Mol. Biol. Evol. 10: 660–676.

■

Phylum Echinodermata

Members of the phylum Echinodermata [ee-KI-no-DER-mah-tah; G., *echinos*, a hedgehog + G., *derma*, skin] are among the most familiar invertebrates observed by people visiting the seashore. Echinoderms get their name from the fact most have **spines** and **tubercles** that project from the surface of the animal. The endoskeleton is composed of calcareous **ossicles** or **plates** that vary in shape and their connection with each other. The skeleton is covered with a thin ciliated epidermis or is embedded in a rather thick, leathery body wall.

There are about 6,500 living species of echinoderms including sea lilies, sea stars (or starfishes), brittle stars, sand dollars, sea urchins, and sea urchins, and sea cucumbers. Approximately 13,000 fossil species are known. Echinoderms are distributed widely, being especially abundant in the Indo-Pacific areas. This strictly marine phylum lacks parasitic and colonial members. None are segmented. The echinoderms are the major invertebrate phylum of **enterocoelous deuterostomes**.

A major characteristic of these largely bottom-dwellers is the basic **pentamerous**, radial symmetry of the adults. This means that an echinoderm's body is arranged typically into five (or multiples thereof) similar and equal parts that radiate from a central axis. The radiating areas, which bear the **tube feet** (or **podia**) of the unique **water-vascular** (or **ambulacral**) **system**, are called **arms** (or **ambulacra, rays**, or **radii**). Between two ambulacra is an **interambulacral** (or **interradial**) area. The oral surface contains the mouth and the aboral surface is away from the mouth.

The water-vascular system is the most distinguishing feature of echinoderms. It is basically an organ system of canals or tubes derived from the spacious, ciliated **enterocoelom**. The water-vascular system serves in locomotion, food gathering, respiration, chemosensation, and excretion.

Unlike most adult echinoderms, the microscopic larvae are bilaterally symmetrical. The first echinoderms were not radially symmetrical either; this results in a primary bilateral symmetry upon which the dominant radial symmetry is superimposed. This is why the echinoderms are placed in the Bilateria and not in the Radiata. The origins of this group including their bilateral beginnings lies deep in the early **Paleozoic Era** (Fig. 22.1).

Classification

The phylum Echinodermata is divided into six classes. The most recently discovered group of echinoderms, class Concentricycloidea (composed of a single species), will not be examined here.

1. **Class Asteroidea.** Unattached, flattened, star-shaped echinoderms with a mouth that is centrally located on the underside of the body and with hollow arms that are not demarcated sharply from the central disc. Asteroids are commonly called sea stars or starfishes. An aboral madreporite and numerous pedicellariae are present in many species. Open ambulacral grooves,

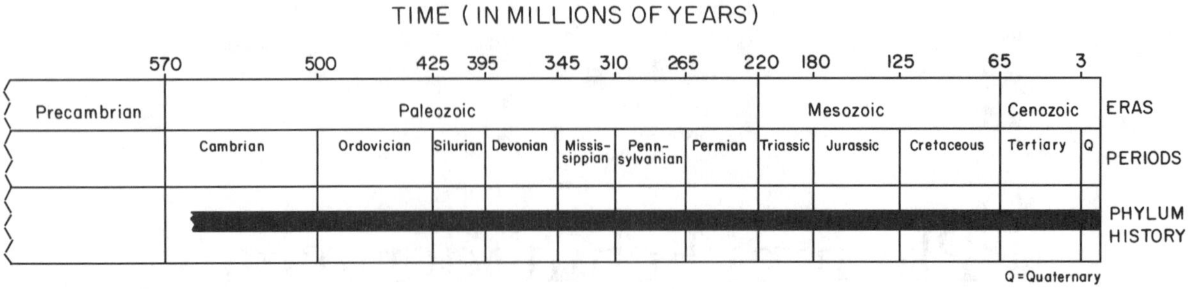

TIME (IN MILLIONS OF YEARS)

570		500		425	395		345	310		265	220	180	125	65	3

Precambrian	Paleozoic							Mesozoic			Cenozoic	ERAS
	Cambrian	Ordovician	Silurian	Devonian	Missis-sippian	Penn-sylvanian	Permian	Triassic	Jurassic	Cretaceous	Tertiary	Q	PERIODS
													PHYLUM HISTORY

Q = Quaternary

Figure 22.1. Geologic history of the phylum Echinodermata.

located on the oral side of the arms, contain tube feet with or without suckers. Examples: *Asterias, Pisaster, Astropecten, Luidia,* and *Solaster.*

2. **Class Ophiuroidea.** Unattached, flattened, star-shaped echinoderms with a mouth that is centrally located on the underside of the body and with elongated and movable, solid arms that are demarcated sharply from the central disc. Ophiuroids are commonly called brittle stars or serpent stars and basket stars. The madreporite is located on the oral side. Pedicellariae and hepatic ceca are lacking. Ambulacral grooves are closed. Examples: *Ophioderma, Ophiothrix* (brittle stars) and *Gorgonocephalus* (basket star).

3. **Class Echinoidea.** Discoid, globoid, or cylindrical echinoderms lacking arms. Echinoids are commonly called sea urchins, heart urchins, and sand dollars. The madreporite is located on the aboral surface. Ambulacral grooves are closed. Most have a mastica-tory apparatus called Aristotle's lantern. Skeleton (test) is rigid with movable long and short spines. Examples: *Strongylocentrotus, Arbacia* (sea urchin),

Echinocardium, Moira (heart urchins), and *Mellita* and *Clypeaster* (sand dollars).

4. **Class Holothuroidea.** Echinoderms with a cylindri-cal and elongated body stretched on an oral-aboral axis. Holothuroids are commonly called sea cucum-bers. Arms are absent and the madreporite is located internally. Tube feet at the anterior end are modified to form oral tentacles. Endoskeleton consisting of microscopic, separate ossicles are embedded in a usu-ally leathery body wall. No spines or pedicellariae present. Ambulacra represented by closed canals. Examples: *Cucumaria, Thyone,* and *Leptosynapta.*

5. **Class Crinoidea.** Echinoderms with a cup-shaped body (theca) to which are attached 5, 10, or more long, hollow arms used to gather food. Crinoids are com-monly called sea lilies and feather stars (comatulids). The oral surface, bearing the mouth and anus, faces upwards, while the crown or calyx contains all the vis-ceral organs. Tube feet lack suckers. No madreporite, spines, or pedicellariae are present. Examples: *Florometra* and *Cenocrinus* (sea lilies), *Neometra,* and *Antedon* (feather stars).

A. Class Asteroidea

The Asteroidea [AS-ter-OI-de-ah; G., *aster,* star + G., *eidos,* form] is perhaps the best known group of echino-derms, containing the sea stars or starfishes. The 1,800 species of sea stars are distributed widely throughout coastal waters of the world. Some sea stars are voracious carnivores feeding on oysters, barnacles, clams, and other marine animals; other are detritus feeders, mud swallowers, or opportunists. The digestive system is well developed. The ciliated enterocoelom is filled with a fluid containing free-moving, phagocytic amoebocytes called **coelomocytes**. These cells, as well as the tube feet of the water-vascular system and **dermal branchiae (papulae)**, aid in the removal of the nitrogenous wastes (ammonium). Gaseous exchange also occurs through the tube feet and dermal branchiae. The skeleton is made up

of calcareous ossicles bound together by connective tis-sue and muscle fibers. Outer circular and inner longitudi-nal fibers lie beneath the dermis. The tube feet contain well-developed longitudinal muscles.

The common North Atlantic sea star, *Asterias forbesi,* will be studied as our representative asteroid. Other gen-era such as *Pisaster* may be available for observation, and the following account will apply to either.

■ Observational Procedure:

External Anatomy
Obtain a preserved and/or dried specimen of the sea star. For the dried specimen, one that has been opened and cleaned is preferred.

1. The radial symmetry is very apparent with all of the five rays radiating from the **central disc** (Fig. 22.2). Although adult sea stars and other echinoderms are radial, they retain vestiges of bilateral symmetry. To appreciate this, a brief consideration will be made of Carpenter's method of designating the radii and interradii. The system can be applied to other adult echinoderms; however, we will use the sea star for instructive purposes.

2. Place the dried (or preserved) specimen on a piece of construction paper with the calcareous, button-shaped **madreporite** on the aboral surface facing upwards and closest to you (Fig. 22.2). Trace the outline of the sea star on the paper and indicate the approximate position of the madreporite. Draw an imaginary line from the madreporite through the opposite **ray**. Continue this line by drawing it on the paper. This ray is designated A and the remaining rays (B, C, D, and E) are lettered counterclockwise from ray A when viewing the sea star from the aboral surface. Add these other lines to the paper.

3. The interradii are designated by the letters of the rays that enclose them (i.e., AB, BC, CD, DE, and EA). Add these labels to the figure. If the animal were cut beginning at the madreporite and extending through the center of ray A, the two halves would be mirror images of one another. Therefore, the animal would conform to a bilateral symmetry. If the animal were cut at right angles to the plane of symmetry, the body consists of three rays (the **trivium**) and the two, opposite rays, C and D, comprise the **bivium**. The madreporite lies on the center of

the bivium and in the CD interradius. Add these labels to the figure. Save this figure for examination when you study the sea cucumber (class Holothuroidea).

Aboral Surface

Take a preserved sea star and place it in a dissection pan with the oral surface facing down.

1. Observe the aboral surface (Fig. 22.3). Note that the aboral surface is spinose with short, calcareous spines covering the area throughout. There is a thin, ciliated epidermis covering the sea star's body including the spines. Observe the structural details of the outer surface with a dissection microscope. Around the bases of the spines are finger-like, fleshy, grayish sacs. These are the **dermal branchiae** (or **dermal papulae**). They are ciliated, hollow projections continuous with the body cavity, serving as respiratory and excretory organs (Fig. 22.4A).

2. Around the spines and throughout the sea star's surface are the minute, pincer-like **pedicellariae** that keep the body surface free of debris and small organisms. Pedicellariae are modified spines, covered with epidermis, and unique to the echinoderms. The pedicellariae should appear as tiny, whitish specks in your specimen, especially on a dried one. Pedicellariae may be stalked or unstalked. Gently scrape the aboral surface of the preserved sea star with your scalpel. Spread thinly a small amount of the scrapings on the slide. Add a drop of water and cover with a cover slip. Observe under low power and then under high power. Note that the pedicellariae are pincer-like and each is opened and closed by adductor and abductor muscles (Fig. 22.4B). Examine a prepared slide showing the pedicellariae. Spines and pedicellariae are also found on the oral surface. Examining the pedicellariae for birefringence will highlight the calcium carbonate ossicles.

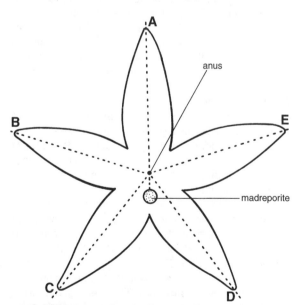

Figure 22.2. Aboral view of a sea star showing Carpenter's system of designating the radii and interradii. See text for discussion of the system. (After Meglitsch 1972.)

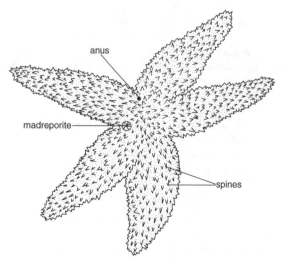

Figure 22.3. Aboral view of sea star.

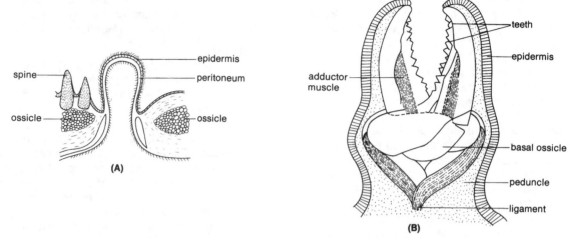

Figure 22.4. Surface structures on sea stars. (A) Section through the body wall of a sea star showing a papula or dermal branchia. (After Cuenot, *Traite de Zoologie*, Vol. XI, Grasse, Ed.) (B) Pedicellaria of a sea star.

3. Two openings occur on the aboral surface (Fig. 22.3). The madreporite, mentioned previously, is the hard, button-shaped plate. If viewed under the dissection microscope, the madreporite's grooves can be seen. The madreporite is part of the water-vascular system to be studied later. The second opening, the anus, is located in the BC interradius and opposite to the madreporite. It is very small and difficult to locate, some sea stars lack the anus.

Oral Surface

1. Turn the sea star over so that the oral surface is facing up. Locate the mouth (Fig. 22.5); this opening is surrounded by a thin, **peristomial membrane**. Note the paired, **oral spines** that extend from each interradius and around the mouth; they protect the mouth and help move food into the opening.
2. Running from the mouth to the tip of each ray is an open, **ambulacral groove** with protective **ambulacral spines** covering each groove. In life these, and the oral spines, are movable. In each ambulacral groove are the fleshy **tube feet** (or **podia**), elements of the water-vascular system. These highly contractile tubes terminate in sucker-discs. Not all sea stars have suckers on their tube feet.
3. At the distal end and on the oral surface of each arm is an eyespot formed by numerous, pigment-cup ocelli. These are sensitive to light. Near the eyespot is a small modified tube foot (**tentacle**), which lacks a sucker.

Internal Anatomy

To study the internal anatomy of the sea star, you will remove the specimen's aboral body wall.

1. Place the preserved sea star with the oral surface down. With scissors, cut off the tips of each arm. Insert the scissors inside the tip of one arm and cut along the edges (i.e., dorsolateral margins) of each arm. Do not cut too deeply. Make sure the body wall at each interradius is cut fully. Cut around the madreporite to leave it attached to the lower part of the body.
2. Carefully lift the aboral surface off the oral side. As you remove the aboral surface, detach all fleshy material adhering by mesenteries to the inner surface of the aboral covering. If the dissection is done properly, you should have all fleshy materials intact and in the lower half of the sea star's body. None of the organs should be adhearing to the aboral part of the body wall. The exposed body cavity is the spacious enterocoelom.

Digestive System

Refer to Figures 22.6 and 22.7 as you read these instructions on examining the digestive system.
1. Locate the short **esophagus** which leads internally from the mouth. The esophagus opens into the large, muscular **cardiac** portion of the stomach. When the sea star feeds, it everts the stomach through the mouth and presses the fleshy tissue against the food. After digestive enzymes have acted upon the food, the stomach and partially digested contents are retracted into the body by **retractor muscles** attached to the ambulacral ridge (Fig. 22.7).
2. Aborally the cardiac stomach leads into the smaller, thin-walled **pyloric stomach**. This pentagonal stomach receives enzymes from five pairs of **pyloric ceca** (or **digestive glands** or **hepatic ceca**). In each arm locate the paired digestive glands. The pyloric stomach

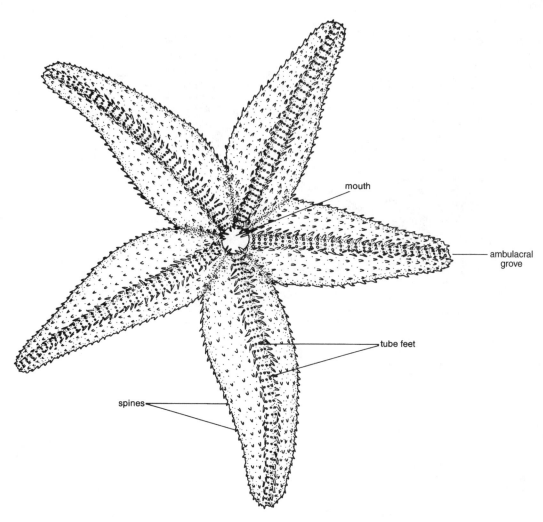

mouth

ambulacral
grove

tube feet

spines

Figure 22.5. Oral view of a sea star.

leads into a very short, inconspicuous **intestine**. A branched sac, the **rectal ceca** (or **intestinal ceca**), of unknown function, lies on top of the pyloric stomach, but has a connection with the intestine. Beyond the opening of the rectal ceca, the short **rectum** leads to the **anus**.

Skeletal System
1. To study the skeleton of the sea star's arm, remove the hepatic ceca in one ray. Note the distinct **ambulacral ridge** composed of middle and lateral rows of ambulacral ossicles (Figs. 22.8 and 22.9). Examine the ridge on your dried specimen. The large, ovoid ossicle at the end of the ridge in the central disc area is the **odontophore**. *Speculate*: Are sea stars segmented? As you study the other classes of echinoderms, consider this question again.

2. Locate the two whitish, threadlike **retractor muscles** that attach the ambulacral ridge and the cardiac stomach. Note that the lateral ambulacral ossicles form the sides of the ambulacral groove on the oral surface, and the fleshy tube feet of the water-vascular system extend through these ossicles. The lower ends of the lateral ossicles rest on a series of **adambulacral ossicles** (Fig. 22.9). The latter bear the long, slender, movable **ambulacral spines** observed along the groove on the oral surface. The **inferomarginal ossicles** compose the bottom, lateral row of strutlike ossicles. Lateral to these, and forming the upper lateral row along the length of the arm, are the **superomarginal ossicles**.

3. Make a cross section of a piece of the arm and observe it under the dissection microscope to see more clearly the above-mentioned ossicles and their

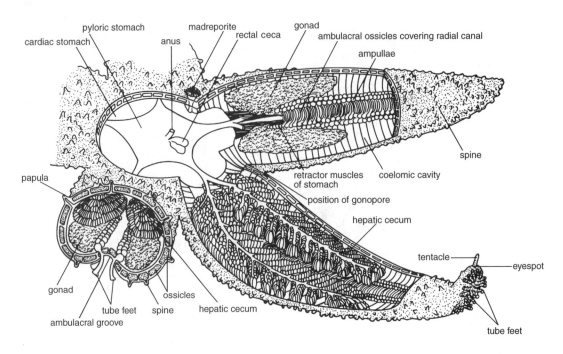

Figure 22.6. Internal aboral view of a sea star showing the digestive system.

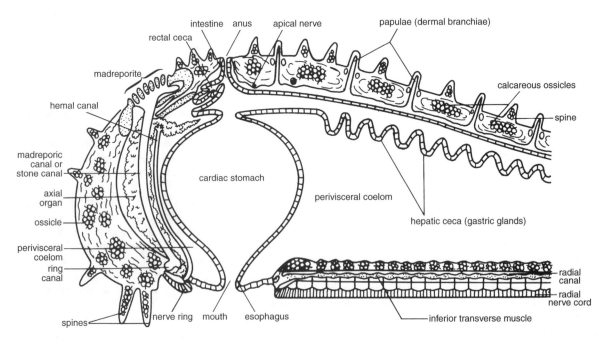

Figure 22.7. Diagrammatic vertical section through a ray and the central disc of *Asterias*. (After Chadwick.)

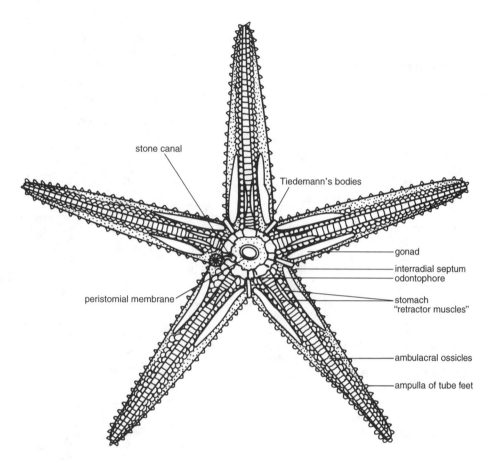

Figure 22.8. Internal aboral view of a sea star after the digestive system has been removed.

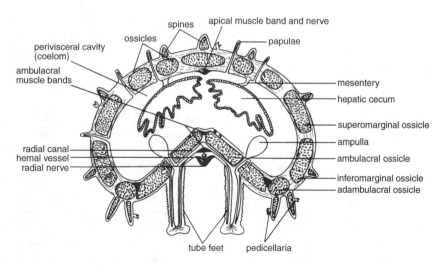

Figure 22.9. Cross section of a sea star's arm.

relationship to each other. If you have a cross section of an arm from a dried specimen, or prepared slide, examine these using a microscope. Study the inner face of the aboral covering and note its ossicles.

Reproductive System
Asterias is dioecious.

1. Locate the two gonads at the base of each arm (Figs. 22.6 and 22.8). These resemble a cluster of grapes. During the reproductive season the gonads nearly fill each arm. At other times, they are small and occupy a small space. Speculate on whether your specimen was taken during its reproductive season or at another time.

2. Two gonads are connected by a single gonoduct that terminates as a single gonopore or gonopore cluster located between the base of the arm in each interradius.

3. Examine a prepared slide of sea star development and locate the large, unfertilized egg showing the nucleus and nucleolus (Fig. 22.10A). Fertilization occurs after the gametes are shed in the water. The zygote undergoes cleavage and soon it develops into a free-swimming, bilaterally symmetrical larva, the **bipinnaria** (Fig. 22.10B). Initially the surface is covered with cilia which eventually become confined to definite ciliated bands used in locomotion and feeding.

4. The bipinnaria in many asteroids, including *Asterias*, becomes a **brachiolaria larva** with the appearance of three additional arms near the sucker at the anterior end (Fig. 22.10C). The brachiolaria later settles down and metamorphoses into a young sea star. Examine whole mounts preparations of these microscopic larvae, locating the ciliated bands, mouth, and anus.

Water-vascular System
Remove the overlying organs and locate the following structures of the water-vascular system (Fig. 22.11): (1) the **madreporite** and (2) the **circumoral canal** (or **ring** or **water canal**) are connected by (3) the calcified **stone canal**; (4) five **radial canals**, one in each arm on the oral side, lead off the **circumoral canal**; (5) short **transverse** (or **lateral**) **canals** arranged in a staggered fashion connect the radial canals and the (6) **tube feet**. This system operates as a hydraulic system. On the inner wall of the circumoral ring and in an interradial position are nine, fleshy **Tiedemann's bodies** in which amoebocytes are formed. An individual tube foot consists of a closed cylinder with a sucker at the distal end and a bulblike **ampulla** at its proximal end. In many sea stars and other echinoderms, but not in *Asterias forbesi*, one to five muscular, saclike **Polian vesicles** connect to the ring canal. Their precise function is unknown.

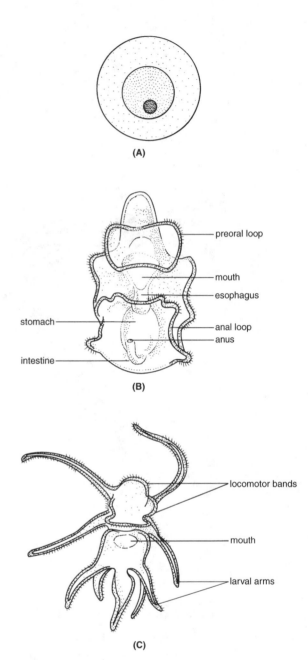

Figure 22.10. Sea star development. (A) Unfertilized egg. (B) Bipinnaria larva. (C) Brachiolaria larva.

Nervous System
The nervous system consists of a plexus of fibers and neurons underlying the epidermis and in the gastrodermis. Remove the tube feet from one ambulacral groove. Observe the minute, yellowish or brown cord that runs along the bottom of the groove along the entire length of the arm. This is the **radial nerve cord** (Fig. 22.9). This constitutes the **ectoneural** portion of

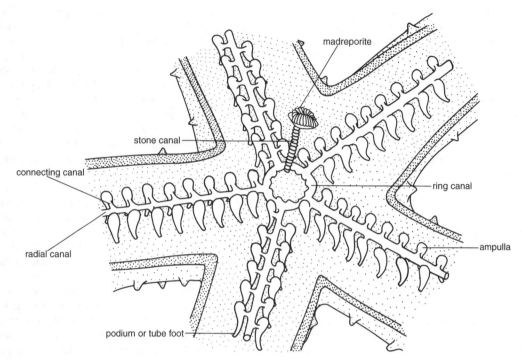

Figure 22.11. Water-vascular system of a sea star.

the nervous system. The **endoneural** portion may be seen in the oral walls of the cardiac and pyloric portions of the stomach and in the median folds of the hepatic ceca. An apical portion of the nervous system occurs medially on the inner surface of the aboral body wall. This single cord is fused with the other four cords in the center of the disc. The apical portion is difficult to find and should not be confused with the large, very apparent, apical longitudinal muscle bands of the arms (Fig. 22.9).

Microscopic Examination

Obtain a prepared slide of a cross section of the arm of a sea star. Using Fig. 22.9 locate the following structures: epidermis, dermis, peritoneum, the various ossicles, enterocoelom, dermal branchiae, pyloric ceca, tube feet, transverse canal, radial canal, radial nerve, and ambulacral groove. If your slide was made from a young sea star, you will not see the gonads or spines. Correlate the structures you see in the slide with those of the dissected specimen.

Live Specimens

When studying live sea stars, exercise care so that the animal is given sufficient time to release its hold before it is lifted from the substratum. If removal is done too quickly, the tube feet may be ripped apart. Also while working with the animals, return them to the seawater every few minutes to prevent drying out.

1. Obtain a sea star and place it into a small dish filled with seawater. Observe the aboral surface of a sea star using low power of a dissection microscope. Are the dermal branchiae visible? Identify the pedicellaria? How large are these structures? Are dermal branchiae and pedicellaria distributed evenly over both sides of the animal?

2. The entire surface is ciliated. To visualize the cilia place an amount of a dilute suspension of carmine powder, carbon black, or fine sand grains on the aboral surface and watch what happens.

3. Carefully turn the animal over and observe its tube feet. Inspect the sucker discs at the distal ends of the feet. How do these suckers function?

4. Remove the sea star from the dish and place it, oral surface down, onto a wet Plexiglas plate. Keep the plate level and observe movement from above and below. Do all the tube feet appear to work in unison? Does there appear to be a pattern to the action of a single tube foot? Does one arm seem to lead and the others follow? What happens when the animal comes upon an obstacle? As you watch the sea star move attempt to determine whether the leading arm ever changes. Describe the locomotory activity of a sea star.

5. Adjust the plate so that it is at an angle of 30° or more. How does the sea star react? Experiment by changing the angle of the tilt. Does the sea star change its locomotory behavior? Slowly turn the plate upside down. What does the sea star do now? Does the sea star move when the plate is held upside down, but at an angle? Summarize your observations and answer the following questions. When moving over a surface do sea stars use their tube feet to pull themselves, push themselves, or do they use a combination of both? What method do you use when you walk?

6. To test your conclusions about sea star locomotion, try the following exercise. Using string, secure a sea star onto a wire mesh screen with its oral surface up. (Do not tie the arms too tightly as this may damage the animal.) Now place a cover glass on to the ambulacra at the distal end of one arm. What happens to the cover glass? Describe the stepping action of the tube feet? Which of the tube feet in the ambulacra are active: only the ones in contact with the glass or all of them? While the glass is being moved, observe the tube feet of the other arms. Are they moving? In what direction are they moving? Is there an overall pattern of action of tube feet in all five ambulacra? Change the position of the cover glass to a different ambulacra. Does this change the pattern of movement of the tube feet? When do arms act as leading arms and when do they act as trailing arms?

What do these observations indicate about motor control by the central nervous system (CNS) of sea stars? *Speculate*: Based on these observations would you say that the CNS of sea stars is sophisticated or unsophisticated? (More advanced studies on the control of locomotion are offered in steps 10 and 11.)

7. Remove the animal from the wire mesh restraint, place it oral side down on a flat surface, and observe righting behavior. How long does it take and how is this accomplished? How do the arms work in coordinating this activity? How do the tube feet aid in righting? How many of the arms become anchored to the surface before the animal can right itself?

8. Carefully lift a sea star up and place a small piece of shrimp (1–2 cm long) directly under the mouth and return the animal to the holding tank. Over the course of the next 30–60 minutes examine the animal to see if you can observe feeding. Notice how the stomach **everts** during this process. (NB: disturbing the animal while it is feeding may cause feeding to cease.)

9. If a burrowing sea star is available (e.g., *Astropecten*), observe burrowing behavior. (To observe burrowing, you will need to provide a tub or aquarium filled with seawater and with a substratum of sand at least 5 cm deep.) Describe how burrowing is accomplished by *Astropecten*.

10. The next series of observations is **optional** as it requires surgery to be done on the specimen while it is alive. (Although the damage will look severe, sea stars have strong powers of regeneration. If the animal is well cared for after the operation, it should be able to repair itself.)

This procedure may be done with the animal free or restrained as discussed in step 6. At the base of the interradius DE, cut the arm into the ambulacra so that the radial nerve is severed (Fig. 22.7). Determine how this effects locomotion. Make another cut, this time into the interradius BC. How do the arms act now in locomotion? Is there a change in the pattern in which arm leads and which trails? Cut the remaining interradii one by one, each time determining how this effect locomotion.

11. Obtain another animal and sever one arm entirely so that the arm still contains a portion of the nerve ring. How does this arm move? Does it act as a leading or a trailing arm? Sever another arm, but this time do so to exclude the nerve ring. How does this arm move? Does it act as a leading or a trailing arm? Consider how the presence (or absence) of a connection to the nerve ring coordinates locomotion.

B. Class Ophiuroidea

Members of class Ophiuroidea [O-fee-ur-OID-de-ah; G., *ophis*, a serpent + G., *oura*, tail + G., *eidos*, form] are called brittle stars because of their tendency to fragment when disturbed, or serpent stars because their long slender arms wriggle in a snakelike motion. Included in this class are the largest echinoderms, the basket stars; these forms have highly branched arms whose ends tend to coil. *Gorgonocephalus* is a basket star that is often available for classroom observation (Fig. 22.12).

Because of the basic star-shaped body and pentamerous condition, ophiuroids resemble the asteroids. As you study these organisms you will see distinct differences between the two groups. Many zoologists consider ophiuroids to be the most successful of echinoderms because they are the largest group, with 2,000 living species. Ophiuroids often occur in dense aggregations on the sea floor. Certainly the brittle stars are the most mobile members of the phylum. There are burrowing ophiuroids and a few can swim. The only commensal echinoderms are ophiuroids. For our laboratory study, *Ophioderma* or some similar brittle star, will be used.

Figure 22.12. *Gorgonocephalus*, a basket star.

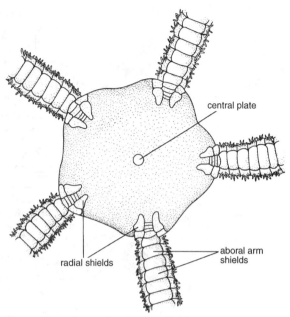

Figure 22.13. External aboral surface of a brittle star.

■ Observational Procedure:

External Anatomy

Obtain dried and preserved individuals for your study.

1. Note that the central disc is composed of calcareous plates or shields arranged in a distinct fashion (Fig. 22.13). Are these plates arranged in some pattern or are they random? The small disc is flattened and rather smooth, lacking the conspicuous and numerous spines observed in the sea star. The disc of some species is spiny, and there is variation in the pattern of the shields. Using a dissection microscope, examine the dorsal side for the madreporite. Is it present there? Is an anus evident?

2. Observe the oral surface containing the mouth (Fig. 22.14). Use the dissection microscope to observe the plates and other external features better. The plates on the oral surface of the central disc consist of the large, interradial **oral shields** (or **buccal plates**), two **aboral** (or **angular**) **plates** lying on either side of the oral shield, and the inward-pointing jaws that extend toward the mouth (Fig. 22.14). Note that the jaws bear teeth at their apices and **oral papillae** along their sides.

3. One of the oral shields is modified as the madreporite of the water-vascular system. The madreporite's oral position differs from that observed in the sea stars where it is located aborally. Examine the arms of your specimen and note the lack of ambulacral grooves. Are dermal branchiae or pedicellariae present?

4. The arms are very flexible because they are composed internally of a series of disc-shaped, calcareous

vertebrae connected by large, intervertebral muscles (Fig. 22.15). Break off a section of an arm from your specimen and using a dissection microscope examine both ends of the arm. Locate both sides of a vertebral osscile. On one side of an ossicle you should observe the **sockets** that articulate with the **processes** of an adjacent socket (Fig. 22.15). What type of joint is this? Where else in the animal kingdom are these joints found? How do such joints permit the snakelike move-

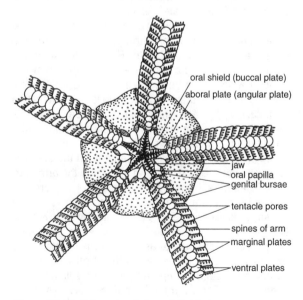

Figure 22.14. External oral view of *Ophioderma*.

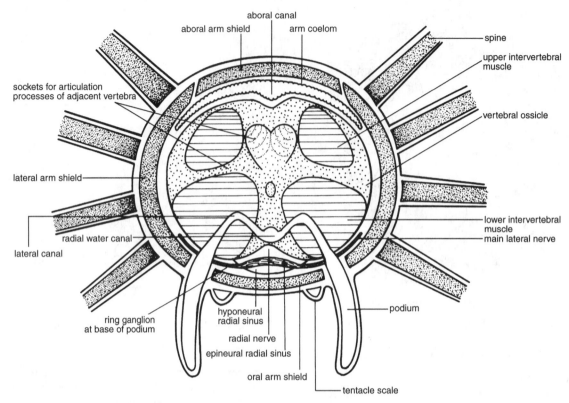

Figure 22.15. Cross section of a generalized ophiuroid arm. (After Barnes.)

ment of the arms? If these organisms possess vertebrae, why aren't they considered to be members of Phylum Chordata, Subphylum Vertebrata (Exercise 25)? The vertebrae represent the ambulacral ossicles that have become located internally.

5. Four longitudinal rows of plates surround the vertebrae externally: one aboral arm plate, two lateral arm plates, and one oral arm plate (Fig. 22.15). Along the outer margins of the lateral arm plates are holes through which the tube feet extend. Note the vertical row of spines protruding between the lateral arm plates (Fig. 22.14). How does their position compare to that of the podia? Unlike the condition in most asteroids, ophiuroid tube feet lack suckers.

6. Locate the small, external openings on the oral surface within the central disc and alongside the arm margins. These are **bursal slits** (Fig. 22.14). There are five pairs of them and each slit leads inward to a **bursa**, which is an internal sac formed by an invagination of the disc's oral surface. Gaseous exchange occurs across the bursae as seawater is pulled in and expelled. In *Ophioderma* and other brittle stars, the gametes are stored temporarily in these slits. In some species (not *Ophioderma*) the bursae serve as brood chambers for the developing brittle stars as well as in

respiration. In addition to gas exchange and reproduction, the bursae may be the principal sites for nitrogenous waste removal.

Internal Anatomy

1. Using a preserved specimen, cut away the surface of the aboral disc. The ophiuroid's enterocoelom is much reduced from that of the sea star. Locate the large, saclike stomach with 10 **stomach pouches** (Fig. 22.16). Alternating in position with these pouches are the gonads and bursae. The stomach is connected to the mouth by a short esophagus. An intestine, hepatic ceca, and anus are lacking. Unlike asteroids, the digestive system in ophiuroids is confined to the central disc. Ophiuroids are scavengers, filter feeders, or suspension feeders. The tube feet are used to capture food.

2. The water-vascular and nervous systems are similar to those of asteroids, with some minor differences. These systems will not be studied in your dissections.

3. Although most ophiuroids are dioecious, hermaphroditic species are not uncommon. Ophiuroid reproduction is similar to that of asteroids. The bilaterally symmetrical larva, when present, is an **ophiopluteus**, characterized by eight arms (Fig. 22.17). Compare this

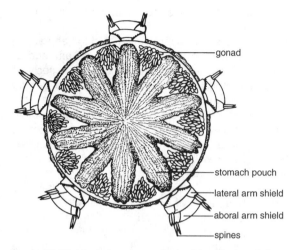

Figure 22.16. Internal anatomy of an ophiuroid, aboral view.

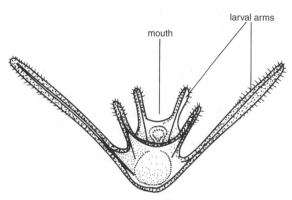

Figure 22.17. Ophiopluteus of a brittle star.

larval form to the brachiolaria larva of the sea star (Fig. 22.10C). How are they similar? How are they different? The ophiopluteus develops gradually into an adult; there is no sessile stage. Examine slide preparations of the larva. Look for the mouth and anus. Are there ciliated bands similar to those found in asteroids?

Live Specimens

Remember to return the animals to the seawater every few minutes to prevent drying out.

1. Place the brittle star in an aquarium and observe its locomotion. Does one arm always lead the others? Determine whether the animal pulls or pushes itself along. How do the arms react when they make contact with objects? Do brittle stars move differently over smooth and rough surfaces? Why are these animals also called serpent stars? How does locomotion in the sea stars compare to that of brittle stars?

2. Remove the brittle star from the aquarium and place the animal on a Plexiglas plate. How does it move now? What happens when the plate is tilted to an angle of 30° or more? Provide the animal with a surface that has some relief (e.g., holes and bumps). Is the animal better able to keep its purchase on this surface then the smooth one? Examine the ambulacral region of the animal. How do the tube feet of the brittle star differ from that of the sea star? How are brittle stars like free-hand rock climbers?

3. Turn the animal over and observe righting behavior. How is it accomplished? Is righting done differently in the sea stars and the brittle stars?

4. If a brittle star that has been starved is available, attempt the following exercise on feeding. Place the animal in a small aquarium of clean (clear) seawater. Add a small amount of very finely-ground clam (or another mollusk) to the water. What is the brittle star's response? After several minutes add a few drops of a dilute solution of toluidine blue to the seawater near to the arms of the animal. What do you see? Illumination from the side may aid you in making these observations. Correlate your observations to what you know about brittle star feeding. ■

C. Class Echinoidea

Sea urchins, sand dollars, heart urchins (spatangoids) belong to the class Echinoidea [ECH-in-OI-de-ah; G., *echinos*, hedgehog + G., *eidos*, form]. There are approximately 1,000 living species of these free-moving echinoderms found in the littoral and benthic zones of the oceans. The body is globular to disclike in shape with movable spines covering the exterior. The spines function in protection, burrowing, keeping the body surface clean, and helping to right the body should it become upended. The gonads of some regular sea urchins are used for food by some people.

Echinoids have a rigid endoskeleton called the **test**. The test and spines are covered with a ciliated epidermis. The ossicles of the test are flattened and sutured together firmly so that in most echinoids the individual ossicles are immobile, unlike the condition in most echinoderms. A body musculature is lacking on the inner surface of the test. The interlocking ossicles radiate in rows from the aboral apex to the mouth that is located in the center of the oral surface. The ambulacra and interambulacra are formed by ossicles arranged in 10 alternating double columns that converge at the two poles. The ambulacral plates have paired holes or pores through which the tube feet extend; holes are lacking on the interambulacral ossicles. Echinoids lack projecting, free-moving arms similar to those of asteroids and ophiuroids.

Although the dominant symmetry is radial, as observed in sea urchins, some echinoids have a secondary, bilateral symmetry. Because of this difference in symmetry, echinoids are divided into two groups: (1) the **regular** echinoids (sea urchins) with radial symmetry and (2) the **irregular** echinoids (heart urchins and sand dollars) with bilateral symmetry superimposed on the radial one.

■ Observational Procedure: Sea Urchin

External Anatomy

Obtain a cleaned specimen (one with the spines removed) and a preserved, complete specimen of a sea urchin, such as *Arbacia* or *Strongylocentrotus*. The following instructions will use both.

1. Note the globular or ovoid shape with the oral surface being flatter than the aboral one. Locate on the oral surface the five ambulacra composed of rows of double ossicles and the five double rows that constitute the interambulacra (Fig. 22.18). The ambulacra

have pores for the tube feet, whereas the interambulacra lack the holes.

2. The anus is at or near the center of the aboral surface; a membrane (the **periproct**) with embedded, minute ossicles bears the anus. In a cleaned specimen the periproct will probably be missing. Five **genital plates** surround the anus; one is modified as the madreporite. How does the madreporite plate differ from the genital plates? On each genital plate locate the **gonopore**. This is where the gametes exit. Does the madreporite possess a gonopore? How many gonopores are present?

3. Alternating with the genital plates locate the five **ocular plates** each of which coincides with an ambulacral region. On each ocular plate find the tiny pore through which passes a light-sensitive tube foot in life.

4. Observe the pattern of the numerous bumps (**tubercles**) to which spines are attached. Do they occur in both the ambulacral and interambulacral regions? Are all the tubercles the same size? Is there a pattern to the distribution of the 1° and 2° tubercles? Using a hand lens observe the smallest of the tubercles. These are the bases to which the pedicellaria attach. How numerous are these structures?

5. The base of each tubercle forms a joint with the concave end of a spine. This attachment permits and movement of the spines. Take a disarticulated spine

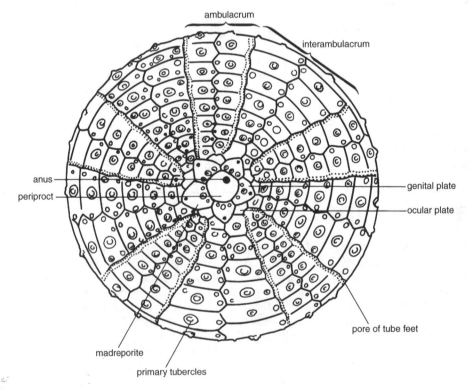

Figure 22.18. Test of *Arbacia*, aboral view.

and fit it onto a tubercle and determine for yourself how this joint works. What type of joint is this? Where else in the phylum and in the animal kingdom are ball-and-socket joints found? Do they all work in the same way?

6. Study the structure of the spines and pedicellariae under the dissection microscope. The latter in most echinoids are stalked and three-jawed. A number of sea urchins have poisonous pedicellariae, but not *Arbacia.* Are the spines a uniform length all over the body? Besides the ball-and-socket joint are spines jointed anywhere else? Break a dried spine in half and examine the broken side using a dissection microscope. Are spines solid or hollow?

7. Using a dissection microscope or hand lens examine the surface of the cleaned test closely. What does the zig-zag pattern of interlocking plates represent? These are the ossicles, but they are all fused. Are all of the ossicles the same size? Describe how the ossicles fit together to produce the test. This type of architectural structure has been around since the 1950s. Who invented it and what name was it given? What other animals have you studied that possess this domelike structure? In carbon chemistry there is an unusual arrangement of 60 carbon atoms that gives this same sort of structure? What name is given to that?

8. If you have extra cleaned specimens try the following exercise to determine the weight to strength ratio of the echinoid version of the geodesic dome. Weigh your specimen to the nearest one tenth of a gram. This is the mass (M) of the test. Now arrange a system whereby weights may be added to the specimen in a manner so that they do not fall off. Add weight a few grams at a time to increase the crushing force (F). How much weight did you end up adding before the test failed? How did the test fail; was it crushed completely or did it just crack at one point? From the class data determine the weight-to-strength ratio for these sea urchins tests. To do this plot on an X-Y axis all values of F as a function of M. Is there a strong correlation? Consider how such a light-weight structure can carry so much weight. How is F redistributed by the test?

9. In the preserved specimen locate the mouth on the oral surface (Fig. 22.19). The opening is surrounded by a peristomial membrane that is thickened along the inner edge to form a lip. Find the five pairs of short, **buccal podia** and the five pairs of bushy, **peristomial gills** in the membrane. These gills are evaginations of the body surface. Exchange of respiratory gases occurs in these gills as well as across the tube feet. Do you see pedicellariae and small spines in the peristomial membrane? You should see protruding through the mouth five teeth, the chewing parts of **Aristotle's lantern.** This complex structure, unique to most echinoids, will be studied later.

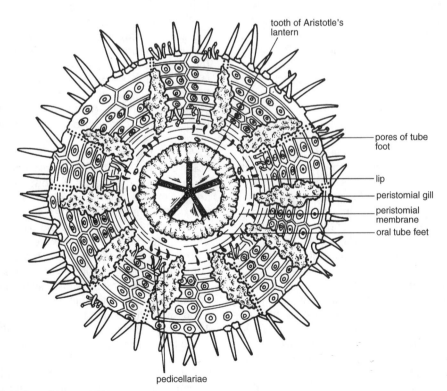

tooth of Aristotle's lantern

pores of tube foot

lip

peristomial gill

peristomial membrane

oral tube feet

pedicellariae

Figure 22.19. Oral view of a sea urchin.

Internal Anatomy

This dissection requires that you do not cut too deeply into the test.

1. Place the urchin oral side down in the dissection pan. Take a pair of scissors or a fine saw and cut through and around the equator of the test. You may remove spines to help in the cutting process, but keep the two halfs together while cutting. Carefully open the two halves, and as you do this, detach or cut the mesenteries holding the gut together.

2. In the oral half locate Aristotle's lantern (Fig. 22.20). The main components of this complex structure are five, radially arranged, calcareous plates called **pyramids**, a large assortment of other plates and muscles, and five oral teeth projecting at the tip of each pyramid. The lantern can be protruded or retracted through the mouth. Sea urchins use their sharp teeth to scrape, tear, and pull food materials. Most sea urchins are herbivores, feeding primarily on algae growing on rocks and similar surfaces.

3. Surrounding the lantern, the **ring canal** with its associated **Pollan vesicles** may be seen. The vesicles are opposite each interambulacrum. The **stone canal** and **axial organ** ascend aborally off the ring canal. The former terminates at the madreporite on the aboral surface. The axial organ is part of the hemal system. This system, found also in other echinoderms, is composed of a complex of channels that are difficult to follow in preserved specimens; the system will not be studied. The exact function of the axial organ is unknown. The **radial canals** leave the ring canal. At that juncture you may have difficulty locating the canals, but they, along with ampullae of the tube feet, may be observed along the inner surface of the test. Note the lack of body muscles attached to the inner test wall. What would account for the lack of this musculature?

4. The digestive system consists of the mouth, a buccal cavity, and a pharynx (the interior of Aristotle's lantern) that leads to a long **esophagus** (Fig. 22.20). Trace the esophagus to the **small intestine** that curves around the inside of the test in a counterclockwise direction as viewed from the aboral surface. The small intestine leads to a **large intestine** that makes an almost complete circuit in the opposite direction. Some workers call the two circuits of the intestine the **stomach**. The large intestine leads to the short **rectum** that opens at the **anus**.

5. Running along the inner surface of the oral part of the intestine locate the ciliated **siphon**. This tube connects the esophagus at one end and the intestine at the other end. The precise function of the siphon is unknown. It may remove water from the intestinal contents, thus concentrating enzymes and food during the initial stages of digestion.

6. Examine the inner surface of the aboral half of the test and locate the five **gonads**. Are these located in the ambulacral or in the interambulacral areas? Each gonad opens to the outside by a genital pore, observed when the external anatomy was studied.

7. All echinoids are dioecious. Fertilization is external in the sea water and a bilateral larva, the **echinopluteus**, forms. The larva metamorphoses into a juvenile urchin without settling down. Study slides of the echinopluteus (Fig. 22.21). Note that the echinopluteus larva of echinoids superficially resembles the ophiopluteus (Fig. 22.16). Compare the larval forms of the echinoderms examined so far. How are they similar? How are they different?

Live Specimens

When studying sea urchins, exercise care so that the animal is given sufficient time to release its hold before it is lifted from the substratum. If removal is done too quickly, the tube feet may be ripped apart. Also while working with the animals, return them to the seawater every few minutes to prevent drying out.

1. Place a live sea urchin such as *Arbacia, Lytechinus*, or *Strongylocentrotus* in a small dish of seawater that will restrict its movements. Observe the animal under low power of a dissection microscope. Are dermal branchiae present? Determine whether the epidermis is ciliated. Turn the animal over and observe the oral side, locating the peristome, buccal tube feet, gills, and **sphaeridia**. The later are short, bean-shaped spines that function as organs of equilibrium.

2. Observe the action of the tube feet on both sides of the animal. Examine the function of the tube feet closely. Do all possess suckers? Do those on the aboral side have different behaviors than those on the oral surface? How do the tube feet of the sea urchin differ from those of sea stars and brittle stars? Find the five light-sensitive tube feet that exit the test from the genital plates. Are they any different in form or behavior from the other tube feet?

3. Identify the pedicellaria and compare them to those of the sea star. Are they the same size? Take a probe and gently prod the animal. How do the spines and pedicellaria respond to the disturbance? What direction do the surrounding spines and pedicellaria point when one spine is handled? Slowly push on a spine. What happens? Does the spine seem to lock into place? This phenomenon is accomplished by **catch** or **cog muscles** that are located at the base of the spine. These muscles may be anesthetized by blowing carbon dioxide over the animal. What happens to the spines when this is done?

4. Attempt to observe the action of the jaws of Aristotle's lantern. How does it function in obtaining food?

5. Place the sea urchin in an aquarium and observe its movements. How do these animals use their spines in locomotion? Turn the urchin upside down and observe

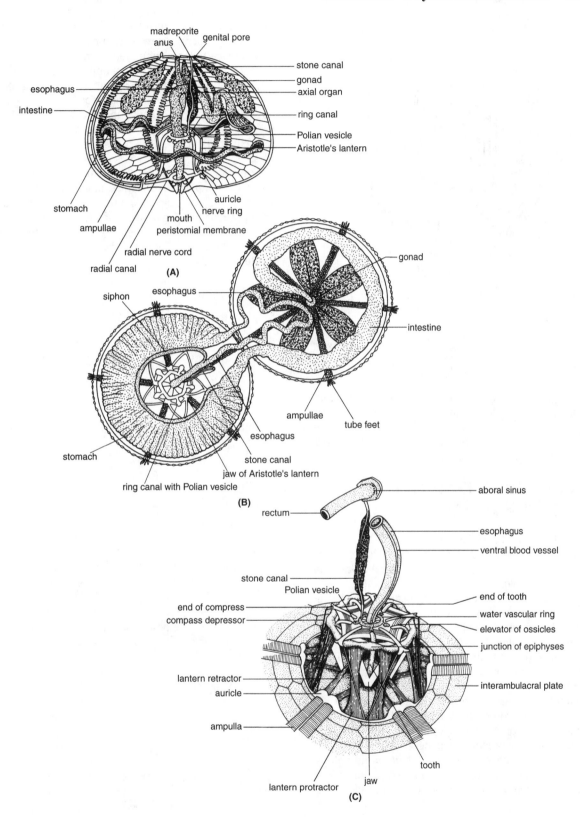

Figure 22.20. Internal anatomy of a sea urchin. (A) Lateral view. (B) Oral-aboral view. (C) Details of Aristotle's lantern.

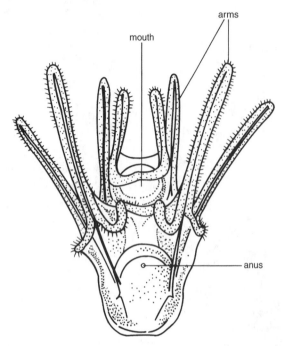

Figure 22.21. Early echinopluteus of an echinoid.

righting behavior. Is it different from what you observed in the sea star or brittle star? At the discretion of your instructor, you will remove tube feet and sphaeridae from the region of the peristome. How do these ablations influence righting behavior?

Observational Procedure: Irregular Urchins

Sand Dollar and Sea Biscuit
Obtain preserved and cleaned sand dollar such as *Mellita* and cleaned tests of the sea biscuit *Clypeaster* and a sea urchin.

1. Compare the tests of the sand dollar and sea biscuit to that of the sea urchin. Are the tests of the sand dollar and sea biscuit similar to that of the sea urchin? How do they differ? Are fused ossicles evident on the external surface? In answering this question you may find that using a hand lens or dissection microscope will bring too much magnification. To see the ossicles, try rotating the test to observe the surface from different angles. Fusion of the ossicles in the test reaches an extreme degree in these forms.

2. Examine the preserved specimen of the sand dollar, noting the covering of small, numerous spines. How do these spines differ in size and distribution from that of the sea urchin? Determine whether they func-

tion as a ball-and-socket joint. Using a dissection microscope find the minute tubercles that cover the outer surface of the cleaned tests. How many of these ball-and-socket joints are present on the tests of these organisms? Are they located on both sides? If tube feet are not used in locomotion by these forms, how do they move? (Compare you answer to the observations that you will make later using live sand dollars.)

3. Locate the mouth on the oral surfaces of these animals. Both sand dollars and sea biscuits possess a modified Aristotle's lantern that cannot be protracted through the mouth. (NB: These structures are often sold as novelties, called "doves," or "angles," in seaside gift shops.) Unlike regular sea urchins, the anus of irregular echinoids has moved from the aboral center to the posterior edge of the body or into the posterior **lunule** (Fig. 22.22).

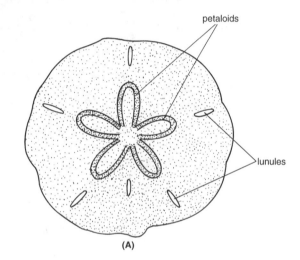

(A)

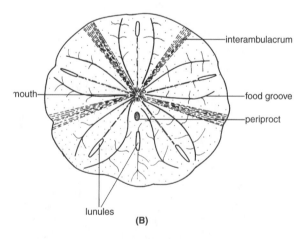

(B)

Figure 22.22. Sand dollar. (A) Aboral view. (B) Oral view of a sand dollar.

4. Lunules are the elongated notches or holes on the sand dollar and their numbers vary among different species. At least in some forms, the lunules aid the sand dollar in burrowing in the sand, although spines are the principal structures for that purpose. Lunules are absent in the sea biscuit (*Clypeaster rosaceus*) of Florida and the West Indies (Fig. 22.23).

5. On the clean specimens, examine the aboral surface under the dissection microscope and observe the five **petaloids** (the ambulacra) that resemble a flower. Find the holes that contained the tube feet. These aboral podia function in gas exchange, but not movement. Peristomial gills like those of sea urchins do not occur in the irregular echinoids.

6. At each interradius and near the base of the petaloids are the **gonopores**, each surrounded by a **genital plate**. Can you determine which of the genital plates functions as the madreporite? To do this you may need to examine several different specimens and/or species to see the pattern. At the base of each petaloid locate the **ocular plate** and the hole that bore a light-sensitive tube foot in life.

7. The internal anatomy of both the sand dollar and sea biscuit is similar to that of sea urchins. Because of the difficulty in dissection, their internal anatomy will not be studied. However, their internal architecture is interesting and should be examined. Obtain a cleaned dried specimen of a sand dollar that has been broken open and examine the inside. How does the inside of the sand dollar compare to that of the sea urchin? Suggest a function for these structures in the living animal? Now examine the internal anatomy of a sea biscuit test that has been cut in half. Does this organism have internal supports as in the sand dollar? Is the test wall thick or thin? What other features are present in the test wall of the sea biscuit? Speculate on the function of the perforations and cavities.

8. If sufficient specimens are available, you may wish to analyze the weight-to-strength ratio of sand dollars (and/or sea biscuits) as was done for the sea urchin. Replot the data from the sea urchin study along with that of the sand dollar. What does this information indicate about the strenght of the tests of these organisms?

9. If a live sand dollar is available for study, place the animal in a small dish of seawater and examine it using a dissection microscope. Is the epidermis ciliated? Are dermal branchiae, pedicellaria, spines, or tube feet visible on either the oral or aboral surface? How large are the spines and tube feet? What are they doing? Sprinkle a small amount of sand on the aboral surface and watch what happens. Observe

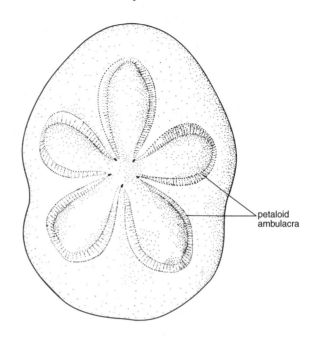

Figure 22.23. Sea biscuit (*Clypeaster*).

burrowing behavior. How do these animals bury themselves? How is the sand dollar like an arthropodian milliped?

Heart Urchins (Spatangoids)
Heart urchins are similar to the sand dollars and sea biscuits just examined. Obtain a cleaned test of an irregular echinoid for study.

1. Note that the body of the heart urchin is egg- or heart-shaped and on the aboral surface there are petaloid ambulacra containing tube feet for gaseous exchange (Fig. 22.24).

2. The mouth and peristome are located toward the anterior end of the flat, aboral surface, and the anus is at the posterior or more pointed end. There are three anterior ambulacra and two posterior ambulacra on the aboral surface. These ambulacra contain pores through which the tube feet protrude.

3. Small, dense spines cover the body surface; the spines are similar to those of sea urchins and sand dollars. Heart urchins lack the complex Aristotle's lantern. Both sand dollars and heart urchins feed on minute organic materials.

4. How do irregular echinoids compare to the others members of their class? Is the construction of the test more similar to that of the sea urchin or of the sand dollar? ■

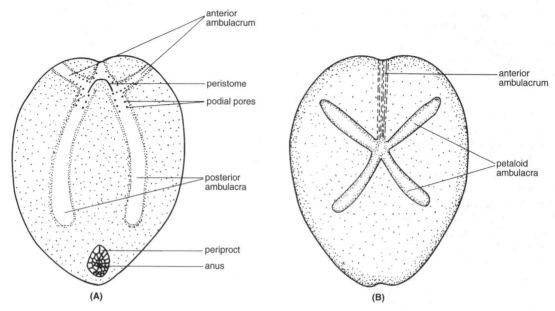

Figure 22.24. Heart urchin. (A) Oral view. (B) Aboral view.

D. Class Holothuroidea

Sea cucumbers are rather bizarre-looking echinoderms. Indeed they are unlike other members of the Echinodermata in a number of features. There are approximately 1,000 living species of the class Holothuroidea [HOL-o-thur-OI-de-ah; G., *holothourion*, a kind of zoophyte + G., *eidos*, form]. They are common in the littoral areas of the sea and also constitute an important fauna of the marine abyss.

The body of a sea cucumber is elongated on an oral-aboral (anteroposterior) axis (Fig. 22.25). The oral end contains the mouth that is surrounded by a number of food-gathering **tentacles**. These mucus-covered structures are modified tube feet and usually can be retracted within the sea cucumber's body. The opposite end bears the anus. The spineless body wall of most sea cucumbers is leathery and contains an embedded endoskeleton of microscopic, calcareous ossicles that are not fused with each other. Tube feet are scattered over the body surface (*Thyone*), grouped in five ambulacral areas (*Cucumaria*), or totally lacking (*Leptosynapta*). Sea cucumbers move about by muscular contraction of their body, aided by the tube feet when present.

Sea cucumbers are typically dioecious and fertilization occurs in the water. During development many species pass through two larval stages: first, an **auricularia** and second, a **doliolaria**. These microscopic larvae resemble those found in other members of the Echinodermata.

Sea cucumbers are noted for their great powers of regeneration. When disturbed or under stress, the animal is unusual in being able to **eviscerate** the internal organs. This is perhaps a means of distracting predators; *Holothuria* and a few others, can eject a sticky mass of **Cuvierian tubules**, that entangle the predator. Regeneration of the lost organs then occurs after the sea cucumber crawls to safety.

■ Observational Procedure:

External Anatomy

Obtain a specimen of *Cucumaria* and/or *Thyone*, or a similar species suitable for dissection.

1. How is the body surface of the sea cucumber obviously different from the other echinoderms examined? Determine the oral-aboral axis of the specimen. Feel the body surface. How does it compare to the other enchionderms examined so far? Is there any evidence of ossicles?

2. In *Cucumaria* the tube feet are present in five distinct ambulacral areas, three on the ventral surface and in two on the dorsal surface. Retrieve your sketch of the symmetry of the sea star and relate it to *Cucumaria*. Ray A is on the midventral line according to Carpenter's designation of the radii, discussed in the exercise on asteroids. The tube feet on the somewhat

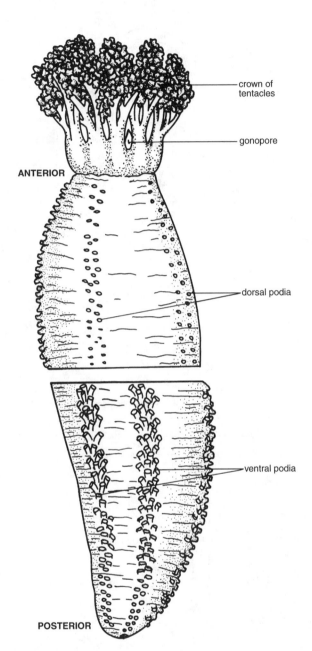

crown of
tentacles

gonopore

ANTERIOR

dorsal podia

ventral podia

POSTERIOR

Figure 22.25. External view of a sea cucumber.

flattened, ventral surface (the **sole**) are better developed than are those on the dorsal surface. Do all the sea cucumbers available for study have the same arrangement of tube feet.

3. Isolate the endoskeletal ossicles for examination by removing a section of the body wall and boiling it in a small quantity of bleach (sodium hypochlorite). (NB: Be sure to do this under safe conditions: e.g., in a fume hood with the window guard down.) During this process the

body wall will be dissolved. Examine the ossicles using a compound microscope (Fig. 22.26). Do these ossicles resemble in anyway those of other echinoderms?

Internal Anatomy

1. Beginning at the posterior end, make a ventral incision through the body wall and extend this incision through the anterior end. Open the body wall laterally and pin it to the dissection pan. Cover the specimen with water. As you study the internal anatomy, refer to Fig. 22.27.

2. The enterocoelom in sea cucumbers is spacious. In life the body cavity is filled with coelomic fluid containing various types of amoebocytes. The peritoneal cilia produce a current that moves the coelomic fluid throughout the enterocoelom.

Digestive System

1. This system consists of the mouth, expanded pharynx, short esophagus, muscular stomach, and long intestine that leads to the muscular rectum (Fig. 22.27). In *Cucumaria*, but not in *Thyone*, the esophagus is absent and the stomach is poorly developed. Note the muscles that attached the rectum to the inner surface of the body wall.

2. Find the pair of highly branched **respiratory trees** that extend off the rectum (Fig. 22.27). The muscular rectum pumps water into the trees and expels it back to the sea. Exchange of respiratory gases occurs within the trees, which are unique to seawater cucumbers. However, the burrowing forms, such as *Leptosynapta*, lack them and exchange of respiratory gases occurs across the body surface.

Water-vascular System

1. Surrounding the pharynx is a hard structure called the **calcareous ring** to which **retractor muscles** are attached posteriorly. These muscles retract the entire oral region inside the body. The calcareous-ring complex may be homologous with the Aristotle's lantern of echinoids.

2. At the base of the pharynx just behind the calcareous ring is the **ring canal**. The dorsal madreporite is connected to the ring canal by the stone canal. The internal position of the madreporite is unlike that of other echinoderms. One to several (often two) saclike, **Polian vesicles** arise from the ring canal. Follow these where they join the canal (Fig. 22.27). Some specimens may have small vesicles, whereas others may have greatly expanded vesicles. The Polian vesicles may maintain pressure within the water-vascular system.

3. Locate the five **radial canals** off the ring canal that run anteriorly to the oral tentacles and then descend to the aboral end of the body. The canals follow the large, **longitudinal muscles** that are attached to the inner surface of the body wall. These lie under the five

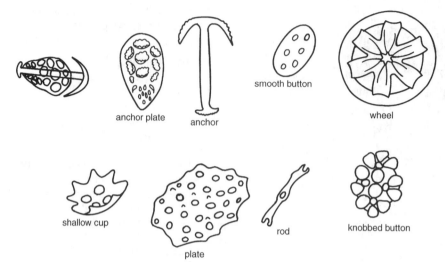

Figure 22.26. Ossicles of several types of sea cucumbers.

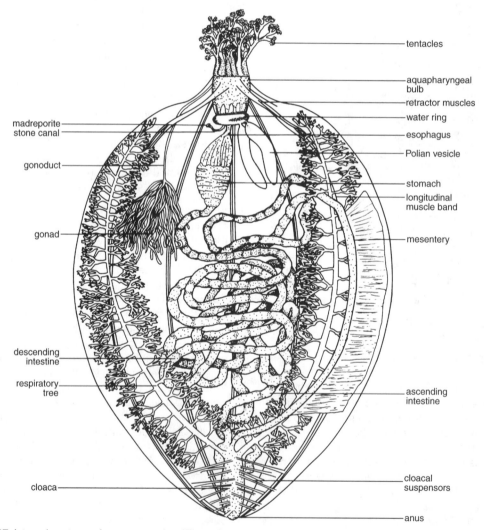

Figure 22.27. Internal anatomy of a sea cucumber (*Thyone*).

ambulacral areas. The canals lead into numerous side canals that connect with the tube feet in the ambulacra. The longitudinal muscles also join the retractor muscles at the anterior end. In addition to these muscles, note that the inner body wall is lined with **circular muscles**. The circular and longitudinal muscles allow the sea cucumber to expand and contract its body.

Reproductive System

The reproductive system is simple in these dioecious animals (Fig. 22.27). Find the moplike **gonad** that consists of a tuft of fine filaments. The gonad increases in size with the approach of sexual maturity. Do you think that your specimen was sexually mature when it was taken? Holothurians are the only echinoderms with a single gonad. A gonoduct carries eggs or sperm anteriorly to a genital papilla located between two dorsal tentacles. Microscopic examination of crushed gonadal filaments will reveal the sex of your specimen. The round head and long tail of the sperm can be seen when stained with methylene blue and examined under oil immersion. Eggs of the female are large and ovoid.

Live Specimens

1. Focus your attention on the cloacal opening of a live animal. Is it slowly opening and closing? Using a Pasteur pipette, place a small amount of seawater that was colored with methylene blue stain near the opening of the cloaca. What happens to the stain? What does this indicate is going on?
2. Try rotating the animal onto a different side. Does it turn back around (right itself)?
3. Observe burrowing behavior in a sea cucumber by placing the animal in a large glass bowl filled with seawater and a sufficient quantity of sand to permit burrowing. How does the sea cucumber bury itself? Is this behavior anything like the burrowing observed in other echinoderms you studied?
4. How do sea cucumbers react to stimuli? Take care in these investigations; some sea cucumbers will eviscerate if abused. If it is appropriate to do so, your instructor will demonstrate this phenomenon. Examine the ejected organs to determine which ones were expelled. What possible advantage can this behavior provide the sea cucumber? What other animals ablate body parts when attacked? ■

E. Class Crinoidea

The class Crinoidea [cri-NOI-de-ah; G., *krion*, a lily + G., *eidos*, form] contains about 100 species of sea lilies and an additional 600 species of feather stars or comatulids [KO-mat-u-lids; L., *comatus*, hairy] (about 600 living species). Crinoids have a worldwide distribution in warm and frigid waters and live at depths ranging from a few meters to over 5,000 m. Some species are gregarious, often forming huge local populations.

The living species represent a mere remnant of a once-thriving group in the geologic past. The crinoid fossil record dates back to the **Cambrian Period** and over 5,000 fossil species have been described from the **Paleozoic Era** (Fig. 22.1).

Most crinoids living today are the stemless comatulids, whereas in the past the majority of species were attached permanently to the sea floor by a **stem** (or **stalk**). Comatulids are attached to a substrate by a stem in their early development, but break away and move above on their jointed **cirri**, somewhat in a walking fashion, or swim by lashing their arms (Fig. 22.28). Sea lilies are similar in structure to the comatulids, but have a stalk for attachment (Fig. 22.29).

■ Observational Procedure:

1. Obtain a specimen of a stemmed crinoid and a stemless comatulid such as *Antedon*. Observe that the body consists of the cup-shaped **calyx** (**theca**) housing the internal organs, the arms constituting a food-gathering apparatus, and the attachment device (**stem** and/or **cirri**) for clinging to the substratum (Fig. 22.28 and 22.29). The combination of the calyx and arms is called **crown**. Feather stars lack the stalk, but have cirri for grasping the substratum.
2. The cup-shaped calyx is composed of external calcareous plates and is attached by its aboral surface. The membranous covering of the calyx is the **tegmen**; it is the oral surface of the organism (Fig. 22.30). Embedded in the tegmen are minute, calcareous plates. Also in the tegmen are tiny, ciliated canals that open into the enterocoelom. These openings may function as a madreporite. Note that the tegmen is divided radially into five ambulacral and five interambulacral areas (Fig. 22.30). The mouth is situated at or near the center of the tegmen and the anus occurs in

Figure 22.28. External view of a feather star (comatulid).

one of the interambulacral areas. The anus is on an elevated **anal cone** that directs feces away from the mouth. Unlike other echinoderms, crinoids have the oral surface is directed upwards.

3. Extending from the periphery of the crown are the five arms. Each arm, along with its ambulacral groove, may branch once or several times, forming 10 or more arms. How many branches does your specimen have? What advantage might a large number of branches offer a sedentary, suspension feeder?

4. Note the rows of spherical, yellow-brown dots along the sides of each ambulacral groove (Fig. 22.31B). These bodies are **saccules** and may serve as waste deposit areas. To see the saccules, observe them with a dissection microscope. The margins of the ambulacral grooves bear movable plates called **lappets**. These can close over the ambulacral groove. Within the grooves are the tentacle-shaped tube feet bearing mucus-secreting papillae used for food capture. Ampullae are lacking on the tube feet. Cilia within the ambulacral groove propel the food to the mouth. Crinoids are suspension feeders.

5. Extending laterally off each arm find the numerous jointed **pinnules** (Fig. 22.30). The ambulacral grooves of the pinnules are continuous with the main arm grooves; both podia and lappets occur in the pinnules. Gametes develop from the germinal epithelium in those pinnules located along the proximal half of the arm length. Both the arms and pinnules are composed of ossicles of the endoskeleton.

6. The stem is composed of a series of stacked, disc-shaped ossicles, each perforated by a central, axial canal for passage of components to certain organs. The discs are known as **columnals**. Because of the internal skeletal ossicles, the stalk has a jointed appearance. View these structures on a fossil crinoid (Fig. 22.29).

7. Crinoids are dioecious. The gametes rupture through the pinnule walls. The fertilized eggs remain attached to the pinnules as in *Antedon* or they are shed in the seawater. The free-swimming larva is a nonfeeding **vitellaria**, similar to that of sea cucumbers (Fig. 22.32). The larva settles to the bottom of the sea floor and develops eventually into an adult.

Due to the difficulty involved in the dissection, the internal anatomy of the crinoid will not be studied. ■

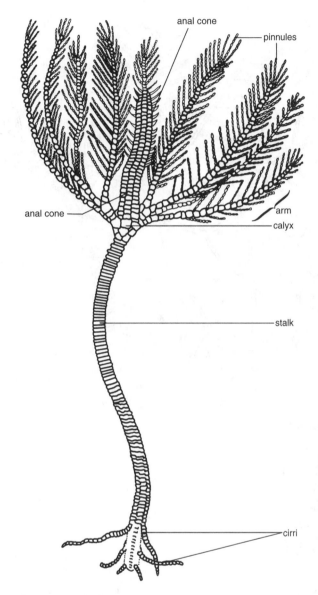

Figure 22.29. External view of a sea lily.

Fossil Echinoderms

Because of their hard, calcareous endoskeleton echinoderms fossilize well and their fossil record is rich. According to many zoologists, this ancient group dates to the lower **Cambrian Period** (Fig. 22.1). Despite a rich fossil record, the exact relationship of the echinoderms to the other invertebrates and to the chordates (Exercise 25) is unknown.

Most of the fossil species were stalked, sessile, and had the oral surface along with the ambulacral groove directed upwards. Many possessed extensions of the groove in a pinnule-like projection called **a brachiole** (Fig. 22.33A). These fossils were similar to modern-day sea lilies. More than likely the ambulacral system of tube feet was first used for feeding, rather than for locomotion. Later the podia served for locomotion and several independent lines toward bilateral symmetry ensued.

Three of the oldest groups (**Cambrian Period**) recognized currently are Helicoplacoidea, Carpoidea, and Blastozoa (Fig. 22.33).

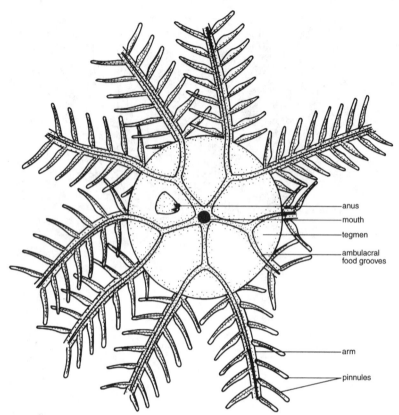

Figure 22.30. Oral surface of a crinoid.

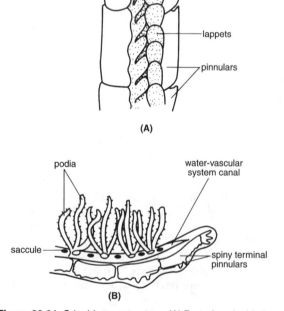

Figure 22.31. Crinoid arms structure. (A) Part of a crinoid pinnule showing lappets and ambulacral groove. (B) Section through the tip of a crinoid pinnule showing the podia and saccules.

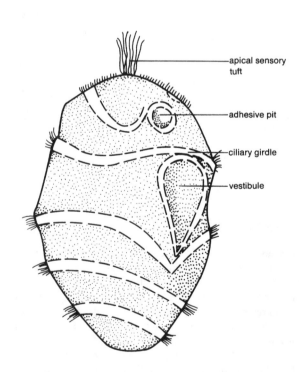

Figure 22.32. Crinoid vitellaria larva.

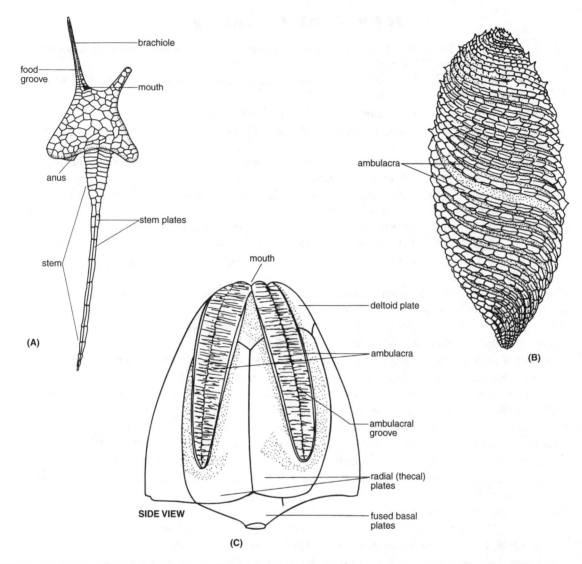

Figure 22.33. Three fossil echinoderms. (A) *Dendrocystites*, a carpoid with one brachiole. (B) *Helicoplacus* showing the spiral ambulacrum. (C) *Pentremites*, a blastoid showing ambulacra. (After various sources.)

■ Observational Procedure:

1. Observe the fossil echinoderms available for your study. Whenever possible determine the oral-aboral axis and recognize the radial symmetry. Do the specimens all exhibit a pentamerous body plan? Locate ambulacra and the interabulacra. Do they resemble the ambulacra and the interabulacra of the extant specimens?

2. If possible relate each specimen to a present-day form. Do any of these specimens resemble modern echinoderms sufficiently for you to assign them to a subphylum or to a class? What evidence did you focus on to do this simple classification?

3. If whole fossil crinoids are available for study compare their form to Fig. 22.29. Examine the columnals. Are they hollow? What sorts of patterns grooves or embossments are present?

4. If the calyx of a blastoid is on display, locate the following features: ambulacra, interabulacra, mouth, anus, and various plates (deltoid, radial, basal). How does this specimen qualify as an echinoderm? ■

Supplemental Readings

Ausich, W.I., and D.J. Bottjer. 1982. Tiering in suspension-feeding communities on soft substrate throughout the Phanerozoic. Science 216: 173–174.

Binyon, J. 1964. On the mode of functioning of the water vascular system of *Asterias rubens*. J. Mar. Biol. Assoc. U.K. 44: 577–588.

Binyon, J. 1972. Physiology of Echinoderms. Pergamon Press, Elmsford, New York.

Broom, D.M. 1975. Aggregation behaviour of the brittle-star *Ophiothrix fragilis*. J. Mar. Biol. Assoc. U.K. 55: 191–197.

Burnett, A.L. 1960. The mechanism employed by the starfish *Asterias forbesi* to gain access to the interior of the bivalve *Venus mercenaria*. Ecology 41: 583–584.

Byrne, M. 1985. Evisceration behavior and the seasonal incidence of evisceration in the holothurian *Eupentacta quinquesemita* (Selenka). Ophelia 24: 75–90.

Christensen, A.M. 1970. Feeding biology of the sea star *Astropecten irregularis*. Ophelia 8: 1–134.

Clark, A.M. 1962. Starfishes and Their Relations. British Museum, London.

Coe, W.R. 1972. Starfishes, Serpent Stars, Sea Urchins, and Sea Cucumbers of the Northeast. Dover Publishers, New York.

Cottam, G., and J. T. Curtis. 1955. On the methods used by the starfish *Pisaster ochraceus* in opening three types of bivalve molluscs. Ecology 36: 764–767.

Duggins, D.O. 1980. Kelp beds and sea otters: an experimental approach. Ecology 61: 447–453.

Durham, J.W. 1966. Evolution among the Echinoidea. Biol. Rev. 41: 368–391.

Durham, J.W., and K.E. Caster. 1963. Helicoplacoidea, a new class of echinoderms. Science 140: 820–822.

Fell, H.B. 1963. The phylogeny of sea stars. Phil. Trans. R. Soc. Lond. Ser. B. 246: 381–485.

Ferguson, J.C. 1969. Feeding, digestion, and nutrition in Echinodermata. *In*: M. Florkin, and B.T. Scheer (eds.). Chemical Zoology, Vol. 3. Echinodermata. Academic Pres, New York, pp. 71–100.

Ferguson, J.C. 1989. Rate of water admission through the madreporite of a starfish. J. Exp. Biol. 145: 147–156.

Haugh, B.N., and B.M. Bell. 1980. Fossilized viscera in primitive echinoderms. Science 209: 653–657.

Howarth, R.W. 1979. Coral reef growth in the Galapagos: limitation by sea urchins. Science 203: 47–50.

Keller, B.D. 1983. Coexistence of sea urchins in seagrass meadows: an experimental analysis of competition and predation. Ecology 64: 1581–1598.

Lawrence, J.M. 1987. A Functional Biology of Echinoderms. Croom Helm, London.

Lovtrup, S. 1975. Validity of the Protostomia-Deuterostomia theory. Syst. Zool. 24: 96–108.

Macurda, D.B., Jr. 1978. These reef animals blossom at night. Smithsonian 9:8 6–88.

McClintock, J.B. 1983. Escape response of *Argopecten irradians* (Mollusca; Bivalvia) to *Luidia clathrata* and *Echinaster* sp. (Echinodermata: Asteroidea). Florida Sci. 46: 95–100.

Millot, N. (ed.). 1967. Echinoderm Biology. Academic Press, New York.

Motokawa, T. 1984. Connective tissue catch in echinoderms. Biol. Rev. 59: 255–270.

Nichols, D. 1969. Echinodermata, 4th ed. Hutchinson University Library, London.

Nichols, D. 1986. A new class of echinoderms. Nature 321: 808.

Paul, C.R.C., and A.B. Smith (eds.). 1988. Echinoderm Phylogeny and Evolutionary Biology. Claredon Press, Oxford.

Seilacher, A. 1979. Constructional morphology of sand dollars. Paleobiology 5: 191–221.

Smith, A.B. 1984. Classification of the Echinodermata. Palaeontology 27: 431–459.

Smith, A.B., and J. Ghiold. 1982. Role of holes in sand dollars (Echinoidea): a review of lunule formation. Paleobiology 8: 242–253.

Smith, G.N., Jr., and M.J. Greenberg. 1973. Chemical control of the evisceration in *Thyone briareus*. Biol. Bull. 144: 421–436.

Strathmann, R.P. 1975. Larval feeding in echinoderms. Am. Zool. 15: 717–730.

Swan, E.F. 1961. Seasonal evisceration in the sea cucumber *Parastichopus californicus*. Science 133: 1078–1079.

Ubaughs, G. 1967. General characteristics of Echinodermata. Treatise on Inverterbrate Paleontology, Part S(1): 53–560.

Vadas, R.L. 1977. Preferential feeding: an optimization strategy in sea urchins. Ecol. Monogr. 47: 337–371.

Vine, P.J. 1973. Crown-of-thorns plagues: the natural causes theory. Atoll Res. Bull. 166: 1–10.

Wilkie, I.C. 1978. Arm autonomy in brittle stars. J. Zool. 186: 311–330.

Wilkinson, C.R., and I.G. Macintyre. (eds.). 1992. The *Acanthaster* debate. Coral Reefs 22(2).

■

Phylum Chaetognatha

Chaetognatha [KEY-to-NATH-ah; G., *chaeto*, bristle + G., *gnathos*, jaw] is a small phylum comprising about 70 species of unsegmented, bilaterally symmetrical, predatory worms. Chaetognaths are commonly called arrow worms because they resemble short arrows or darts. They range in size from a few millimeters to no more than 15 cm long; most are less than 4 cm. The arrow worm body consists of three regions: (1) a rounded head armed with grasping spines, (2) an elongate trunk with one to two pairs of lateral fins, and (3) a postanal tail with spatulate tail fin (Fig. 23.1). The epidermis is covered by a thin, transparent cuticle. Sensory hair-fans, important in detecting vibrations produced by potential prey (e.g., copepods), are arranged in a pattern of several rows. In the neck is the ciliary loop, consisting of two concentric rings of ciliated epidermal cells whose function is unknown. Internally, arrow worms possess several coelomic compartments. Traditionally, chaetognaths have been considered to be deuterostomes, especially because of the presence of a tripartate coelom. However, this intrepation has been challenged and their phylogenetic position is uncertain. Some researchers place them with the aschelminths (pseudocoelomates), while others suggest a protostome relationship.

Arrow worms are exclusively marine. Found in all oceans, they are especially common in tropical and subtropical waters. Most arrow worms are planktonic, except for species of the genus *Spadella* which inhabit the benthos of shallow waters.

Classification

There is a single class, Sagittoidea, with two orders based on presence or absence of ventral transverse muscles.

■ Observational Procedure:

Observe several specimens of an arrow worm, such as *Sagitta*, using a dissection microscope. Identify the three body regions: head, trunk, and tail (Fig. 23.1). In the head region, locate the large, chitinous spines used to seize prey. Anterior to the spines are two rows of small teeth that aid in prey capture. The head and spines may be covered by a protective fold of the body wall called the **hood**; the hood may be folded back. Locate the vestibule on the ventral side of the head. In this small chamber prey are held before being swallowed. Occasionally, specimens are found with prey still present in the vestibule. Note the eyes on the dorsal side of the head.

Locate the two pairs of lateral fins and the tail fin. Illuminating the specimen from above (epi-illumination) may aid in finding these nearly transparent structures. Note that internally the fins possess raylike supports. The fins are delicate structures and often are distorted or lost in preserved specimens. Some species (e.g., *Spadella*) have one pair of lateral fins. The sensory hair-fans are difficult to see, but sometimes may be seen in stained specimens. Note the longitudinal muscle bands that run the length of the animal.

301

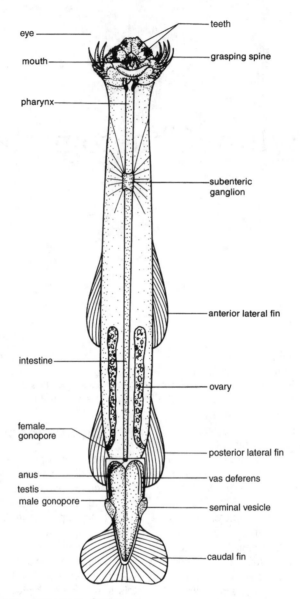

Figure 23.1. General anatomy of the chaetognath, *Sagitta*.

The gut is a simple straight tube. At the anterior end is a muscular pharynx that leads to the long, straight intestine. An anus is located on the ventral side in the region of the trunk-tail septum. Find the paired ovaries in the posterior part of the trunk. Carefully manipulate your specimen so that you can see the female **gonopores** just before the level of the trunk-tail septum. In the tail, locate the paired testes and seminal vesicles. The latter are embedded laterally in the body wall. ■

Supplemental Readings

Alvariñõ, A. 1965. Chaetognaths. Oceanogr. Mar. Biol. Ann. Rev. 3: 115–194.

Bieri, R., and E.V. Thuesen. 1990. The strange worm Bathybelos. Am. Sci. 78: 542–549.

Bone, Q., H. Kapp, and A.C. Pierrot-Bults (eds.). 1991. The Biology of Chaetognaths. Oxford University Press, London.

Feigenbaum, D. 1982. Feeding by the chaetognath, *Sagitta elegans*, at low temperatures in Vineyard Sound, Massachusetts. Limnol. Oceanogr. 27: 699–706.

Feigenbaum, D., and M.R. Reeve. 1977. Prey detection in the Chaetognatha: response to a vibrating probe and experimental determination of attack distance in large aquaria. Limnol. Oceanogr. 22: 1052–1058.

Ghirardelli, E. 1968. Some aspects of the biology of the chaetognaths. Adv. Mar. Biol. 6: 271–375.

Ghirardelli, E. 1995. Chaetognaths: two unsolved problems: the coelom and their affinities. *In*: G. Lanzavecchia, R. Valvassori, and M.D. Candia Carnevali (eds.). Body Cavities: Function and Phylogeny. Selected Symposia and Monographs Unione Zoologica Italiana, 8, Mucchi, Modena.: pp. 61–67.

Grant, G.C. 1992. The Phylum Chaetognatha. A Bibliography. Virginia Institute of Marine Science. School of Marine Science. The College of William and Mary. Special Papers in Marine Science 7: 1–166.

Moreno, I. (ed.). 1993. Proceedings of the II International Workshop on Chaetognatha. Universidad de les Illes Baleares, Palma.

Newbury, T.K. 1972. Vibration preception by chaetognaths. Nature 236: 459–460.

Telford, M.J., and P.W.H. Holland. 1993. The phylogenetic affiniities of the chaetognaths. A molecular analysis. Mol. Biol. Evol. 10: 660–676.

■

Phylum Hemichordata

Hemichordata [hem-ee-KOR-dah-tah; G., *hemi*, half + G., *chorda*, string] contains about 85 species of burrowing or sessile, enterocoelous, marine animals. Their bodies are divisible into an anterior **proboscis**, middle **collar**, and a long posterior **trunk** (Fig. 24.1). **Pharyngeal gill slits** for gas exchange occur in most hemichordates. Other features include a complete digestive system, open circulatory system, nervous system combining invertebrate and vertebrate features, and separate sexes. Two classes of this widely distributed phylum are represented by acorn worms and pterobranchs.

Pterobranchia [ter-o-BRAN-chi-ah; G., *pteron*, wing + G., *branchia*, gills] consists of fewer than 15 small (1–5 mm) species. Most of these uncommon, deep-water invertebrates live in secreted tubes forming colonies or aggregations. Extending dorsally from the collar region are one or more arms bearing several food-gathering, hollow, ciliated tentacles. Pterobranchs may be more closely related to phoronids (Phoronida; Exercise 18) and brachiopods (Brachiopoda; Exercise 21) then to enteropneusts. Graptolites, an enigmatic group of fossils, may be pterobranchs (Armstrong et al. 1984). Because of the rarity, unavailability, and small size of pterobranchs, they will not be studied in this exercise.

There are approximately 70 species of acorn worms in the class Enteropneusta [en-TER-o-NEUS-tah; G., *enteron*, gut + G., *pnein*, to breathe]. These slow-moving worms range from a few millimeters to greater than 1 m in length. They are found in shallow waters of intertidal and littoral zones beneath shells and rocks, in seaweed, or burrows dug in mud or sand. The burrows are generally U-shaped and mucus-lined. Enteropneusts are either suspension feeders or deposit feeders. Those of the former type trap minute food particles in mucus on the proboscis moving that material to the mouth by ciliary action. Deposit feeders extract organic materials from the vast amount of sand and mud that they consume. Undigested remains are deposited on the surface of the sediment as coiled **fecal casts**.

Acorn worms are dioecious. Gametes from the gonads in the trunk region escape via a pore into the water. Fertilization is external and development is either direct or indirect. Many species produce a free-swimming, ciliated **tornaria larva** that is similar to the bipinnaria larva of sea stars (Echinodermata: Asteroidea; Exercise 22.A). Asexual reproduction after fragmentation has been reported from some enteropneusts.

Classification

1. **Class Pterobranchia.** Hemichordates possessing one or more pairs of tentacular arms arising from the dorsal area of the collar. One pair of gills present or gills absent. Examples: *Rhabdopleura* and *Cephalodiscus*.

2. **Class Enteropneusta.** Hemichordates with numerous gill slits. Examples: *Saccoglossus* and *Balanoglossus*.

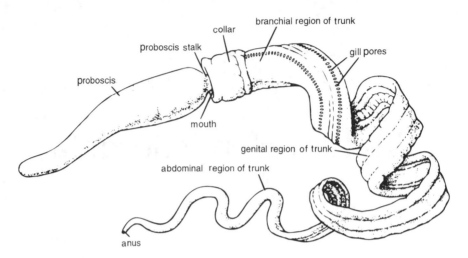

Figure 24.1. External view of the hemichordate.

■ Observational Procedure:

Whole Acorn Worm

1. Obtain an acorn worm such as *Saccoglossus* (*Dolichoglossus*) *kowalevskii* which lives along the North American and European Atlantic coasts. Place in a small dish containing water and examine it using the dissection microscope. Locate the three body regions: proboscis, collar, and trunk (Fig. 24.1). Each region corresponds to the body divisions of a typical deuterostome (e.g., **protosome, mesosome**, and **metasome**). Internally, each region has its own coelomic cavity: a single protocoel of the proboscis, paired mesocoels of the collar, and paired metacoels of the trunk. These and other internal structures will be examined later when you study a longitudinal section of the worm.

2. Locate the proboscis of your worm and the others available in the lab. What shape is it? What sorts of characteristics would you predict it possesses for this activity? A narrow isthmus, the **proboscis stalk**, connects the proboscis to the collar region (Fig. 24.1). Be very careful should you attempt to find this shaft of tissue, for in doing so you may damage the specimen.

3. Focus your attention on the collar. Describe the shape is this structure. Is the collar a uniform shape? Attempt to locate the mouth on the anterior margin of the ventral side of the collar. (Be careful here.) It should be found just below the proboscis stalk (Fig. 24.1). The mouth marks the ventral side of the animal. Using a dissection microscope, examine the dorsal side of the proboscis and attempt to locate the small **proboscis pore** on the stalk. This pore allows water to enter and leave the protocoel during burrowing. Why might this be important to the functioning of the pro-

boscis? If you cannot find this structure, you may be able to see it when you examine a longitudinal section of the worm.

4. Most of the animal's body consists of the trunk which has a midventral and middorsal **longitudinal ridge** (Fig. 24.1). The trunk can be divided into three regions; locate each region on your specimen as it is discussed. (1) The anterior **branchial** region contains numerous U-shaped gill slits. Locate the **gill pores** in paired rows along the dorsolateral area of the branchial trunk region. Each pore leads to an internal branchial sac that communicates with the slit. (2) The middle **genital** region contains the gonads. (3) The posterior **abdominal** region contains intestine with the paired **intestinal diverticula**. After passing through the gut, undigested materials are eliminated by the terminal anus.

5. If you are to dissect the specimen, your instructor will provide you with the appropriate instructions. Otherwise, set the specimen aside but do not return it to the museum jar just yet. You may need to refer back to the specimen while you study some of the internal structures of the animal.

Slides—*Balanoglossus*

Longitudinal and Sagittal Sections

1. Obtain longitudinal and sagittal sections of an acorn worm such as *Balanoglossus*. Examine each section under low power of a compound microscope to become familiar with the orientation of the specimens on the slides. Locate the proboscis, collar, mouth, and pharyngeal region with gill slits. As you proceed through this portion of the exercise, it will be best if you are able to examine more than one slide of each kind.

Although all of these slides are longitudinal sections, each may show a slightly different aspect of the animal, and thus provide a different view of important anatomical features.

2. Locate the proboscis and describe how it is constructed. The outer most layer is the epidermis, a region rich in glandular cells bearing both cilia microvilli. Switch to medium magnification and examine the entire edge of the proboscis. Are the glandular cells evenly distributed in the epidermis? If not, where are they concentrated? What function does it serve to have gland cells concentrated in one region or another? Using some advanced lighting conditions (i.e., high power and phase contrast) you may be able to see the cilia, which otherwise show up as a thin fuzzy region.

3. Beneath the epidermis locate the **basal lamina** and a layer of connective tissue **dermis**. Below those two layers find the regions of circular and longitudinal muscles. How can you tell the circular and longitudinal muscles apart? What is their function in the proboscis? Locate the protocoel (Fig. 24.2). This body cavity is not extensive and may seem to be subdivided into chan-

nels due to the presence of intervening tissue. On the dorsal side of the animal, attempt to find the proboscis pore on the proboscis stalk (Fig. 24.2). This channel connects the protocoel to the outside. How do the circular and longitudinal muscles, protocol, and proboscis pore function in the burrowing action of the proboscis?

4. Using low power, scan the basal region of the proboscis where it connects to the stalk. In this area locate the narrow **buccal diverticulum** or **stomochord**. In the past, this feature was thought to be a **notochord** (Fig. 24.2), and because of that, hemichordates were placed in **Phylum Chordata** (Exercise 25). Locate the **proboscis skeleton** (Fig. 24.2). Switch to high power and examine this structure for cells; are any present? The proboscis skeleton is cartilaginous and wishbone in shape with the wings extending into the collar. What function(s) might this structure serve?

5. At the base of the proboscis, just anterior to the proboscis skeleton, locate a network of vessels. This is the **glomerulus**, a structure believed to have excretory functions. Behind the glomerulus and dorsal to the stomochord is the heart. Where is the protocoel in relation

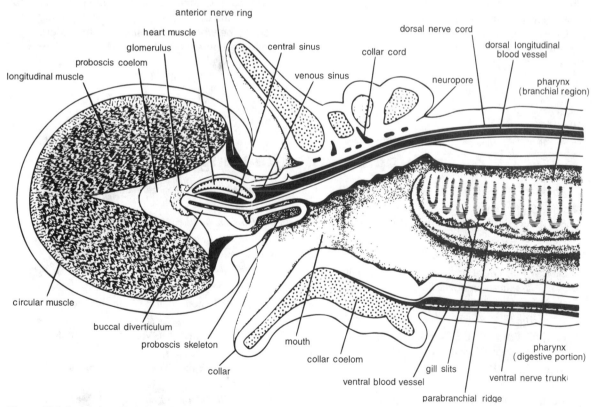

Figure 24.2. Internal sagittal view of the anterior region of a hemichordate.

to these other structures? Given this spatial relationship, how might excretion be accomplished? Locate the anterior nerve ring; it should be found above the heart almost in the wall of the proboscis.

6. Focus your attention to the region of the collar, noting the presence of folds. Is the epidermal layer of the collar as richly supplied with glands as was the proboscis? The largest anatomical features to be seen in the collar region are the mouth and buccal cavity. Locate cavities of the mesocoel in the dorsal and ventral sides of the collar. Also try to find the dorsal and ventral nerve cords. Extensions of the collar nerve cord may appear as a series of cavities (Fig. 24.2). When these cavities open to the outside they are referred to as **neuropores**. Are any neuropores present in the sections available for your study? The dorsal and ventral blood vessels also may be found. Are any structures resembling the proboscis skeleton present in the collar region? What distinguishes the proboscis skeleton from the other structures in this region?

7. Finally, examine the pharyngeal or branchial portion of the upper trunk. The most obvious feature present here is the large empty space of the pharynx. Locate the cavities of the metacoel in the dorsal and ventral sides of the collar. Locate the gills and their support structures, the gill arches. Note that the gill tissue is very thin and, of course, cellular. Observe that their skeletal supports (the **gill arches**) are thicker and acellular. You should be able to trace the blood vascular system as a series of winding tubes that lead to the gills. Other structures that you may see at this level include mucous glands and ovaries or testes.

Cross Section (trunk, through level of the intestine)

1. Obtain a cross section of an acorn worm such as *Balanoglossus*. Examine the section under low power of a compound microscope to become familiar with the orientation of the specimen. As noted in the previous section, when you proceed through this exercise, it will be best if you examine more than one slide.

2. Locate the large open space in the center of the section; this is the pharynx. Determine the dorsal and ventral side of the animal. The structures that will aid you the most in making this determination are the **skeletal supports** of the gills. They originate on the dorsal side and pass ventrally around the pharynx. Locate the gill skeletal support and examine them under high power. How are they similar to the proboscis skeleton?

3. Switch back to a lower power and find the dorsal nerve cord and beneath it the dorsal blood vessel. On the opposite side of the animal, locate the ventral nerve cord and beneath it the ventral blood vessel.

4. Locate the metacoel; it is the space between the pharynx and the longitudinal muscle. Now focus you attention to the gill slits; these are openings in the pharynx that communicate to the outside via the external gill pores. Find the following structures: (1) gill slits; (2) the space between the gill slits and the external gill pores; (3) gill pores; (4) gill bars. ■

Supplemental Readings

Armstrong, W.G., P.N. Dilly, and A. Urbanek. 1984. Collagen in the pterobranch coenecium and the problem of graptolite affinities. Lethaia 17: 145–152.

Barrington, E.J.W. 1940. Observations on feeding and digestion in *Glossobalanus minutus*. Q. J. Micros. Sci. 82: 227–260.

Barrington, E.J.W. 1965. The Biology of Hemichordata and Protochordate. W. H. Freeman, San Francisco, CA.

Duncan, P.B. 1987. Burrowing structure and burrowing activity of the funnel-feeding enteropneust *Balanoglossus aurantiacus* in Bogue Sound, North Carolina, USA. Mar. Ecol. 8: 75–95.

Hadfield, M.G. 1975. Hemichordata. *In*: A.C. Geise and J.S. Pearse (eds.), Reproduction of Marine Invertebrates, Vol. II. Academic Press, New York, pp. 185–240.

Harrison, F.W., and E.E. Ruppert (eds.). 1996. Microscopic Anatomy of Invertebrates, Vol. 15. Hemichordata, Chaetognatha, and The Invertebrate Chordates. Wiley-Liss, New York.

Knight-Jones, W.E. 1952. On the nervous system of *Saccoglossus cambrensis*. Philos. Trans. R. Soc. Lond. B 246: 315–354.

Stebbing, A.R.D., and P.N. Dilly. 1972. Some observations on living *Rhabdopleura compacta*. J. Mar. Biol. Assoc. U.K. 52: 443–448.

PHYLUM CHORDATA

Nearly 50,000 extant species comprise the phylum Chordata, but most of them are members of the subphylum Vertebrata [VER-ta-BRA-tah; L., *vertebratus*, jointed or articulated]. The two other subphyla. Urochordata [U-ro-kor-DA-tah; G., *uro*, tail + G., *chordata*, string] and Cephalochordata [SEF-a-lo-kor-DA-tah; *cephala*, head], are of some interest to both invertebrate and vertebrate zoologists as members may be similar to the ancestral vertebrates. All chordates possess paired pharyngeal gill slits, a dorsal tubular nerve cord, and a supportive notochord during some portion of their lives.

Jefferies, R.P.S. 1986. The Ancestry of Vertebrates. Cambridge University Press, Melbourne.

Løvtrup, S. 1977. The Phylogeny of Vertebrates. Wiley, London.

Maisey, J.G. 1986. Heads and tails: a chordate phylogeny. Cladistics 2: 201–256.

Romer, A.S. 1967. Major steps in vertebrate evolution. Science 158: 1629–1637.

Phylum Chordata

Most of the nearly 50,000 species comprising phylum Chordata [kor-DAH-tah; G., *chorda*, string] are vertebrates (subphylum Vertebrata). However, this phylum contains two nonvertebrate subphyla (Urochordata and Cephalochordata) with a total of some 1,400 species. All three subphyla are united because they possess **gill slits, notochord,** and a **dorsal tubular nerve cord** at some time in their life. Although some instructors prefer to

leave part or all of this phylum to courses in vertebrate zoology, the nonvertebrate chordates are valid subjects for study by students of invertebrate zoology. The nonvertebrate chordates, together with the hemichordates (Exercise 24), are known as the **protochordates**. The subphylum Cephalochordata has a fossil beginning in the mid-**Cambrian Period** (Fig. 25.1).

TIME (IN MILLIONS OF YEARS)

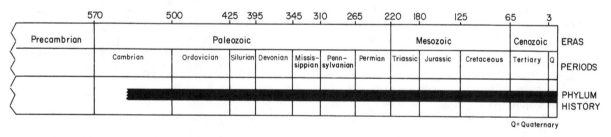

Figure 25.1. Geological history of the Cephalochordata.

A. Subphylum Urochordata (Tunicata)

Subphylum Urochordata [URO-chor-DA-tah; G., *uro*, tail, + G., *chorda*, string] is comprised of small to medium-sized individuals that scarcely resemble the chordate description. The three chordate characteristics are present only during the larval stage. In the adult the gut is U-shaped, with a spacious pharynx perforated by numerous gill slits. Because the body is covered with a tough test or **tunic**, urochordates are commonly called **tunicates**.

Classification

Exclusively marine, free-living, filter-feeding chordates lacking segmentation and hard parts. Three (sometimes four) classes are recognized. In this exercise only members of classes Ascidiacea and Thaliacea will be studied.

1. **Ascidiacea:** Shallow to deep-sea, sessile or sedentary, barrel or bag-shaped urochordates. Adults are either solitary or colonial in jelly masses containing a few to many minute individuals. Ascidians are often found as members of the marine fouling community. The tunic, which may be massive, is a composite of protein and a polysaccharide resembling cellulose. Most species are monoecious. The larval stage resembles a tadpole. Examples: *Botryllus, Ciona, Corella, Ecteinascidia, Molgula,* and *Styla.*

2. **Larvacea (Appendicularia):** Small, solitary, planktonic urochordates in which the adult appears to have conserved the organization of the ascidian larval stage (**neoteny**): i.e., they resemble the tadpole larva of an ascidian. The larvacean body is ovoid in shape with a long thin tail. They live within gelatinous cases or houses that they secrete. Example: *Oikopleura.*

3. **Thaliacea:** Planktonic, colonial urochordates (or with a life cycle in which colonial and solitary forms alternate). Individuals are more-or-less barrel-shaped with a large open pharynx. Colonies range in size from a few centimeters to several meters. Locomotion is accomplished by a rhythmic contraction and expansion of the tunic that results in a jet propulsion. Examples: *Pyrosoma* and *Salpa.*

■ Observational Procedure: Class Ascidiacea

Whenever possible, observe living animals. Place the specimens to be examined in a small bowl filled with water sufficient to cover the animals. Use sea water if live animals are used, allowing them sufficient time to relax and extend their siphons. Try not to disturb the animals once they have relaxed. If your specimens are small and transparent, you may wish to make your observations with the aid of a dissection microscope. Lighting from the side or beneath the animals will work best for making most of the observations necessary for detailed study. Obviously, students examining preserved specimens will ignore directions established for live specimens.

Solitary Ascidians

Begin your observations with a solitary ascidian such as *Corella, Ciona, Molgula* and/or *Styela* (= *Cynthia*) (Fig. 25.2).

1. What features do these animals possess that would permit you to recognize them in the field? Are your live specimens producing inhalent and exhalent currents?

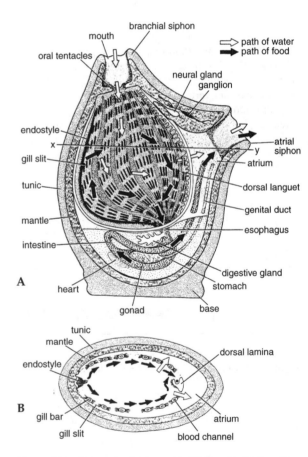

Figure 25.2. Generalised sea squirt. (**A**) Longitudinal section of a sea squirt depicting movement of water and food particles. (**B**) Cross section of sea squirt at the level of line X-Y. (After Sherman and Sherman and other sources.)

Which siphon is inhalent and which is exhalent? How strong are these currents? What produces them? The following directions will aid you in answering these questions.

Using a Pasteur pipette, gently deposit a few drops of a dilute suspension of carmine power or carbon black near each siphon. How do the particles move; are they taken into the animal; where do they enter? One aperture, the inhalent, oral, or **branchial** (G., *brachium*, gill) siphon takes water into the animal, while the other, the exhalent or **atrial** (L., *atrium*, vestibule) siphon directs water out of the animal (Fig. 25.2).

If your specimen is sufficiently transparent, it may be possible to track the movement of particles into the digestive tract. Do the particles accumulate within the gut? Over time you may be able to see the entire gut fill with this colored food. Are any of the particles rejected immediately, or do they all go into the animal? Do any particles exit the animal via the atrial siphon before passing through the gut?

2. Feel the covering or tunic that encases the animals. What descriptive terms would you use to describe the tunic (hard, soft, firm, flexible)? If you felt the tunic of a live animal, what happened when you touched the animal? Why are these ascidian commonly called **sea squirts**? Do they squirt from only one siphon? How is expulsion of water accomplished?

3. The branchial siphon is located at the anterior end and is directed parallel to the long axis of the body. The atrial siphon indicates the dorsal side of the animal (Fig. 25.2). Determine the anterioposterior axis of your specimen. How are these animals bilaterally symmetrical? Which side is left and which is right? Tunicates are usually attached to a substratum. To what sorts of substrata are your specimens attached?

4. Examine the body of *Ciona* or *Molgula* more closely. It is said to be divided into two regions: thorax or **branchial** and abdomen or **visceral**. To differentiate these regions you must be able to observe the internal structures of your specimen (Fig. 25.2). This may not be possible unless the tunic is transparent enough to permit light to pass through. If the specimens available for study do not permit internal observations, obtain a whole mount slide of a specimen such as *Ecteinascidia* and make the following observations. [NB: *Ecteinascidia* is a colonial ascidian (see below), but may be examined here to get a good idea of ascidian morphology. The study of more than one slide may be necessary for you to complete these observations.]

5. To start, identify the branchial and atrial siphons, then locate the pharynx. The pharynx marks the branchial portion of the animal. How large is the branchial portion of the body with respect to the whole animal? Just inside the pharynx locate the oral tentacles and between the branchial and artial siphons, the

subneural gland. The **endostyle** runs down the ventral side of the pharynx. It secretes a mucus that aids in food capture. On the side opposite the endostyle, locate the **dorsal lamina**. Examine the pharynx and identify the gill slits (stigmata) and the tissue that surrounds them, the gill bars. How numerous are the gill slits?

6. At the base of the pharynx is the esophagus. This marks the beginning of the visceral portion of the animal. Trace the esophagus from the pharynx to the stomach and then to the intestine. The intestine travels up the dorsal side of the animal, becoming the rectum before terminating at the anus. On which side of the body does the anus empty? Locate the genital duct and the genital openings. Where are they located? Why might the atrium be considered to be a kind of a **cloaca**?

7. The following directions are to be used if you are to dissect a specimen. The purpose of the dissection is to open the pharynx so that gill slits may be examined and so that both the endostyle and dorsal lamina will be exposed. To do this, begin cutting from the branchial siphon down the left side of the body. This cut should be near, but not on, the ventral margin (Fig. 25.2).

Insert the point of a pair of scissors into the branchial siphon and carefully cut through the tunic and beneath it, the mantle. Continue the cut down to the base of the animal and then around the base and up through the atrial siphon. Complete the cut by inserting the scissors into the atrial siphon and cutting up to the branchial siphon.

Carefully trim any extra tissue holding the left and right halves of the animal together and remove the left side of the animal exposing the pharynx. Pin the sea squirt to the base of a dissection pan and cover it with water. Identify the structures previously examined in steps 5 and 6. Remove a piece of the pharyngeal wall, cover it with water, and examine it using a compound microscope. Locate the gill slits and gill bars. If a live specimen was used, note the activity of the cilia.

Colonial Ascidians
There are two kinds of colonial ascidians: **social** and **compound**. The former are connected by a **stolon**, while the latter are embedded in a common tunic.

1. *Social Ascidians*—If available, observe the colony organization of a social ascidian such as *Ecteinascidia* or *Perophora* (Fig. 25.3). How are the individuals arranged on the stolon? Are all individuals in the colony the same size? Is there a pattern to the arrangement? What does this indicate? Identify the branchial and atrial siphons. Identify the internal structures in the individuals. The directions given above (steps 3 through 5) may be followed for *Ecteinascidia*. In addition, note the presence of stalks at the posterior (aboral) end of the animal. What do they represent?

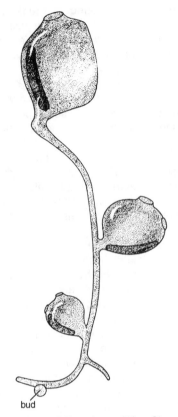

Figure 25.3. A colony of *Perophora*. (After Sherman and Sherman.)

2. *Compound Ascidians*—Observe a compound ascidian such as *Aplidium* (= *Amaroucium*) or *Botryllus* (Fig. 25.4). These specimens tend to resemble a slab of hardened jelly, and may contain hundreds of individuals within the colony. What features do these animals possess that would permit you to recognize them in the field? How does the tunic of a colonial tunicate feel? Is it the same as that of the solitary ones you examined?

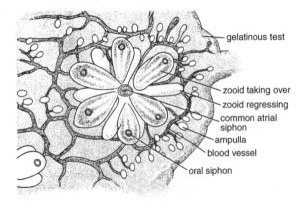

Figure 25.4. Portion of a colony of *Botryllus*. (After Sherman and Sherman.)

Determine whether a live colonial ascidian will squirt when disturbed?

3. Observe your specimen using either a hand lens or a dissection microscope and locate the clusters of animals (**zooids**) with their individual buccal siphons and their single, common atrial siphon. Note how groups of zooids form a collective unit with a common atrial siphon. This functional unit is called a **rosette** (Fig. 25.4). Which of these openings is larger, those of the branchial siphons or the common atrial siphon? In relaxed specimens, attempt to measure the siphon diameters and determine whether the cross-sectional area of the branchial siphons of a rosette equals or exceeds that of the common atrial siphon. (To do this you will need to use a dissection microscope equipped with an ocular micrometer). Consider the significance of the relationship of total cross-sectional area to the hydrodynamics of water movement through the rosette.

4. Using a sharp, single-edged razor blade, carefully dissect out one or more zooids from the common tunic. Be careful to get an entire animal for study. Place the zooids in a depression slide, add some water, and cover

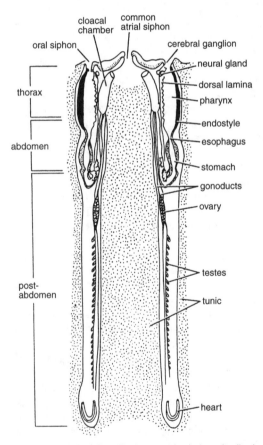

Figure 25.5. *Aplidium* (= *Amaroucium*). Longitudinal section of a portion of two zooids in a colony. (After Sherman and Sherman.)

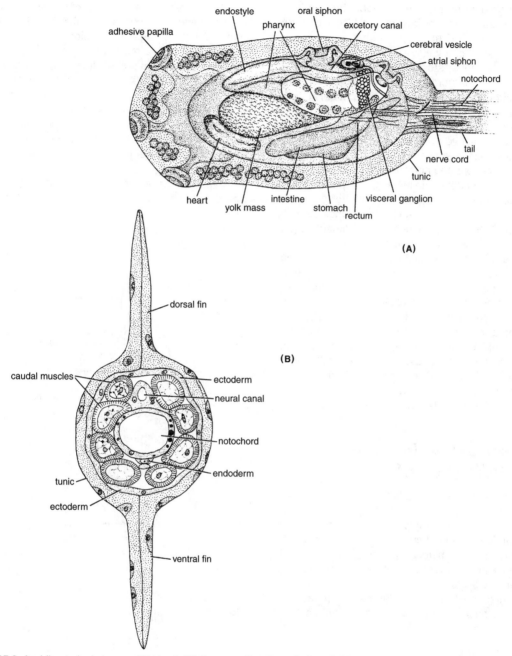

Figure 25.6. Ascidian tadpole larvae. (A) Head. (B) Cross section through the tail. (Sherman and Sherman and other sources.)

the preparation with a coverslip. Examine the zooids with the aid of a dissection microscope. Note that the body form is more elongate than in the solitary tunicates. The branchial (thorax) and visceral (abdomen) regions are again present, but colonial forms also possess a **postabdomen**. This region contains the gonads and heart (Fig. 25.5). Locate these organs in your specimens.

Tadpole Larva

1. Examine a whole-mount slide of ascidian tadpole larva (Fig. 25.6). (You may need to study more than one slide to complete these observations.) Identify the head and tail regions. Describe the body shape. Why is this larval stage given the name tadpole? Differentiate the tunic from the rest of the body tissues.

2. At the anterior end locate the **adhesive papillae**. What do these structures resemble? Also locate the epidermis, branchial and atrial siphons, endostyle, branchial basket (pharynx with gill slits), and digestive system (esophagus, stomach, intestine, and anus). Just posterior to the branchial siphon locate the **sensory vesicle** (cerebral vesicle) with **cerebral ganglion, statolith**, and **ocellus** (eyespot). Locate the dorsal nerve cord that runs toward the posterior end of the animal just above the notochord. Find the longitudinal muscle bands (dorsal and ventral) that surround both the nerve cord and notochord. Dorsal and ventral fins extend the length of the tail (Fig. 25.6). Is the notochord birefringent? What does this tell you about its composition? What component of the tadpole larva is birefringent? What is the biochemical composition of the tunic that would make it birefringent?

3. Compare the general organization of the siphons, branchial basket, and internal organs of the larval stage to that of the adult (Fig. 25.6). How are they similar? What portion of the tadpole larva will become the oral siphon? Where will the aboral region develop?

4. Compare the structures found in adults with those of the larvae. To do this, make a list of those structures that are present in both stages? Although differing in size, how are these structures similar? Also inventory the structures unique to the tadpole stage and note their specific function. ■

■ Observational Procedure: Class Thaliacea, Salps

Live thaliaceans are not readily available from commercial supply houses and unfortunately these delicate animals do not preserve well. Nevertheless, a general understanding of salp morphology is possible even with preserved materials. Place a salp in a small bowl filled with water and examine it with the aid of a dissection microscope.

1. The anterioposterior axis may be determined by locating the bulbous gut near the atrial siphon (Fig.

25.7). This lies in the posterior region of the body. At the other end is the buccal or oral siphon. Locate each end and the openings of each siphon.

2. Note the transparent gelatinous tunic. Gently nudge the tunic with a blunt probe. How does it differ from the tunic of the sea squirts you have just studied? Beneath the tunic, located in the body wall, are the muscle bands. How would these bands be used in creating a jet propulsion of the animal through the water.

3. Most of the body is taken up with the spacious atrium, but you should also find the endostyle and the comblike **gill bar**.

4. Given the overall body plan of salps, contemplate how they are adapted for a planktonic existence ■

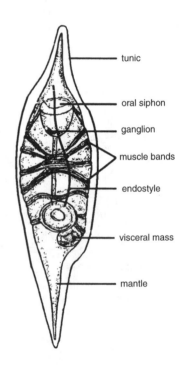

Figure 25.7. Generalized thaliacean.

B. Subphylum Cephalochordata

Cephalochordata [CEPH-a-lo-chor-DA-tah; G., *cephala*, head + G., *chorda*, string], also known as Acrania [ah-CRAY-ne-ah; G., *a*, without + G., *crania*, skull], are a very small group of benthic, marine, fish-like chordates. Because of their bladelike appearance, cephalochordates are commonly known as **lancelets**; individuals are sometimes referred to by the old genus name **amphioxus**. Although adults only range in size from 4 to 8 cm, these organisms can be very significant in the benthic ecology of certain regions as their populations can reach enormous numbers. Up to 5,000 individuals per square meter have been reported in shallow Caribbean waters. In Asia lancelets are harvested for food.

Classification

A total of 23 described species. Example: *Branchiostoma* (= *Amphioxus*).

■ Observational Procedure:

Whole Mount

Obtain a prepared whole mount of *Branchiostoma*.
1. Using a dissection microscope, identify the anterio-posterior axis and the dorsal and ventral sides (Fig. 25.8). The animal is compressed laterally and pointed at both ends, the posterior end being thinner and more pointy. How do the etymons of amphioxus and lancet fit the shape of this animal? Note the absence of a distinctive head region; the blunt anterior end encloses a cavity, the **vestibule**, by a **buccal** or **oral hood**. Is your specimen lying on its right or left side? What other features can you make out at this level (Fig. 25.8)?
2. Switch to a compound microscope to continue your study. At the anterior end locate the rostrum and below that the oral hood (Fig. 25.9). Lining the edges of the oral hood are the tentacles or **buccal cirri** that project outside. How do they resemble a cage? What function might the cirri serve? Inside of the buccal cavity, locate the **wheel organ**. In the living animal this ciliated structure creates a current that brings food to the mouth. Just in front of the wheel organ on the roof of the buccal cavity, locate **Hatschek's pit**. How is this structure well-positioned for the production of mucus which aids in feeding? Posterior to the wheel organ is the **velum** with **velar** or **oral tentacles**; these are sensory in nature.
3. Behind the velar tentacles, locate the large pharynx composed of gill bars alternating with gill slits (Fig. 25.9). These structures slope at an angle of about 45° to the long axis of the body. Identify the **endostyle** on the ventral margin of the pharynx. Food and water separate in the pharynx, with the water passing through the gill slits, into the **atrium**, and out the **atriopore**.
4. Posterior to the pharynx the gut narrows into an esophagus that widens into the stomach. Anteriorly a blind diverticulum, the **digestive** or **hepatic cecum**, extends into the atrium along the right side of the pharynx. From the stomach the gut again narrows into the intestine. At the junction of the stomach and intestine is the **iliocolonic ring**. This structure usually stains more darkly than the rest of the gut. Near the caudal fin, the intestine opens to the outside through an anus. Note that the anus does not empty into the atrium as in the ascidians (Fig. 25.8).
5. Above the gut locate the notochord, dorsal nerve cord, and dorsal fin ray (Fig. 25.8). Note that the fin ray appears to be constructed as a series of boxes. These are the **fin-ray boxes**. Compare the structure of the fin ray with that of the notochord; how can they be distinguished? The ventral fin also contains fin-ray boxes. In both dorsal and ventral fins, these compartments taper off before the beginning of the tail. Fin-ray boxes are believed to be food storage chambers.
6. Note how the notochord is composed of a series of compacted disc-shaped structures; each of these is a modified muscle cell, specialized for skeletal support. Amphioxus swims using a sinusoidal motion. Speculate on how a notochord, consisting of a stack of discs, might be better able to handle the shear forces generated during sinusoidal swimming than a solid flexible bar.
7. Examine the myotomes (Fig. 25.8). What is their shape and how are they arranged along the body? What is the functional significance of this arrangement? Is amphioxus segmented? Of what significance is this to amphioxus? How many times has true segmentation evolved in the invertebrates?
8. Examine the nerve cord from its anterior extreme to the posterior end. Is it the same thickness the entire length? On the dorsal side of the anterior extremity, attempt to locate the **neuropore** (Kölliker's pit), an opening to the outside. What other features are evident within the nerve cord? At the anterior-most end is a small, dark **pigment spot** and scattered down its length along the ventral margin are the **cup ocelli**. Examine the cup ocelli using high power magnification. Why were they given their name? Cup ocelli tend to be concentrated at the anterior end of the nerve chord.

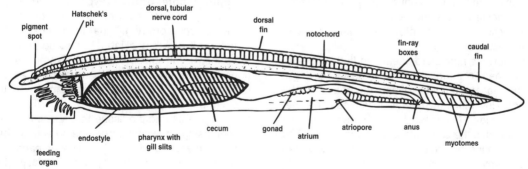

Figure 25.8. Lateral view of *Branchiostoma*.

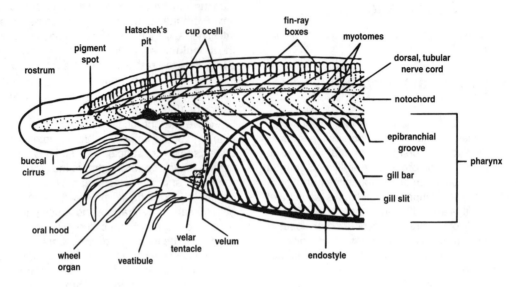

Figure 25.9. Enlargement of the anterior end of *Branchiostoma*.

They are more numerous in the region of the buccal cavity and the end of the pharynx (Fig. 25.8).

Cross Section

Obtain a prepared slide with cross sections of an amphioxus through several levels, at least one of which is through the pharynx. As you find the structures described below, note their location in the whole organism.

1. Begin your observations with a section through the level of the pharynx (Fig. 25.10). At the dorsal surface find the dorsal fin and fin-ray box. (At this level the ventral fin will not be present.) Below the dorsal fin locate the nerve cord. What evidence indicates that it is tubular? Can you see any nerves exiting the nerve cord in your section? (You may have to examine all the sections on the slide to find a ventral root nerve.) Directly below the nerve cord is the notochord. Is there anything distinctive about the notochord in this or any of the other sections?

2. Just below the notochord attempt to find the paired dorsal aortae. To either side of the nerve cord and notochord, locate a stack of muscle masses, the myotomes. How many myotomes do you see on either side of the body? Are these muscle masses paired? Do all pairs have the same cross-sectional area? Why not? Re-examine a whole mount of amphioxus and determine why the myotomes look like this in cross section (Fig. 25.8). Note how these muscle units

appear to be stacked one upon the other up the side of the body.

3. Locate the sheets of connective tissues (**myocommas** or **myosepta**) that separate the myotomes. Examine the sheath that surrounds the notochord. This is the **notochord sheath**; its fibrous nature is seen best using phase contrast lighting. Note how it connects to the myosepta. What function might this network of sheaths serve in locomotion (i.e., the myocommas interconnecting with the notochord sheath)? Scan the body wall of the entire organism. Are other muscles evident? In cross sections at the level of the pharynx, locate the transverse muscle layer. What function might it serve?

4. Examine the pharyngeal region (Figs. 25.8 and 25.10). At the most dorsal part of the pharynx, locate the **epibranchial groove** (hyperpharyngeal groove) and at the ventral-most part, locate the endostyle (hypobranchial groove). What is the shape of these structures? Are they ciliated? In some preparations the pharynx will still contain food. Examine your sections to determine what food amphioxus eats. How is food gathering in amphioxus similar to that in ascidians?

5. The gill slits and the tissue that supports them, the gill bars, slope downward from the anterior to the posterior at an angle of about 45°. However, each gill slit is really two slits separated by a **secondary gill bar**, extending between the **primary gill bars**. The two may be distinguished by the presence of fine cross bars or **synapticula**. The synapticula run perpendicular to the gill bars, connecting primary bars, but bypassing the secondary bars (Fig. 25.10).

6. Protonephridia (tubular flagellated solenocytes) are present dorsally, just below the notochord. They connect the small coelom to the artium. Locate the ventral aorta just below the endostyle. These structures are usually difficult to find, so you may need to examine several slides to locate them.

7. To either side of the pharynx find the atrial cavity. Inside the artium locate the hollow digestive or hepatic cecum (liver). (Remember that the digestive cecum is found only on the right side of the body. What does it mean if you were to find this structure on the left side of the body in your slide?) Examine the cells that line the lumen of the digestive cecum. What features does the cecum possess that is indicative of it having a function in digestion?

8. Amphioxus is dioecious. Small, dense cells comprise the testes, while cells of the ovaries are larger and fewer in number. Is your specimen male or female? Be sure to examine specimens of both sexes so that you can tell them apart.

9. Examine the body wall and locate the epidermis and dermis layers. Two metapleural folds may be seen in the body wall in sections of amphioxus through the pharynx. If your cross section is a composite of several sections from different levels, locate the pharynx (or gut), notochord, nerve cord, atrium, and myotomes in each. As you locate these structures attempt to determined the approximate level from which the section was cut in the organism. ■

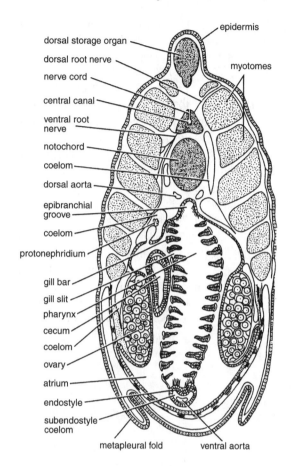

Figure 25.10. Cross section of *Branchiostoma* at the level of the pharynx. (After Sherman and Sherman.)

Supplemental Readings

Abbott, D.P., and W. Trason. 1968. Two new colonial ascidians from the west coast of North America. Bull. S. Calif. Acad. Sci. 67: 143–153.

Alldredge, A.L. 1976. Appendicularians. Sci. Am. 235(1): 94–102.

Alldredge, A.L. 1977. House morphology and mechanisms of feeding in the Oikopleuridae (Tunicata, Appendicularia). J. Zool. Lond. 181: 175–188.

Alldredge, A.L., and L.P. Madsin. 1982. Pelagic tunicates: unique herbivores in the marine plankton. Bio. Sci. 32: 655–663.

Alexander, R.McN. 1975. The Chordates. Cambridge University Press, Cambridge, UK.

Barham, E.G. 1979. Giant larvacean houses: observations from deep submersibles. Science 205: 1129–1131.

Barrington, E.J.W. 1965. The Biology of Hemichordata and Protochordata. W.H. Freeman and Co., San Francisco, CA.

Barrington, E.J.W., and R.P.S. Jefferies (eds.). 1975. Protochordates. Symp. Zool. Soc. Lond. 35.

Berrill, N.J. 1950. The Tunicata. Ray Society, London.

Berrill, N.J. 1961. *Salpa*. Sci. Am. 204: 150–160.

Bone, Q. 1972. The Origins of Chordates. Oxford University Biology Readers, Number 18. Oxford University Press, London.

Cloney, R.A. 1978. Ascidian metamorphosis: review and analysis. *In*: F-.S. Chia and M.E. Rice (eds.), Settlement and Metamorphosis of Marine Invertebrate Larvae. Elsevier, New York, pp. 255–282.

Flood, P.R. 1991. A simple technique for preservation and staining of the delicate houses of oikopleurid tunicates. Mar. Biol. 108: 105–110.

Gans, C., and R.G. Northcutt. 1983. Neural crest and the origin of vertebrates: a new head. Science 220: 268–274.

Goodbody, I. 1974. The physiology of ascidians. Adv. Mar. Biol. 12: 1–149.

Gutmann, W.F. 1981. Relationships between invertebrate phyla based on functional-mechanical analysis of the hydrostatic skeleton. Am. Zool. 21: 63–81.

Hirose, E., T. Maruyama, L. Cheng, and R.A. Lewin. 1996. Intracellular symbiosis of a photosynthetic prokaryote, *Prochloron* sp., in a colonial ascidian. Invert. Biol. 115(4): 343–348.

Jørgenson, C.B., and E.D. Goldberg. 1953. Particle filtration in some ascidians and lamellibranchs. Biol. Bull. 105: 447–489.

Løvtrup, S. 1977. The Phylogeny of Vertebrates. Wiley, London.

Madin, L.P. 1974. Field observations on the feeding behavior of salps. Mar. Biol. 25: 143–147.

Millar, R.H. 1971. The biology of ascidians. Adv. Mar. Biol. 9: 1–100.

Stokes, M.D., and N.D. Holland. 1996. Reproduction of the Florida lancelet (*Branchiostoma floridae*): Spawning patterns and fluctuations in gonad indexes and nutritional reserves. Invert. Biol. 115(4): 349–359.

Wickstead, J.H. 1967. Chordata sub-phylum Acrania (= Cephalochordata) family Branchiostomidae. Fiches d'Identification du Zooplankton 11.

General References

While this cannot be considered to be an exhaustive list, below are some of the important pieces of literature on invertebrates. After each exercise we list additional literature that deals specifically with the phylum under consideration (Supplemental Readings).

Abbott, D.P. 1987. Observing Marine Invertebrates: Drawings from the Laboratory. G.H. Hilgard (ed.). Stanford University Press, Standford, CA.

Alexander, R.McN. 1979. The Invertebrates. Cambridge University Press, Cambridge.

Allen, K., and D. Briggs (eds.). 1990. Evolution and the Fossil Record. Smithsonian Institution Press, Washington, DC.

Anderson, J.N., A.D.A. Rayner, and D.W.H. Walton. 1984. Invertebrate-microbial Interactions. Cambridge University Press, New York.

Ax, P. 1987. The Phylogenetic System. Wiley, New York.

Backeljav, T., B. Winnepenninckx, and L. de Brujn. 1993. Cladistic analysis of metazoan relationships: a reappraisal. Cladistics 9: 167–181.

Barnes, R.S.K. (ed.), 1984. A Synoptic Classification of Living Organisms. Sinauer Associates, Sunderland, MA.

Barnes, R.S.K., P. Calow, and P.J.W. Olive. 1988. The Invertebrates: A New Synthesis. Blackwell Scientific Publications, Palo Alto, CA.

Barrington, E.J.W. 1979. Invertebrate Structure and Function, 2nd ed. Wiley, New York.

Barth, R.H., and R.E. Broshears. 1982. The Invertebrate World. Saunders, New York.

Bayer, F.M., and H.B. Owre. 1968. The Free-Living Lower Invertebrates. Macmillan, New York.

Becklwemishev, W.N. 1969. Principles of Comparative Anatomy of Invertebrates. Vol. 1: Promorphology. Vol. 2: Organology. University of Chicago Press, Chicago.

Bengtson, S. (ed.). 1994. Early Life on Earth. Nobel Symposium No. 84. Columbia University Press, New York.

Bereiter-Hahn, J., A.G. Matoltsy, and K.S. Richards (eds.) 1984. Biology of the Integument. Vol. 1. Invertebrates. Springer-Verlag, New York.

Boardman, R.S., A.H. Checthmam, and A.J. Rowell. 1986. Fossil Invertebrates. Blackwell Scientific Publications, Oxford, UK.

Boucher, D.H. (ed.). 1985. The Biology of Mutualism. Oxford University Press, New York.

Britton, J.C., and B. Morton. 1989. Shore Ecology of the Gulf of Mexico. University of Texas Press, Austin, TX.

Brown, F.A., Jr. (ed.). 1950. Selected Invertebrate Types. Wiley, New York.

Brusca, R.C., and G.J. Brusca. 1990. Invertebrates. Sinauer Associates, Sunderland, MA.

Bullough, W.S. 1950. Practical Invertebrate Anatomy. Macmillan, New York.

Cable, R.M. 1977. An Illustrated Laboratory Manual of Parasitology, 5th ed. Bugress Publishers, Minneapolis, MN.

Cairns, S.D. 1991. Common and Scientific Names of Aquatic Invertebrates from the United States and Canada: Cnidaria and Ctenophora. American Fisheries Society, Bethesda, MD.

Calow, P. 1981. Invertebrate Biology. Halsted Press, Wiley, New York.

Conway Morris, S. 1993. The Fossil Record and the Early Evolution of the Metazoa. Nature (Lond.) 361: 219–225.

Casanova, R., and R.P. Ratkevich 1981. An Illustrated Guide to Fossil Collecting. Naturegraph Publishers, Happy Camp, CA.

CDC (Centers for Disease Control and Prevention). Morbidity and Mortality Weekly Report. U.S. Department of Health and Human Services/Public Health Service. Available over the World Wide Web at the following URL address. http://www.cdc.gov/epo/mmwr/mmwr.html

Cheng, T.C. 1986. General Parasitology, 2nd ed. Academic Press, New York.

Chia, F.-S., and Mary E. Rice (eds.). 1977. Settlement and Metamorphosis of Marine Invertebrate Larvae. Elsevier, New York.

Clark, R.B. 1964. Dynamics in Metazoan Evolution. Oxford University Press, Oxford, UK.

Clarkson, E.N.K. 1986. Invertebrate palaeontology and evolution, 2nd ed. Allen & Unwin, London.

Cloud, P., and M.F. Glaessner. 1982. The Ediacarian period and system: Metazoa inherit the Earth. Science 218: 783–792.

Connor, J. 1993. Seashore Life on Rocky Coasts. Monterey Bay Aquarium, Monterey, CA.

Conway Morris, S., J.D. George, R. Gibson, and H.M. Platt (eds.) 1985. The Origins and Relationships of the Lower Invertebrates. Systematics Association Special Vol. No. 28. Clarendon Press, Oxford.

Corning, W.C., and S.C. Ratner (eds.). 1967. Chemistry of Learning: Invertebrate Research. Plenum Press, New York.

Cox, F.E.G. 1982. Modern Parasitology. Blackwell Scientific Publications Oxford, UK.

Crawford, C.S. 1981. Biology of Desert Invertebrates. Springer-Verlag, New York.

Dales, R.P. 1981. Practical Invertebrate Zoology, 2nd ed. Wiley, New York.

Delly, J.G. 1985. Narcosis and preservation of freshwater animals. Am. Lab. (April): 31–40.

Diehl, F.A., J.B. Feeley, and D.G. Gibson. 1971. Experiments Using Marine Animals. Aquarium Systems, Inc. Eastlake, OH.

Droser, M.L., R.A. Fortey, and X. Li. 1996. The Ordovician radiation. Am. Sci. 84: 122–131.

Edmondson, W.T. (ed.). 1959. Fresh-water Biology, 2nd ed. Wiley, New York.

Evans, S.M. 1968. Studies in Invertebrate Behaviour. Heinemann Educational, London.

Fautin, D., and G.R. Allen. 1992. Field Guide to Anemonefishes and Their Host Sea Anemones. Western Australian Museum Press, Perth, WA, Australia.

Field, K.G., G.J. Olsen, D.J. Lane, S.J. Giovannoni, M.T. Ghislin, E.C. Raff, N.R. Pace, and R.A. Raff. 1988. Molecular phylogeny of the Animal Kingdom. Science 239: 748–753.

Fotheringham, N., and S.L. Bruenmeister. 1989. Beachcomber's Guide to Gulf Coast Marine Life (Florida, Texas, Louisiana, Mississippi, Alabama (2nd ed.). Gulf Publishing Co., Houston, TX.

Freeman, W.H., and B. Bracegirdle. 1982. An Atlas of Invertebrate Structure. Heinemann Educational Books, London, UK.

Gage, J.D., and P.A. Tyler. 1991. Deep Sea Biology—A Natural History of Organisms at the Deep-Sea Floor. Cambridge University Press, New York.

Garstang, W. 1985 (reprinted). Larval Forms and Other Zoological Verses. The University of Chicago Press, Chicago.

Giese, A.C., and J.S. Pearse. 1974/1979. Reproduction of Marine Invertebrates, Vols. 1–5. Academic Press, New York.

Glaessner, M.F. 1984. The Dawn of Animal Life. Cambridge University Press, New York.

Gosliner, T.M., D.W. Behrens, and G.C. Williams. 1995. Coral Reef Animals of the Indo Pacific. Sea Challengers, Inc. Monterey, CA.

Gosner, K.L. 1971. Guide to Identification of Marine and Estuarine Invertebrates. Wiley, New York.

Gotshall, D.W. 1994. Guide to Marine Invertebrates. Sea Challenges, Monterey, CA.

Gould, S. Jay. 1989. Wonderful Life: the Burgess Shale and the Nature of History. W.W. Norton, New York.

Harrison, F.W., and E.E. Ruppert (treatise eds.). 1991–continuing. Microscopic Anatomy of Invertebrates. Wiley-Liss, New York.

Harrison, F.W., and R.R. Cowden (eds.). 1982. Developmental Biology of Freshwater Invertebrates. Liss, New York.

Heatwole, H. 1996. Energetics of Desert Invertebrates. Springer-Verlag, New York.

Hickman, C.P. 1973. Biology of the Invertebrates. Mosby, St. Louis, MO.

House, M.R. (ed.) 1979. The Origin of Major Invertebrate Groups. Systematics Association Special Vol. No. 12. Academic Press, New York.

Humann, P. 1992. Reef Creature Identification—Florida, Caribbean, Bahamas. New World Publications, Jacksonville, FL.

Hyman, L.H. 1940/1967. The Invertebrates, Vols. I–VI. McGraw-Hill, New York.

Inglis, W.G. 1985. Evolutionary Waves: Patterns in the origin of Animal Phyla. Aust. J. Zool. 33: 153–178.

Jensen, G.C. 1995. Pacific Coast Crabs and Shrimps. Sea Challengers, Inc., Monterey, CA.

Kaestner, A. 1967/70. Invertebrate Zoology, Vols. I–III. Wiley, New York.

Kaplan, E.H. 1988. Southeastern and Caribbean Seashores (Peterson Field Guide Series). Houghton Mimin. Co., Boston, MA.

Kerfoot, W.C. 1980. Evolution and Ecology of Zooplankton Communities. University Press New England, Hanover, HN.

Klaus, R. 1982. Ecology of Marine Parasites. University of Queensland Press, New York.

Kobayashi, M., M. Takahshi, H. Wada, and N. Satoh. 1993. Molecular Phylogeny Inferred from Sequences of Small Subunit Ribosomal RNA Supports the Monophyly of the Metazoa. Zool. Science 10: 827–833.

Kozloff, E.N. 1983. Seashore Life of the Northern Pacific Coast. University of Washington Press, Seattle, WA.

Kozloff, E.N. 1987. Marine Invertebrates of the Pacific Northwest. University of Washington Press, Seattle, WA.

Kozloff, E.N. 1990. Invertebrates. Saunders, Philadelphia, PA.

Lanzavecchia, G., R. Valvassori, and M.D. Candia Carnevali (eds.). 1995. Body cavities: Function and Phylogeny. Selected Symposia and Monographs Unione Zoologica Italiana, 8, Mucchi, Modena.

Laufer, H., and R.G.H. Downer (eds.). 1988. Endocrinology of Selected Invertebrate Types. Liss, New York.

Laverack, M.S., and J. Dando. 1987. Lecture Notes on Invertebrate Zoology, 3rd ed. Blackwell Scientific Publications, Oxford, UK.

Lehmann, U., and G. Hillmer. 1983. Fossil Invertebrates. Cambridge University Press, Cambridge, UK.

Lincoln, R.J., and J.G. Sheals. 1979. Invertebrate Animals. Collection and Preservation. British Museum (Natural History). Cambridge University Press, Cambridge, UK.

Little, C., and J.A. Kitching. 1996. The Biology of Rocky Shores. Oxford University Press, New York.

Lunt, G.G., and R.W. Olsen (eds.). 1988. Comparative Invertebrate Neurochemistry. Cornell University Press, Ithaca, NY.

Lutz, P.E. 1985. Invertebrate Zoology. Addison-Wesley, Reading, MA.

MacGinitie, G.E., and N. MacGinitie. 1968. Natural History of Marine Animals, 2nd ed. McGraw-Hill, New York.

MacInnis, A.J., and M. Voge. 1970. Experiments and Techniques in Parasitology. W.H. Freeman, San Francisco, CA.

Marquardt, W.C., and R.S. Demaree, Jr. 1985. Parasitology. Macmillan, New York.

Marshall, A.J., and W.D. Williams (eds.) 1972. Textbook of Zoology: Invertebrates. Elsevier, New York.

Martinez, A.J., and R.A. Harlow. 1994. Marine Life of the North Atlantic-Canada to New England. Marine Life, Wenham, MA.

McKinney, F.K. 1991. Exercises in Invertebrate Paleontology. Blackwell Scientific Publications, Boston, MA.

Meglitsch, P.A., and F.R. Schram. 1991. Invertebrate Zoology, 3rd ed. Oxford University Press, London, UK.

Meinkoth, N.A. 1981. Audubon Society Field Guide to North American Seashore Creatures. Audubon Society, New York.

Meyer, M.C., and O.W. Olsen. 1980. Essentials of Parasitology. Wm. C. Brown, Dubuque, IA.

Mitchell, D.H., and T.E. Johnson (eds.) 1984. Invertebrate Models in Aging Research. CRC Press. Boca Raton, FL.

Moffett, S.B. 1996. Nervous System Regeneration in the Invertebrates. Springer-Verlag, New York.

Moore, R.C., et al. (eds.). 1954–continued. Treatise on Invertebrate Paleontology. 32 Parts, Geological Society of America and University of Kansas Press, Lawrence, KS.

Morris, R.H., D.P. Abbott, and E.C. Haderlie. 1980. Intertidal Invertebrates of California. Stanford University Press, Stanford, CA.

Murray, M. 1967. Hunting for Fossils. Collier Books, Macmillan, New York.

New, T.R. 1995. Introduction to Invertebrate Conservation Biology. Oxford University Press, New York.

Nielsen, C. 1995. Animal Evolution. Oxford University Press, New York.

Niesen, T.M. 1994. Beachcombers Guide to California Marine Life—(Covers Common Marine Fauna and Flora from San Francisco to San Diego). 1994. Gulf Publishing Co., Houston, TX.

Noble, E.R., and G.A. Noble. 1982. Parasitology: the Biology of Animal Parasites, 5th ed. Lea and Febiger, Philadelphia, PA.

Nybakken, J.W. 1996. Diversity of Invertebrates: A Laboratory Manual. Wm. C. Brown, Dubuque, IA.

Parker, S.P. 1982. Synopsis and Classification of Living Organisms, Vols. 1–2. McGraw-Hill, New York.

Patterson, D.J. 1992. Free-Living Freshwater Protozoa: A Color Guide. CRC Press, Boca Roton, FL.

Pearse, V. (ed.) 1995–continuing. Invertebrate Biology. This journal continues the traditions of the Transactions of the American Microscopical Society.

Pearse, V., J. Pearse, M. Buschsbaum, and R. Buschsbaum. 1987. Living Invertebrates. Boxwood Press, Pacific Grove, CA.

Pechenik, J.A. 1996. Biology of the Invertebrates, 3rd ed. Wm.C. Brown, Dubuque, IA.

Peckarsky, B.L., P.R. Fraissinet, M.A. Penton, D.J. Conklin, Jr. 1990. Freshwater Macroinvertebrates of Northeastern North America. Cornell University Press, Ithaca, NY.

Pennak, R.W. 1989. Fresh-water Invertebrates of the United States, 3rd ed. Wiley, New York.

Pierce, S.K., T.K. Maugel, and L. Reid. 1987. Illustrated Invertebrate Anatomy. Oxford University Press, New York.

Raff, R.A. 1996. The Shape of Life. University of Chicago Press, Chicago, IL.

Ricketts, E.F., J. Calvin, J.W. Hedgpeth, and D.W. Phillips 1986. Between Pacific Tides, 5th ed. Stanford University Press, Stanford, CA.

Ruppert, E.E., and R.D. Barnes. 1994. Invertebrate Zoology, 6th ed. Harcourt Brace, Fort Worth, TX.

Ruppert, E.E., and R.S. Fox. 1988. Seashore Animals of the Southeast. University of South Carolina Press, Columbia, SC.

Russell-Hunter, W.D. 1968. A Biology of Lower Invertebrates. Macmillan, New York.

Russell-Hunter, W.D. 1969. A Biology of Higher Invertebrates. Macmillan, New York.

Russell-Hunter, W.D. 1979. A Life of Invertebrates. Macmillan, New York.

Schmidt, G.S., and L.S. Roberts. 1989. Foundations of Parasitology, 4th ed. Times Mirror/Mosby, St. Louis, MO.

Sefton, N., and S. Webster. 1986. Caribbean Reef Invertebrates. Sea Challengers, Inc., Monterey, CA.

Shaw, A.C., S.K. Lazell, and G.N. Foster. 1974. Photomicrographs of Invertebrates. Fletcher and Sons Ltd., Norwich, UK.

Sherman, I.W., and V.G. Sherman. 1976. The Invertebrates: Function and Form. Macmillan, New York.

Simonetta, A.M., and S. Conway Morris. 1991. The Early Evolution of Metazoa and the Significance of Problematic Taxa. Cambridge University Press, Cambridge, UK.

Smith, D.L. 1977. A Guide to Marine Coastal Plankton and Marine Invertebrate Larvae. Kendall/Hunt Publishing Co., Dubuque, IA.

Smith, R.I., and J.T. Carlton. 1975. Light's Manual: Intertidal Invertebrates of the Central California Coast (3rd ed.). University of California Press, Berkeley, CA.

Stachowitsch, M. 1992. The Invertebrates: An Illustrated Glossary. Wiley-Liss, New York.

Tasch, P. 1980. Paleobiology of the Invertebrates. Wiley, New York.

Thorp, J.H., and A.P. Covich. 1991. Ecology and Classification of North American Freshwater Invertebrates. Academic Press, San Diego, CA.

Tomlinson, J.T. 1976. Invertebrate Behavior. San Francisco State University, San Francisco, CA.

Trager, W. 1986. Living Together. The Biology of Animal Parasitism. Plenum Press, New York.

Trueman, E.R. 1975. The Locomotion of Soft-Bodied Animals. Elsevier, New York.

Turgeon, D.D. 1988. Common and Scientific Names of Aquatic Invertebrates from the United States and Canada: Mollusks. American Fisheries Society, Bethesda, MD.

Valentine, J.W. 1975. Adaptive strategy and the origin of grades and ground-plans. Am. Zool. 15: 391–404.

Valentine, J.W. (ed.). 1985. Phanerozoic Diversity Patterns. Princeton University Press, Princeton, NJ.

Vermeij, G.J. 1978. Biogeography and Adaptation. Harvard University Press., Cambridge, MA.

Vermeij, G.J. 1987. Evolution and Escalation: An Ecological History of Life. Princeton University Press, Princeton, NJ.

Vernberg, F.J., and W.B. Vernberg (eds.) 1981. Functional Adaptations of Marine Organisms. Academic Press, New York.

Vine, P. 1986. Red Sea Invertebrates. Immel Publishing Limited, New York.

Vogel, S. 1981. Life in Moving Fluids. Willard Grant Press, Boston, MA.

Voss, G.L. 1976. Seashore Life of Florida and the Caribbean. Banyan Books, Miami, FL.

Welsh, J.H., R.I. Smith, and A.E. Kammer. 1968. Laboratory Exercises in Invertebrate Physiology. Burgess Publishing Co., Minneapolis, MN.

Whittington, H.B. 1985. The Burgess Shale. Yale University Press, New Haven, CT.

Williamson, D.I. 1992. Larvae and Evolution: Toward a New Zoology. Chapman and Hall, New York.

Willmer, P. 1990. Invertebrate Relationships: Patterns in Animal Evolution. Cambridge University Press, Cambridge, UK.

Index